PROCEEDINGS OF THE INTERNATIONAL CONFERENCE
ON HYDROLOGY AND WATER RESOURCES,
NEW DELHI, INDIA, DECEMBER 1993

VOLUME 3

WATER-QUALITY HYDROLOGY

Water Science and Technology Library

VOLUME 16/3

The titles published in this series are listed at the end of this volume.

PROCEEDINGS OF THE INTERNATIONAL CONFERENCE ON HYDROLOGY AND WATER RESOURCES, NEW DELHI, INDIA, DECEMBER 1993

Volume 3

WATER-QUALITY HYDROLOGY

edited by

VIJAY P. SINGH

Department of Civil and Environmental Engineering, Louisiana State University, Baton Rouge, U.S.A.

and

BHISHM KUMAR

National Institute of Hydrology, Roorkee, India

SPRINGER-SCIENCE+BUSINESS MEDIA, B.V.

A C.I.P. Catalogue record for this book is available from the Library of Congress

ISBN 978-94-010-4176-8 ISBN 978-94-011-0393-0 (eBook)
DOI 10.1007/978-94-011-0393-0

Printed on acid-free paper

To

Mother India

TABLE OF CONTENTS

SECTION 3: SEDIMENT TRANSPORT IN UPLAND AREAS

SECTION 4: SOLUTE TRANSPORT IN VADOSE ZONE

SECTION 5: SOLUTE TRANSPORT IN GROUND WATER

SECTION 6: WATER QUALITY MODELING

PREFACE

Water is vital to life, maintenance of ecological balance, economic development, and sustenance of civilization. Planning and management of water resources and its optimal use are a matter of urgency for most countries of the world, and even more so for India with a huge population. Growing population and expanding economic activities exert increasing demands on water for varied needs--domestic, industrial, agricultural, power generation, navigation, recreation, etc. In India, agriculture is the highest user of water.

The past three decades have witnessed numerous advances as well as have presented intriguing challenges and exciting opportunities in hydrology and water resources. Compounding them has been the growing environmental consciousness. Nowhere are these challenges more apparent than in India. As we approach the twenty-first century, it is entirely fitting to take stock of what has been accomplished and what remains to be accomplished, and what accomplishments are relevant, with particular reference to Indian conditions.

India has had a long tradition of excellence spanning over several millennia in construction of water resources projects. She has produced civil engineers of highest quality, some of whom rank among the best in the world. In keeping with India's tradition of honoring distinguished engineers at their retirement, it was decided to hold an International Conference on Hydrology and Water Resources, December 20-22, 1993, in New Delhi, to celebrate the retirement of Professor Satish Chandra, Director of the National Institute of Hydrology at Roorkee, who has had a long and distinguished career spanning over three decades in hydrologic education, research, and development. Through this forum, it was hoped to bring together from all over the world researchers, educationists, and practitioners to New Delhi, India, to present the results of their latest work, to share their experiences, to interact with Indian water resources professionals, and to, hopefully, set the stage for future course of hydrology and water resources.

The objectives of this conference were therefore (1) to assess the current state of the art of hydrology and water resources, (2) to discuss applicability of hydrologic and water resources models, (3) to enhance interdisciplinary interaction, (4) to discuss research needs in hydrology and water resources, and (5) to determine the directions for further research.

We received an overwhelming response to our call for papers. The subject matter of the conference was divided into 36 major topics encompassing virtually all facets of hydrology and water resources. Each topic comprised an invited state-of-the-paper and a number of contributed papers. These contributions provided a natural blend to evolve a synthesized body of knowledge on that topic. Extended abstracts of all the invited and contributed papers were assembled in a pre-conference proceedings volume. This helped stimulate discussion and exchange of ideas during the conference.

The papers presented at the conference were reviewed by the Organizing Committee. As a result, some of the papers did not pass the review and were therefore eliminated from inclusion in the final proceedings. The papers contained in this book, WATER-QUALITY HYDROLOGY, represent one part of the conference contributions.

The other parts are embodied in three separate companion books, SURFACE-WATER HYDROLOGY, SUBSURFACE-WATER HYDROLOGY, and WATER RESOURCES PLANNING AND MANAGEMENT, which are being published simultaneously. Arrangement of the contributions in these books under four different titles was a natural consequence of the diversity of topics presented in the papers. These books can be treated almost independently, although some overlap may exist between them.

This book contains six sections encompassing major aspects of environmental hydrology. Each section normally starts with invited state-of-the-art papers,

ACKNOWLEDGMENTS

The International Conference on Hydrology and Water Resources was sponsored and co-sponsored by a number of organizations. The sponsors supported the conference financially without which it might not have come to fruition. Their financial support is gratefully acknowledged. The co-sponsors extended their support by announcing the conference through their journals, periodicals, transactions, newsletters, or magazines. The publicity so gained helped with increased participation in the conference, and is sincerely appreciated. The following is a list of symposium sponsors and co-sponsors.

Sponsors

Wallac Oy
Finland

Pharmacia United Limited
New Delhi, India

Electrotek Instruments Inc.
Madras, India

Elron Instruments Company Pvt. Ltd.
New Delhi, India

Department of Power
Govt. of Arunachal Pradesh
Itanagar, India

Environmental Measurements & Control
Cochin, India

AIC Agro Instruments (P) Ltd.
Calcutta, India

Spectrum Associates
New Delhi, India

Anubhav Printers & Packers
Roorkee, India

Ajay Printers
Roorkee, India

Co-Sponsors

American Geophysical Union
USA

International Association for
Hydraulic Research
The Netherlands

International Water Resources
Association
USA

Soil and Water Conservation Society
USA

Association of American Geographers
USA

International Statistical Institute
The Netherlands

SECTION 1:

CHEMICAL TRANSPORT

NONLINEAR REACTIVE SOLUTE TRANSPORT: A PRACTICAL AND FAST SOLUTION METHOD

D. A. Barry and K. Bajracharya
Department of Environmental Engineering,
Centre for Water Research,
University of Western Australia,
Nedlands, Western Australia 6009

ABSTRACT. Mixing cell models have been used frequently to model solute transport coupled with reactions. The main advantage of these models is their conceptual simplicity. This allows them to be solved sequentially with chemical speciation models to predict chemical concentrations in combined reaction/transport problems. Mixing cell models are actually just explicit finite-difference solutions of the governing advection-dispersion equation. It can be shown that the inclusion of reactions in the "standard" mixing cell model degrades its second-order accuracy. We propose, therefore, an improved model which maintains second-order accuracy regardless of whether reactions are included. In addition, the improved model is unconditionally stable, unlike the standard scheme. We show that nonequilibrium reactions can also be included without difficulty. Next, we show that different boundary conditions can be incorporated into the mixing cell models. In particular, a third-type surface condition is considered. For this case, to maintain second-order accuracy of the improved model, it is necessary to dispense with the explicit nature of the scheme on the boundary. At other locations the scheme is still explicit. Other conditions, e.g., a finite mass of solute available at the surface, as would be the situation if a landfill was considered, can be handled in the same way. The method can be extended to cater for multilayered porous media. The method can also be extended easily to simulate multispecies transport. Laboratory data on the precursor effect has been used to demonstrate the use of the multispecies transport model. Our results demonstrate that nonlinear reactions and transport can be modelled very efficiently and quickly.

1. Introduction

The mixing cell concept has been used frequently to predict migration of reactive and non-reactive solutes in soil and groundwater (Schweich and Sardin, 1981; Van Ommen, 1985; Rao and Hathaway, 1989; Dudley et al., 1991; Bajracharya and Barry, 1993c). Mixing cell models are easy to code and efficient to compute. They can be linked with chemical speciation models to simulate complex reactions coupled with transport.

Explicit finite-difference representations of the transport equation underlie the mixing cell concept. For tracer solutes, the standard model is second-order accurate in the space and time discretization. It can be shown that the inclusion of solute adsorption in the transport equation reduces the accuracy of the mixing cell model. We present an improved mixing cell model which maintains second-order accuracy. In addition, the improved model is unconditionally stable.

V. P. Singh and B. Kumar (eds.), Water-Quality Hydrology, 3–17.

There are two types of predictive models used for analysing transport of reacting solutes (Rubin, 1983). In one, the reaction is sufficiently fast to be described by an equilibrium model. In the other type, the reaction is relatively slow and is quantified by a nonequilibrium model. We show that the improved method can cater for both cases. Also, different surface boundary conditions can be easily incorporated into the improved mixing cell model.

In reality soils are never homogeneous and the soil solutions never contain a single species only. Soil heterogeneity can be partially accounted for by considering layered soils. Such porous media have been considered by a number of researchers (Shamir and Harleman, 1967; Selim, et al., 1977; Kreft, 1981; Barry and Parker, 1987, Barry et al., 1987). Here this problem is treated also using improved mixing cell model. An example of multispecies transport is considered by making use of the laboratory data on the precursor effect considering an ion-exchange reaction (Starr et al., 1982).

Our main purposes in this paper are to:

(i) show that the improved mixing cell model can describe nonequilibrium reactions very efficiently,
(ii) include different types of surface boundary condition in the numerical scheme,
(iii) show the use of the model in the case of a multilayered porous medium, and
(iv) show the use of the model for multispecies transport.

In the following section the governing equations for single species transport with various surface and exit conditions and reactions are first described followed by improved mixing cell model. The governing equations in multispecies transport are then presented with the ion exchange reactions.

2. Governing Models

2.1. SINGLE SPECIES TRANSPORT

The governing one-dimensional solute transport equation with adsorption is given by (e.g., Yortsos, 1987):

$$\frac{\partial(S+C)}{\partial t} = D_s \frac{\partial^2 C}{\partial z^2} - V\frac{\partial C}{\partial z} \tag{1}$$

(symbols are defined in the notation list). This equation describes solute transport subject to adsorption under steady flow conditions. The solid phase solute concentration, S, is assumed to be given as an equivalent fluid phase concentration. Initially, the normalised solute concentration in both the solid and liquid phases is taken as:

$$C(z,0) = S(z,0) = 0, \tag{2}$$

although the right-hand side can be non zero if necessary.

We consider next the different boundary conditions used to solve eq. (1).

2.1.1. *Entrance Boundary Conditions*

One type of boundary condition is a constant concentration of solute at the soil surface,

$$C(0,t) = C_o. \tag{3a}$$

Equation (3a) is based on the assumption of a "well-mixed" solute reservoir at the entrance and assumes that the concentration is continuous across the inlet boundary (van Genuchten and Parker, 1984). Alternatively, eq. (3a) is appropriate for describing the solute flux at $z = 0$ (Kreft and Zuber, 1978).

The other commonly used surface condition is the mixed, or flux boundary condition, given by (Brenner, 1962):

$$VC(0,t) - D_s \frac{\partial C(0,t)}{\partial z} = VC_o. \tag{3b}$$

Equation (3b) results in a macroscopic discontinuity in concentration at $z = 0$.

Finite source boundary conditions are typically imposed in landfill modelling studies. It is assumed that the pollutant concentration in the landfill is initially C_o, and that this concentration decreases with time as leachate is transmitted through the soil by advection and dispersion. If the equivalent thickness of leachate in the landfill is h_f, and n_a is the porosity, then the surface concentration at any time, t, is given by (Rowe and Booker, 1985):

$$C(0,t) = C_o - \frac{n_a}{h_f} \int_0^t VC(0,\tau) - D_s \frac{\partial C(0,\tau)}{\partial z} \, d\tau. \tag{3c}$$

This entrance condition applies to a finite source of solute at the entrance boundary. Note that setting $h_f \rightarrow \infty$ recovers eq. (3a). In the illustrative examples given below, boundary condition (3a) is used to solve the nonequilibrium problem and surface conditions (3b) and (3c) are used to solve the case of equilibrium reactions.

2.1.2. *Exit Conditions*

For a semi-infinite domain, we require that the solute concentration remains bounded throughout the soil profile, i.e. (van Genuchten and Parker, 1984):

$$\left. \frac{\partial C}{\partial z} \right|_{z \rightarrow \infty} = 0. \tag{4a}$$

For a finite column, the commonly used exit condition is given by (Brenner, 1962):

$$\left. \frac{\partial C}{\partial z} \right|_{z = L} = 0. \tag{4b}$$

In other words, eq. (4b) assumes that the solute concentration is continuous across the exit boundary, although solute dispersion is negligible outside the porous medium.

Another case of interest is the exit boundary condition for a soil column that is freely draining (Barry and Sposito, 1988):

$$C(L,0) = 0. \tag{4c}$$

Equation (4c) applies also in case of landfills, where base concentrations can be assumed zero when the groundwater velocity is quite large relative to the infiltrating solute (Rowe and Booker, 1985). This becomes the case of infinite dilution, and is a reasonable assumption given that flow rates in the unsaturated zone are typically much lower than those in the saturated zone. The incorporation of these boundary conditions into the mixing cell model is discussed below.

2.1.3. *Adsorption*

The relationship between the solid and liquid phase solute concentration at equilibrium is given by the adsorption isotherm and is of the form:

$$S = f(C). \tag{5a}$$

A linear isotherm is given by

$$S = (R-1)\,C, \tag{5b}$$

where R is the retardation factor. An S-curve isotherm is (Barry, 1992)

$$S = k_4\left\{1 - \left[1 + (k_2\, C)^{k_1}\right]^{k_3}\right\}. \tag{5c}$$

The term $\partial S/\partial t$ in the left hand side of eq. (1) describes the reaction process. When the reaction is sufficiently fast, the adsorption isotherm is written as an algebraic function of C (eq. 5) and so is used to eliminate the adsorbed phase concentration, S (Rubin, 1983; Clancy and Jennings, 1988). Such an operation gives rise to the equilibrium transport model. When the reaction process is insufficiently rapid, an additional differential rate equation must be used to solve (1). This results in the nonequilibrium transport model. A thorough discussion concerning classes of reactions can be found in the paper by Rubin (1983). Adsorption kinetics can be expressed mathematically as:

$$\frac{\partial S}{\partial t} = \alpha\,[f(C) - S]. \tag{6}$$

It should be noted that, at equilibrium, $\partial S/\partial t = 0$ and $S = f(C)$. Equations (1) and (6), together with appropriate boundary conditions, constitute a nonequilibrium transport model.

2.2. IMPROVED MIXING CELL MODEL FOR SINGLE SPECIES

Upon dropping second derivative term in the advective-dispersive equation, eq. (1), one gets a pure advective transport equation given by:

$$\frac{\partial(C + S)}{\partial t} = -\,V\frac{\partial C}{\partial z}. \tag{7}$$

Finite-difference solutions of (7) produce numerical dispersion which is controlled to simulate solutions to eq. (1).

In the standard mixing cell model, the temporal derivative is approximated by forward differencing and the advective term is approximated by backward differencing. The standard mixing cell solution has a truncation error of $O(\Delta z^2, \Delta t)$ (Bajracharya and Barry, 1993b) and the following scheme is proposed which is still explicit but is more accurate than the standard model. In this scheme the advective term is approximated by the average of the backward difference approximations at the (i,j) and (i,j+1) grid points as in the conventional Crank-Nicolson scheme (e.g., Noye, 1982). After simplfication, the improved scheme is given by

$$(2 + C_r)\,C(i,j+1) + 2S(i,j+1) = 2\,[C(i,j) + S(i,j)] + C_r\,[C(i-1,j) - C(i,j) + C(i-1,j+1)],\tag{8}$$

where we use the notation $C(z,t) = C(i\Delta z, j\Delta t) = C(i,j)$, while

$$C_r = \frac{V\Delta t}{\Delta z}\tag{9}$$

is the Courant number. At equilibrium, the solid phase concentration is given by:

$$S(i,j+1) = f\,[C(i,j+1)].\tag{10}$$

By making use of eq. (10), eq. (8) reduces to an equation with C(i,j+1) as the only unknown. In the case of nonequilibrium adsorption, the finite-difference analog of eq. (6) is given by:

$$S(i,j+1) = \frac{\frac{\alpha\Delta t}{2}\{f\,[C(i,j+1)] + f\,[C(i,j)]\}}{1 + \frac{\alpha\Delta t}{2}} + \frac{\left(1 - \frac{\alpha\Delta t}{2}\right)S(i,j)}{1 + \frac{\alpha\Delta t}{2}}\tag{11}$$

Equations (8) and (11), when combined, again result a single equation with C(i,j+1) as the only unknown. Next we show that eqs. (8) and (11), the improved mixing cell model, are consistent.

By making use of the third-order Taylor series expansion in eq. (8), we get:

$$\frac{\partial(C + S)}{\partial t} = \frac{V\Delta z}{2}\frac{\partial^2 C}{\partial z^2} - V\frac{\partial C}{\partial z} + O(\Delta z\Delta t, \Delta z^2, \Delta t^2),\tag{12}$$

i.e., the error is $O(\Delta z\Delta t, \Delta z^2, \Delta t^2)$. In this scheme the numerical dispersion introduced is $V\Delta z/2$. When the transport equation (1) is to be solved by eq. (8), the relation $D_s = V\Delta z/2$ is imposed, thus fixing the spatial step length.

Similarly, eq. (11) reduces to:

$$\frac{\partial S}{\partial t} = \alpha\,[f\,(C) - S] + O(\Delta t^2),\tag{13}$$

and so eqs. (12) and (13) are consistent with respect to truncation error.

Since a stability analysis is possible only for simple cases, the scheme obtained by combining eqs. (8) and (11) was checked for stability using numerical experiments. In all cases the scheme was found to be unconditionally stable. Unlike the standard scheme, this allows large time steps to be used without the solution being corrupted by roundoff errors. The improved mixing cell is unconditionally stable when adsorption is not considered (Bajracharya and Barry, 1993a).

2.2.1. *Incorporation of Surface Boundary Conditions*

The incorporation of eq. (3a) is straightforward so only conditions (3b) and (3c) are considered below. The entrance condition (3b) is incorporated at z = 0 by the central difference approximation (Noye, 1982). This reduces to a system of equations in two unknowns at the first two grid points. The system of equations is:

$$\left(1 + \frac{C_r}{2} + \frac{C_r V \Delta z}{D_s}\right) C(0,j+1) - \frac{C_r}{2} C(1,j+1) + S(0,j+1) =$$
$$\left(1 - \frac{C_r}{2} - \frac{C_r V \Delta z}{D_s}\right) C(0,j) + \frac{C_r}{2} C(1,j) + \frac{2 C_r V \Delta z}{D_s} C_o + S(0,j) \tag{14a}$$

and

$$-\frac{C_r}{2} C(0,j+1) + \left(1 + \frac{C_r}{2}\right) C(1,j+1) + S(1,j+1) =$$
$$\frac{C_r}{2} C(0,j) + \left(1 - \frac{C_r}{2}\right) C(1,j) + S(1,j). \tag{14b}$$

Equations (10), (14a), and (14b) are used to compute the concentration at z = 0. At the rest of the locations, the scheme reduces to a single equation in one unknown.

In the case of the finite source boundary condition, eq. (3c) is first differentiated with respect to time to yield

$$\frac{\partial C(0,t)}{\partial t} = -\frac{n_a}{h_f}\left[V C(0,t) - D_s \frac{\partial C(0,t)}{\partial z}\right]. \tag{15a}$$

Writing eq. (15a) in finite-difference form with a forward difference in time and a centred difference in space and making use of eq. (8), we get two sets of equations with two unknowns as follows

$$\left[1 + \frac{C_r}{2} + \frac{C_r}{2}\left(\frac{4 h_f \Delta z}{n_a D_s \Delta t} + \frac{2 V \Delta z}{D_s}\right)\right] C(0,j+1) - \frac{C_r}{2} C(1,j+1) =$$
$$\left[1 - \frac{C_r}{2} + \frac{C_r}{2}\left(\frac{4 h_f \Delta z}{n_a D_s \Delta t} - \frac{2 V \Delta z}{D_s}\right)\right] C(0,j) + \frac{C_r}{2} C(1,j) \tag{15b}$$

with eq. (14b) representing the other equation of the set. Note that no adsorption is assumed to take place at z = 0. However adsorption proceeds at $z = 0^+$. Equations (14b) and (15b) are used to compute the concentration at z = 0.

2.2.2. *Boundary Layer Correction*

In the mixing cell model, no exit condition can be specified and the solution obtained is that for a semi-infinite system. At the exit, the boundary condition is often taken as eq. (4b). Parlange and Starr (1975) have shown that the effect of eq. (4b) is to increase solute concentration in the profile in the region $z \approx L$. Solutions found for a semi-infinite column can be corrected to account for this small concentration increase. When P < 4, the relatively short column length produces a significant increase in solute concentration throughout the soil profile and eq. (1) must be solved directly using eq. (4b). For P > 4 it is possible to approximate the effect of eq. (4b) using a boundary layer correction. We find, following a similar procedure to that employed by Barry et al. (1986), that the corrected solution to a finite column is, for $P > 4$

$$C^*(z,t) = C(z,t) - \frac{\exp\left[\beta(z-L)\right]}{\beta}\frac{\partial C}{\partial z}\bigg|_{z=L} \tag{16a}$$

where C is the solution obtained for a semi-infinite spatial domain. The corrected solid phase concentration is

$$S^* = f(C^*). \tag{16b}$$

Since eqs. (16a) and (16b) must satisfy eq. (1), the value of β is closely approximated by

$$\beta = \frac{V}{2D}\left[1 + \sqrt{\left(1 - \frac{4\gamma D}{V}\right)}\right], \tag{16c}$$

where γ is,

$$\gamma = \frac{\dfrac{\partial^2 C}{\partial z^2}\bigg|_{z=L} - \dfrac{D_s}{V}\dfrac{\partial^3 C}{\partial z^3}\bigg|_{z=L}}{\dfrac{\partial C}{\partial z}\bigg|_{z=L}}. \tag{16d}$$

This boundary layer correction is applied when condition (4b) is imposed.

Similarly, for the case of zero concentration at the column exit, another boundary layer correction can be obtained. In this case, the effect of eq. (4c) is to decrease the solute concentration in the profile near $z \approx L$. We find that the corrected solution to a finite column is, for $P > 4$

$$C^*(z,t) = C(z,t) - \exp\left[\beta'(z-L)\right]C(L,t), \tag{17a}$$

where β' is given by

$$\beta' = \frac{V}{2D}\left[1 + \sqrt{\left(1 + \frac{4\gamma' D}{V^2}\right)}\right], \tag{17b}$$

while γ' is

$$\gamma' = \frac{D_s\dfrac{\partial^2 C}{\partial z^2}\bigg|_{z=L} - V\dfrac{\partial C}{\partial z}\bigg|_{z=L}}{C(L,t)}. \tag{17c}$$

This correction is applied when exit condition (4c) is used. Equations (16d) and (17c) are incorporated by using a centred finite-difference scheme for the first, second and third spatial derivatives.

2.3. LAYERED POROUS MEDIA

The same advective-dispersive equation (1) is used to describe transport and reaction of chemical species in layered media. The case of a two-layered medium is treated here. This problem can be easily solved by treating each layer to be semi-infinite, in which case the mixing cell model can be directly applied using the same equation for each layer. The exit concentration in one layer gives the entrance concentration for the next, i.e., continuity of concentration (Shamir and Harleman, 1967). In the exact solution, the additional constraint at the interface between the two layers is the continuity of mass flux. The results of the time-moment analysis given by Barry and Parker (1987) and Barry et al. (1987) were used to obtain the following condition for the mixing cell solution to be close to the exact solution

$$\left| \frac{d_1 d_2 - d_1^2 [1 - \exp(-P_1)] - d_2^2 [1 - \exp(-P_2)]}{d_1 l_1 + d_2 l_2} \right| << 1 \qquad (18)$$

where $d_i = D_i/V_i^2$, $l_i = L_i/V_i$, $P_i = V_i L_i/D_i$, $i = 1, 2$. The condition used to obtain eq. (18) is that the ratio of the difference of the variances of the exact solution and the mixing cell solution to the variance of the exact solution should be small. Note that condition (18) is valid for nonreactive and linearly adsorbing solutes only. However, our numerical checks demonstrate that it is reasonable for nonlinear reactions as well.

2.4. MULTISPECIES TRANSPORT

If the only reactions taking place are adsorption or ion exchange then, for a one-dimensional homogeneous soil column, the basic equations describing the transport of reacting species considering instantaneous and time dependent reaction can be expressed as (e.g., Barry, 1992):

$$\frac{\partial(S_k + C_k)}{\partial t} = D_s \frac{\partial^2 C_k}{\partial z^2} - V \frac{\partial C_k}{\partial z} \qquad k = 1, 2, .., n. \qquad (19)$$

To solve eq. 19, an additional n equations are required. The chemical reaction laws describing the ion exchange process or other surface reaction laws provide these equations. For example, if only ion exchange is considered, then among the n exchanging ions in the system, there are n-1 independent equilibrium expressions of the form (e.g., Valocchi, 1981):

$$K_{lk} = \left(\frac{S_l^*}{C_l}\right)^{v_k} \left(\frac{C_k}{S_k^*}\right)^{v_l}, \qquad (20)$$

where

$$S_k^* = \frac{S_k}{CEC} \qquad (21)$$

The additional equation is provided by the cation exchange capacity (CEC), i. e.,

$$CEC = \sum_{k=1}^{n} S_k. \tag{22}$$

The combination of eqs. 21 and 22 result in the n multispecies exchange isotherms:

$$S_k = f_k(C_1, C_2, .., C_n). \qquad k = 1, 2, .. , n. \tag{23}$$

To solve eq. 19, we specify initial and boundary conditions that correspond to a step function input at the boundary z = 0

$$\begin{aligned} C_k(z, 0) &= C_{kn}, \quad z > 0, \\ C_k(0, t) &= C_{kf}, \quad t \geq 0. \end{aligned} \tag{24}$$

The system of nonlinear equations (19) - (23) with initial and boundary conditions (24) constitute the multispecies equilibrium model and are here solved by improved mixing cell model. The improved mixing cell solution for this is described below.

2.5. IMPROVED MIXING CELL SOLUTION OF MULTISPECIES TRANSPORT

Upon applying the improved mixing cell method to the multispecies transport equation, we obtain,

$$\begin{aligned} \left(1 + \frac{C_r}{2}\right) C_k(i,j+1) + S_k(i,j+1) &= C_k(i,j) + S_k(i,j) \\ + \frac{C_r}{2}\left[C_k(i-1,j+1) + C_k(i-1,j) - C_k(i,j)\right], &\quad k = 1, 2, .. , n. \end{aligned} \tag{25}$$

By making use of eq. (23), eq. (25) reduces to a set of n nonlinear equations with $C_k(i,j+1)$ as the unknowns which is solved by Newton's method.

3. Comparison with Crank-Nicolson method

The results of kinetic nonequilibrium simulations are first discussed followed by the results obtained by incorporating the time-dependent surface boundary conditions. The two layer problem is then discussed considering linear and nonlinear adsorption isotherms.

3.1. KINETIC NONEQUILIBRIUM

The problem solved by the improved mixing cell is eq. (1) with initial condition (2), the boundary conditions (3a), (4a) and an S-curve isotherm given by (5c). The reactive transport problem was also solved by the Crank-Nicolson method [error of $O(\Delta z^2, \Delta t^2)$]. The Crank-Nicolson scheme is a common way to solve eq. (1) numerically. The set of nonlinear algebraic equations, obtained upon applying the method, was solved by iteration. Note that the spatial domain used in the Crank-Nicolson scheme was large enough so that the results were unaffected by the exit condition imposed.

We considered a nonlinear S-curve isotherm to test the accuracy of the mixing cell model. This isotherm represents qualitatively the isotherms likely to be encountered in practice. The breakthrough curve predictions of the Crank-Nicolson scheme and those obtained from the improved mixing cell model are compared in Figure 1. It is clear that the mixing cell model predictions are in excellent agreement with the Crank-Nicolson scheme.

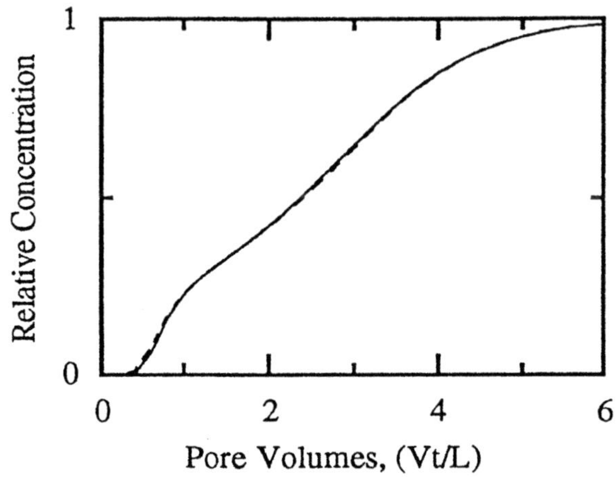

Figure 1. Comparison of the predictions of the improved mixing cell model (dashes) with those of the Crank-Nicolson scheme (line) for the S-curve eq. (5c) ($P = 16$, $\alpha L/V = 1.6$, $k_1 = 3$, $k_2 = 3.375$, $k_3 = -2$, $k_4 = 1.5299$).

3.2. TIME DEPENDENT BOUNDARY CONDITIONS

We show, in Figures 2a and 2b, the mixing cell and the Crank-Nicolson solutions for the transport equation (1), with the S-curve isotherm (same as used in Figure 1) and entrance boundary conditions (3b) and (3c), and exit conditions (4b) and (4c) respectively at a dimensionless time $V^2t/D_S = 40$. Note that, the boundary layer correction, eq. (16a), was applied to the mixing cell solution in Figure 2a whereas eq. (17a) was applied in the mixing cell solution in Figure (2b).

Observe that the mixing cell predictions are close to the Crank-Nicolson solution. It is also found that the mixing cell solution was faster by at least an order of magnitude as compared with the time taken by the Crank-Nicolson scheme.

3.3. TWO LAYERED POROUS MEDIA

Equation (1) is solved with the initial condition (2) and with a step input boundary condition (3a) for the adsorption isotherm given by eq. (5). For the linear isotherm it was found that the expression on the left hand side of the inequality of eq. (18) had to be less than 0.0312 in order to get the accurate solutions using mixing cell models. This corresponds approximately to a layer Peclet number of 16. For the nonlinear isotherm followed by linear adsorption isotherm, the layer Peclet number of 16 also gave results close to that obtained from the exact solution using Crank-Nicolson scheme. Note that, at the interface, continuity of concentration as well as mass flux were assumed in the Crank-Nicolson solution where as only continuity of concentration was considered in the mixing cell solution.

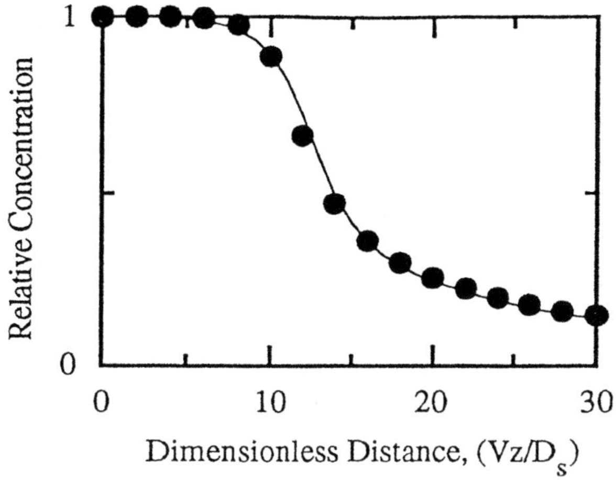

Figure 2a. Concentration profiles simulated using the mixing cell (solid circles) and Crank-Nicolson scheme (line) considering boundary conditions (3b) and (4a) ($P = 30$, $V^2t/D_s = 40$).

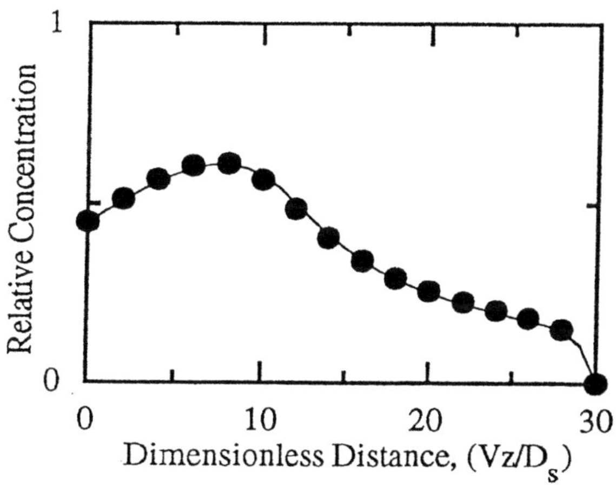

Figure 2b. Concentration profiles simulated using mixing cell (solid circles) and Crank-Nicolson scheme (line) considering boundary conditions (3c) and (4b) ($P = 30$, $Vh_f/D_s = 20$, $n_a = 0.4$, $V^2t/D_s = 40$).

The results for a two layered porous medium are shown in Figure 3 for the nonlinear S-curve followed by a linear isotherm. The constants of the S-curve used were the same as those shown in the figure caption of Figure 1. A retardation factor, R = 1.145 was used for a linear isotherm and $V_1L_1/D_1 = V_2L_2/D_2 = 16$. The input considered was a pulse type with a duration of 2.4 pore volumes. The case of two layers is presented here but the scheme can be applied for any number of layers.

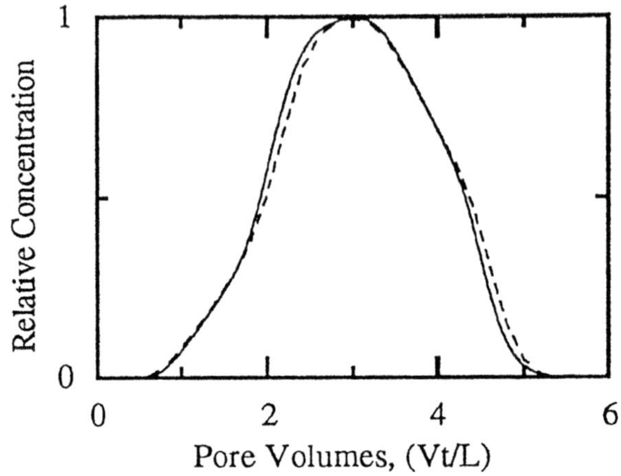

Figure 3. Breakthrough curves simulated using the mixing cell (dashes) and Crank-Nicolson scheme (line) considering boundary conditions (3a) and (4b) (P = 50, R = 1.145).

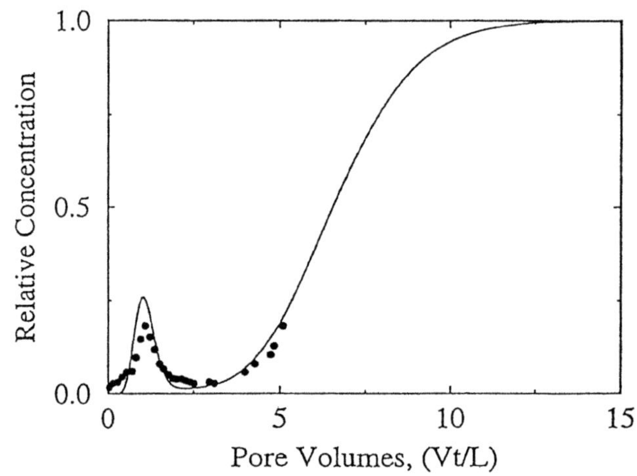

Figure 4. Mixing cell prediction (line) and observed (circle) breakthrough curve of Ca

3.4. APPLICATION OF MULTISPECIES TRANSPORT TO LABORATORY DATA

The precursor effect occurs when a dilute influent solution displaces a high-concentration resident solution. Some of the influent material is transported with the resident solution and appears around 1 pore volume in the breakthrough curve. Barry et al. (1987a) have given the detailed numerical analysis of this effect. Here the same problem is treated using the improved mixing cell model. The labelled and unlabelled Ca is considered as two different species in the ion exchange isotherm. Here the column initially contains unlabelled Ca at the concentration of 2 eq l^{-1}.and the labelled Ca is zero initially in the interstitial liquid. The influent solution contains labelled Ca of concentration 0.01 eq l^{-1} but the concentration of unlabelled Ca is nil. The length of the column was 2.55 cm. The cation exchange capacity of the soil matrix was 0.055 meq ml^{-1} and the selectivity coefficient is assumed to be unity. Figure 4 shows the improved mixing cell prediction of the precursor effect for a Peclet number of 25.5. This result is in excellent agreement with the solution of Barry et al. (1987a) (not shown in figure). The figure also shows the experimental results of Starr et al. (1982).

4. Conclusions

It has been demonstrated that the improved mixing cell model is second-order accurate even when reactions are included. Because the improved mixing cell is unconditionally stable, it allows a larger time step to be used than the standard mixing cell model, with a commensurate reduction in computation time. Not surprisingly, it was found that the Crank-Nicolson scheme requires at least an order of magnitude more time than the mixing cell computation.

The boundary conditions (3b) and (3c) when incorporated in the improved mixing cell model remove the explicit nature of the scheme at the entrance boundary. At the rest of the locations the scheme is still explicit.

Mixing cell models have been used to simulate finite column miscible displacement experimental results without the boundary layer correction. We have demonstrated that mixing cell models can be used to simulate finite columns with different exit conditions by making use of boundary layer corrections. We have shown also that different surface conditions can be incorporated without difficulty and that the case of finite source boundary condition, which is often used in the landfill modelling studies, can be successfully simulated by the mixing cell model. The improved mixing cell can be successfully applied for the case of a multilayered porous medium. For mixing cell models to give accurate results, the layer Peclet number should each be greater than 16. We have demonstrated that the improved mixing cell model can be easily applied to the multispecies transport case.

At lower Peclet numbers, differences between the more accurate Crank-Nicolson scheme and the mixing cell results are generally less than the experimental errors (Villermaux, 1981). Since mixing cell models are simple to code and fast to compute, they are, hence, very attractive tools for simulating reactive transport of solutes. However, since the spatial step size should be close to $2D_s/V$, the method fails when V is very small or D_s is very large.

Notations

C_r	Courant number, $V\Delta t/\Delta z$.
C	normalised liquid phase concentration.
C^*	boundary layer corrected normalized liquid phase concentration.
C_k	liquid phase concentration of kth species.
C_o	normalised influent liquid concentration.
d_i	D_i/V_i^2
D_s	dispersion coefficient, L^2T^{-1}.
f(C)	function of C (adsorption isotherm).
h_f	equivalent leachate thickness in the landfill, L.
k_i	isotherm fitting parameter used in (39) and (40) (i = 1,2,3,4).
K_{lk}	selectivity coefficient.
l_i	L_i/V_i for layer i = 1, 2.

n_a porosity of the landfill liner.
P_i column Peclet number for layer i = 1, 2, V_iL_i/D_i.
R retardation factor.
S normalised solid phase concentration (equivalent liquid phase).
S^* corrected normalised solid phase concentration (equivalent liquid phase).
S_k Solid phase concentration of kth species
t time, T.
V mean pore water velocity, LT^{-1}.
z distance below soil surface, L.
α rate constant, T^{-1}.
β constant given by eq. (16c), L^{-1}.
γ constant given by eq. (16d), L^{-1}.
β' constant given by eq. (17b), L^{-1}.
γ' constant given by eq. (17c), T^{-1}.
Δt time step, T.
Δz spatial step, L.

References

Bajracharya, K., and Barry, D. A., 1993a. Mixing cell models for nonlinear equilibrium single species adsorption and transport. *J. Contam. Hydrol.*, .12, 227 - 243.

Bajracharya, K., and Barry, D. A., 1993b. Mixing cell models for nonlinear nonequilibrium single species adsorption and transport. *Water Resour. Res.*, .29, 1405-1414.

Bajracharya, K., and Barry, D. A., 1993c. Comments on "Sorption of Cadmium and Copper from an acid mine waste extract by two calcareous soils: Column studies" by L. M. Dudley, J. E. Mclean, T. H. Furst, and J. J. Jurinak. *Soil Science,* 1, 68 - 70.

Barry, D. A., 1992. Modelling contaminant transport in the subsurface: Theory and computer programs, in *Modelling Chemical Transport in Soil: Natural and Applied Contaminants.* H. Ghadiri and C. W. Rose (eds.), Lewis Publishers, Chelsea, Michigan, pp. 105 - 144.

Barry, D. A., Parlange, J.-Y., and Starr, J. L., 1986. Interpolation method for solving the transport equation in soil columns with irreversible kinetics. *Soil Sci.,* 142(5), 296-307.

Barry, D. A., and Parker, J. C., 1987. Approximations for solute transport through porous media with flow transverse to layering. *Transp. Porous Media,* 2, 65-82.

Barry, D. A., Parlange, J.-Y., Rand, R., and Parker, J. C., 1987. Addendum to "Approximations for solute transport through porous media with flow transverse to layering". *Transp. Porous Media,* 2, 421-423.

Barry, D. A., Parlange, J.-Y., and Starr, J. L. 1987a. Numerical analysis of the precursor effect. *Soil Sci.* 143, 309-317.

Barry, D. A., and Sposito, G., 1988. Application of the convection-dispersion model to solute transport in finite soil columns. *Soil Sci. Soc. Am. J.,* 52(1), 3-9.

Brenner, H., 1962. The diffusion model of longitudinal mixing in beds of finite length, Numerical values. *Chem. Eng. Sci.,* 17, 229-243.

Clancy, K. M., and Jennings, A. A., 1988. Experimental verification of multicomponent ground water contamination predictions. *Water Resour. Bull.,* 24, 307-316.

Dudley, L. M., Mclean, J. E., Furst, T. H., and Jurinak, J. J., 1991. Sorption of cadmium and copper from an acid mine waste extract by two calcareous soils: Column studies. *Soil Sci.*, 151(2), 121-135.

Kreft, A., 1981. On the boundary conditions of flow through porous media and conversion of chemical flow reactors. *Bull. de l' Acad. Polonaise des Sci.*, Series des Sciences techniques, 29(9-10), 521-529.

Kreft, A., and Zuber, A., 1978. On the physical meaning of the dispersion equation and its solutions for different initial and boundary conditions. *Chem. Eng. Sci.*, 33, 1471-1480.

Noye, J., 1982. Finite difference methods for partial differential equations. In J. Noye (ed.), *Numerical Solutions of Partial Differential Equations*, pp. 3-137, North-Holland Publishing Company, Amsterdam, Netherlands.

Parlange, J.-Y., and Starr, J. L., 1975. Linear dispersion in finite columns. *Soil Sci. Soc. Am. Proc.*, 39, 817-819.

Rao, B. K., and Hathaway, D. L., 1989. Three-dimensional mixing cell solute transport model and its application. *Ground Water*, 27(4), 509-516.

Rowe, R., K., and Booker, J. R., 1985. 1-D pollutant migration in soils of finite depth. *J. Geotech. Eng.*, 111, 479-499, 1985.

Rubin, J., 1983. Transport of reacting solutes in porous media: Relation between mathematical nature of problem formulation and chemical nature of reactions. *Water Resour. Res.*, 19(5), 1231-1252.

Selim, H. M., Davidson, J. M. and Rao, P. S. C., 1977. Transport of reactive solutes through multilayered soils. *Soil Sci. Soc. Am. J.*, 41, 3-10.

Schweich, D., and Sardin, M., 1981. Adsorption, partition, ion exchange and chemical reaction in batch reactors or in columns. *J. Hydrol.*, 50, 1-33.

Shamir, U. Y., and Harleman, D. R. F., 1967. Dispersion in layered porous media. *Proc. Am. Soc. Civil Eng. Hydr. Div.*, 93, 236-260.

Starr, J. L., Parlange, J.-Y., and Braddock, R. D. 1982. Dispersion in soil columns: The precursor effect. *Soil Sci.* 133, 111-115.

Valocchi, A. J., Street, R. L., and Roberts, P. V., 1981. Transport of ion-exchanging solutes in groundwater: Chromatographic theory and field simulation. *Water Resour. Res.* 17, 1517 - 1527.

van Genuchten, M. Th., and Parker, J. C., 1984. Boundary conditions for displacement experiments through short laboratory soil columns. *Soil Sci. Soc. Am. J.*, 48, 703-708.

Van Ommen, H. C., 1985. The "mixing-cell" concept applied to transport of non-reactive and reactive components in soils and groundwater. *J. Hydrol.*, 78, 201-213.

Villermaux, J., 1981. Theory of linear chromatography. In A. E. Rodrigues and D. Tondeur (eds.). pp. 83-140. *Percolation Processes: Theory and Applications*. NATO Advanced Study Institute. Espinho, Portugal, 17-29 July 1978. Sijthoff and Noordhoff, Alphen Aan Den Ryn, The Netherlands.

Yortsos, Y. C., 1987. The relationship between immiscible and miscible displacement in porous media. *AIChE. J.*, 33, 1912-1915, 1987.

HEAVY METAL ADSORPTION IN SOIL: COMPARISON OF BISOLUTE ADSORPTION MODELS AND LABORATORY EXPERIMENTS

K. Bajracharya, D. A. Barry, *Centre for Water Research, Department of Environmental Engineering, University of Western Australia, Nedlands, Western Australia 6009,*
S. Vigneswaran, *School of Civil Engineering, University of Technology, Sydney , NSW 2007,* and
A. Das Gupta, *Division of Water Resources, Asian Institute of Technology, G.P.O. Box 2754, Bangkok 10501, Thailand.*

ABSTRACT. The effect of Zn and NH_4 on the adsorption of Cd has been studied at a pH of 6 for two soils. The experiments were carried out for concentrations of Zn and NH_4 in the ranges present in most industrial wastewaters. The competitive experiments were conducted also with wastewater from a zinc refinery. It was found that both Zn and NH_4 suppressed the ability of Cd to adsorb onto soil. The adsorption of Cd and Zn was determined to be less in the refinery wastewater than that in the single solute batch experiments. Different theories have been proposed to predict the equilibrium concentrations of solutes in the bisolute system using single solute isotherm constants. Four simple mathematical models to predict equilibrium concentrations in the bisolute system have been tested for their applicability, viz., Competitive Freundlich Adsorption Model (CFAM), Modified Competitive Freundlich Adsorption Isotherm Model (MCFAM), Ideal Adsorbed Solution Theory (IAST), and the one given by Sheindorf et al. (1981), (SM). The MCFAM and the SM were found to be satisfactory in predicting the concentrations in the bisolute system and at the same time more easily incorporated into bisolute transport models.

1. Introduction

The fate of the pollutants in the soil-water environment is determined by the environmental conditions with which they are associated. For example, pH is one of the most critical factors in governing the distribution of Cd between solid and fluid phases in the soil. In the pH interval 4 to 7.7, the adsorption capacity of the soils increased 2 to 3 times for a pH increase of one unit (Christensen, 1984). It has also been found that for Cd the predominant process taking place at a pH of 6 is adsorption.

There are numerous compounds in landfill leachates and industrial wastewater effluents. The soil also might contain various cations already adsorbed onto its adsorption sites. The simultaneous presence of all elements and compounds inhibit or accelerate the processes of adsorption, complexation, and precipitation taking place (Griffin et al., 1976).

The main objective of this work is to study the effect of Zn and NH_4 on the adsorption of Cd in soils at a pH of 6. A MCFAM model is developed to predict the equilibrium concentrations of solutes in the bisolute system and is compared with the three other available bisolute models, viz., CFAM, IAST, and SM.

V. P. Singh and B. Kumar (eds.), Water-Quality Hydrology, 19–26.

2. Materials and Method

Experimental soil and sand obtained from Saraburi, Thailand were used in this study. The soil was air dried and the less than 2mm fraction was used in the experiments. The soil characteristics were analysed by standard procedures (Klute, 1986). The proportion of sand, silt and clay were determined by the hydrometer method. The pH of the soil suspensions in water was measured by a glass electrode (HM - 26S, TOA Electronics Ltd.). The pycnometer method was used to determine the particle density. The permeability was determined by a constant head experiment. The ammonium saturation method was used to determine the cation exchange capacity. The approximate method (Jacobs et al., 1971) was used to determine the organic matter in soil. The sand and soil characteristics are given in Table I.

TABLE I. Characteristics of sand and soil.

Characteristic	Sand	Soil
Sand (%) < 2.000 mm	100	67.5
Silt (%) < 0.020 mm		17.5
Clay (%) < 0.002 mm		15.0
pH	8.7	7.9
Particle density (g/cm^3)	2.65	2.58
Permeability (cm/s)	0.212	0.00245
CEC (meq/100g)	1.6	14.2
Organic matter (%)	0.57	3.63

TABLE II. Zinc wastewater characteristics.

Parameters	
COD (mg/L)	12.3
BOD (mg/L)	5.6
pH	6.9
Cd (μg/L)	43.1
Zn (mg/L)	3.0
NH_4 (mg/L)	1.0

It is not known what cations were initially present in the adsorption sites of the soils but the study consisted of conducting adsorption experiments with the experimental soil in single solute (Cd, Zn, NH_4) and bisolute systems (Cd and Zn or Cd and NH_4) treating the system as a black box. It was determined, however, that the soil contained negligible Cd initially.

Chlorides of Cd, Zn and NH_4 were used in the preparation of standard Cd, Zn, and NH_4 solutions respectively. Reagent grade chemicals were used in preparing standard solutions as given in Standard Methods for the Examination of Water and Wastewater (APHA, AWWA, WPCF, 1985).

Zinc refinery wastewater was also equilibrated with sand and soil to study the competitive adsorption process. The wastewater characteristics are shown in Table II.

The single solute equilibrium adsorption experiments involved the equilibration of one gram of the soil samples with 200 ml of Cd, Zn, and NH_4 solutions of known concentrations. The experiments were carried out for different concentrations of Cd, Zn, and NH_4. The pH of 6 was maintained throughout the experiment. The pH was adjusted by adding HNO_3 and NaOH. For the determination of liquid phase concentration, the supernatant was filtered and acidified (pH range 2-4) to enable measurement of the total concentration of Cd and Zn by Atomic Absorption Spectrophotometry. Titrimetric methods were used for determination of NH_4 . The solid phase concentration was computed from the relation (symbols are collected in the notation list):

$$S_{io} = \frac{(C_{oi} - C_{io})\,V}{M} \tag{1}$$

Competitive adsorption experiments were conducted in the sand and soil by varying the initial Cd concentration while keeping the initial concentration of Zn constant. Zn concentrations of 1.2 and 5.0 mg/L were used for sand and soil respectively. When the effect of NH_4 on Cd adsorption was studied, NH_4 concentration was kept at 50 mg/L. The effect of NH_4 was studied only for sand. In all cases the different Cd concentrations used were 100, 200, 400, 600, 800, and 1000 μg/L.

Batch adsorption experiments were conducted with zinc wastewater in sand and soil by equilibrating 1 gram of sand and soil with 200 ml and 100 ml of wastewater respectively. In each of the adsorption experiments, the reactors consisted of 250 ml pyrex glass flasks. Before use, reactors were thoroughly cleaned in nitric acid, then rinsed in distilled water and oven dried.

3. Empirical Competitive Models

The Freundlich isotherm is suitable for highly heterogeneous surfaces of the adsorbent (Fritz and Schlunder 1981). A number of competitive models have been reported by Murali and Aylmore (1983), Crittenden et al. (1985), Fettig and Sontheimer (1987), and Sheintuch and Rebhun (1988). A MCFAM is described below along with the available three other models CFAM, IAST, and SM.

3.1 COMPETITIVE FREUNDLICH ADSORPTION MODEL (CFAM)

For a single solute system, Freundlich equilibrium isotherm is given by:

$$S_{io} = K_i C_{io}^{N}. \tag{2}$$

The empirical competitive Freundlich isotherm for solute 'i' is given by Murali and Aylmore (1983):

$$S_i = \frac{S_{io} K_i C_{io}}{\sum_{i=1}^{n} K_i C_{io}}. \tag{3}$$

The rationale behind eq. 3 is that the distribution of solutes in the ratio $K_i/\Sigma K_i$. For the bisolute case, solid phase concentration of species 1 can be written as:

$$S_1 = \frac{S_{1o}}{1 + \phi \dfrac{K_2 C_{2o}}{K_1 C_{1o}}}, \tag{4}$$

where an arbitrary fitting coefficient, ϕ, has been introduced. This model shall be referred to as Competitive Freundlich Adsorption Model (CFAM). Observe that the equilibrium values of the solid phase concentrations of the solutes in the bisolute system are predicted from the equilibrium liquid and solid phase concentrations of the solutes in the single solute system.

In order to compute the liquid phase equilibrium concentration in the bisolute system, the following mass balance is used:

$$MS_i + VC_i = MS_{io} + VC_{io} = VC_{oi}. \tag{5}$$

Thus, if the initial solute concentrations, amount of adsorbent and amount of solution are known, then the equilibrium solid and liquid phase concentrations of solutes can be computed easily by any standard numerical scheme for solving nonlinear equations.

3.2 MODIFIED COMPETITIVE FREUNDLICH ADSORPTION MODEL (MCFAM)

The model presented above necessitates the computation of both solid and liquid phase single solute equilibrium concentrations in addition to the single solute isotherm constants. The model (MCFAM) described below eliminates this extra step of computation.

In this approach, the competitive bisolute model is approximated as:

$$S_1 = \frac{K_1 C_1^{N_1}}{1 + \phi \dfrac{K_2 C_2}{K_1 C_1}} \tag{6}$$

In eq. 6, both liquid phase and solid phase concentrations are the equilibrium concentrations in the bisolute system. However, the relation makes use of the single solute isotherm constants with the fitting parameter ϕ. In addition to the single solute isotherm constants a few bisolute experiments are necessary to determine the competition coefficient ϕ. Equations 5 and 6 constitute a system of nonlinear equations which can be easily solved numerically.

3.3 IDEAL ADSORBED SOLUTION THEORY (IAST)

Crittenden et al. (1985) used IAST to predict multicomponent adsorption equilibria in activated carbon. The same principle has been used for the case of multisolute equilibria in soil. The final equation for the invariant liquid phase concentration of solute i in the mixture of n solutes in equilibrium is given by:

$$C_i = \frac{S_i \left(\sum_{j=1}^{n} \dfrac{S_j}{N_j} \right)^{\frac{1}{N_i}}}{\left(\sum_{j=1}^{n} S_j \right) \left(\dfrac{K_i}{N_i} \right)^{\frac{1}{N_i}}}. \tag{7}$$

Equations 5 and 7 again form a system of nonlinear equations which can be easily solved numerically.

3.4 Sheindorf et al. (1981) Model (SM)

The multicomponent Freundlich isotherm derived by Sheindorf et al. (1981) is given by:

$$S_1 = K_1C_1\left(C_1 + \phi_{12}C_2\right)^{N_1 - 1}$$
$$S_2 = K_2C_2\left(C_2 + \phi_{21}C_1\right)^{N_2 - 1} \quad (8)$$
$$\phi_{12} = \frac{1}{\phi_{21}}$$

In addition to the single solute constants a few bisolute experiments are required to evaluate the coefficients ϕ_{12} and ϕ_{21}. Equations 5 and 8 again form a system of nonlinear equations which can be solved numerically.

The fixed point iterative method was used to solve all the above competitive adsorption models in this study.

4. Results and Discussion

At a pH of 6, a significant amount of Cd, Zn, and NH_4 was adsorbed on sand and soil. Sand adsorbed less than that adsorbed by the soil. It is evident that the soil contained clay and more organic matter than that in sand. This is responsible for the higher adsorption in the soil. The single solute isotherm parameters are shown in Table III, below.

TABLE III. Single solute Freundlich isotherm parameters.

Solute	Parameter	Sand	Soil
Cd	K	0.90	3.22
	N	0.80	0.84
Zn	K	0.53	5.34
	N	0.83	0.72
NH_4	K	0.22	3.04
	N	0.82	0.59

The effect of Zn on Cd adsorption is shown in figures 1 and 2. In figure 1, the value of the competition coefficient for CFAM, MCFAM, and SM were, respectively, 0.025, 0.12, and 0.1. The equilibrium Cd concentrations in single and bisolute systems for sand and soil are shown in the same figures along with the predictions by the four competitive adsorption models. In figure 2, the value of the competition coefficient for CFAM, MCFAM, and SM were, respectively, 0.0005, 0.05, and 0.2. The adsorption ability of Cd on the soil has been suppressed by the presence of Zn. The extent of suppression is significant in both the sand and soil. The reduction of Cd adsorption was approximately 20 to 30% in each case. Christensen (1984) has also found that increasing the Ca concentration from 10^{-3} to 10^{-2} M reduces the sorption capacity of the sandy loam by a third.

The predictions for the case of Cd adsorption in sand by all four models were of the same order of accuracy. The IAST model cannot be fitted to the data because it does not involve any fitting parameter but involves the isotherm parameters which are fixed as found from the single solute results.

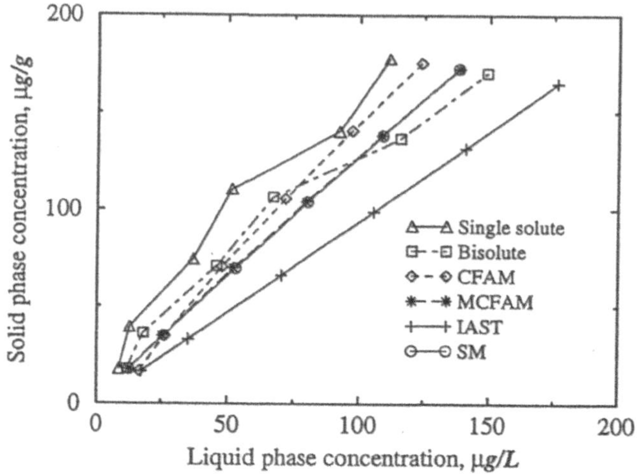

Figure 1. Effect of Zn on adsorption of Cd in sand with model simulations

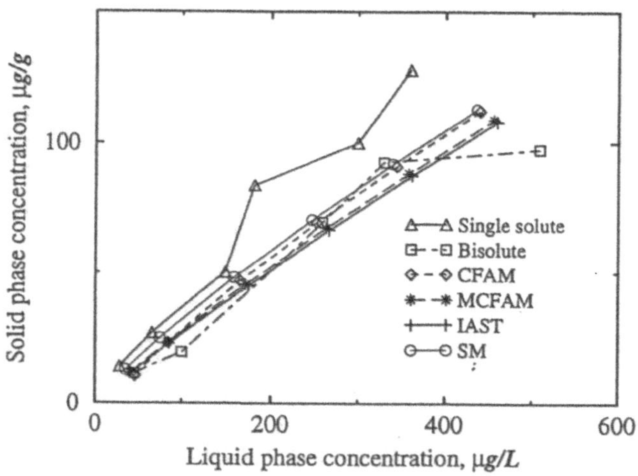

Figure 2. Effect of Zn on adsorption of Cd in soil with model simulations.

The adsorption of Cd was also affected by the presence of NH_4. These experiments were conducted only for sand. The effect of the presence of NH_4 in the adsorption of Cd in sand is shown in figure 3. In figure 3, the value of the competition coefficients for CFAM, MCFAM, and SM were, respectively, 0.012, 0.05, and 0.12. The reduction of total Cd adsorbed was approximately 50%. In this case also the predictions for the case of Cd adsorption in sand by all four models were of the same order of accuracy.

The results of the batch experiments with zinc refinery wastewater are tabulated in Table IV. The adsorption of Cd and Zn was found to be less in Zinc refinery wastewater than that in the single solute batch experiments. In the table also shown is the solid phase concentration for the multisolute system. The solid phase computations are shown based on both competitive and single solute assumptions for comparison purposes. In addition, only predictions from the IAST model are tabulated.

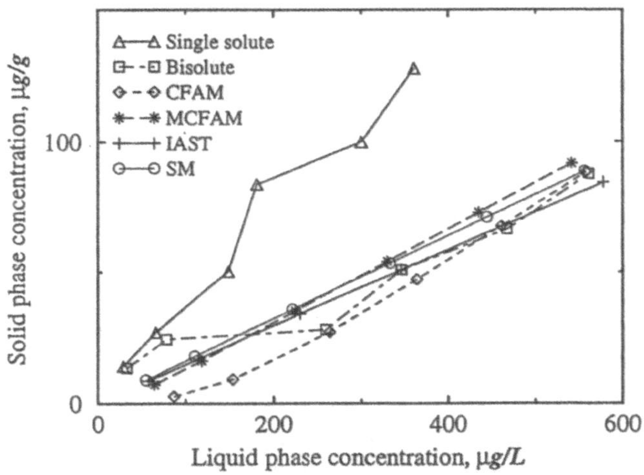

Figure 3. Effect of NH_4 on adsorption of Cd in sand with model simulations.

Of the four models CFAM and IAST are implicit functions of the solid phase concentration whereas the MCFAM and SM are explicit in solid phase concentration. As such the problem is quite simplified when solving the transport equation in conjunction with nonlinear adsorption equations described by such equations of explicit nature.

TABLE IV. Results of batch experiments with Zinc refinery wastewater.

Medium	Solute	Liquid Conc.(µg/L)		Solid Phase Conc. (µg/g)		
		Expt.	Predicted*	Expt.	Predicted*	Computed#
Sand	Cd	26.0	16.5	3.4	5.3	12.2
	Zn	1431.0	865.4	41.8	154.9	220.5
Soil	Cd	29.4	31.9	3.1	5.6	55.1
	Zn	582.0	473.8	31.8	85.2	522.7

* Predicted based on IAST model
\# Computed based on single solute adsorption

5. Conclusions

The effect of Zn and NH_4 on the adsorption of Cd onto sand and soil were found to suppress the adsorption capacity significantly. The four models CFAM, MCFAM, IAST, and SM could predict competitive adsorption process satisfactorily. Of the four models, MCFAM and SM are found to be more easily incorporated in a multispecies transport model because these are explicit function in terms of liquid phase concentration.

Notations

C_i	liquid phase concentration of species 'i' in the bisolute system, µg/L
C_{io}	liquid phase concentration of species 'i' in the single solute system, µg/L
C_{oi}	initial liquid phase concentration of species 'i', µg/L
S_i	solid phase concentration of species 'i' in the bisolute system, µg/g
S_{io}	solid phase concentration of species 'i' in the single solute system, µg/g
K_i	adsorption isotherm parameter, $(\mu g/g)(\mu g/L)^{-N}$
N	adsorption isotherm parameter
M	mass of adsorbent, g
V	volume of solution, L
ϕ	fitting parameter

References

APHA, AWWA, WPCF, 1985. Standard Methods for the Examination of Water and Wastewater, 16th Edition.

Christensen, T. H., 1984. Cadmium soil sorption at low concentrations: I. Effect of time, cadmium load, pH, and calcium. *Water, Air, and Soil Poll.*, 21, 105-114.

Crittenden, J. C., Luft, P., Hand, D. W., Oravitz, J. L., Loper, S. W., and Arl, M., 1985. Prediction of multicomponent adsorption equilibria using ideal adsorbed solution theory, *Env. Sci. Tech.*, 12(11), 1037-1043.

Fettig, J., and Sontheimer, H., 1987. Kinetics of adsorption on activated carbon: II. Multisolute systems, *J. Env. Eng.*, 113(4), 780-794.

Fritz, W. and Schlunder, E. U., 1981. Competitive adsorption of two dissolved organics onto activated carbon., *Chem. Eng. Sci.*, 36, 721-730.

Griffin, R. A., Shimp, N. F., Steele, J. D. Ruch, R. R., White, W. A., and Hughes, G. M., 1976. Attenuation of pollutants in municipal landfill leachate by passage through clay. *Env. Sci. & Tech.*, 10(13), 1262-1268.

Klute, A., 1986. Methods of soil Analysis, Part 1, Physical and Mineralogical Methods, American Society of Agronomy, Inc., Soil Science Society of America, Inc. Madison, Wisconsin, USA.

Jacobs, H. S., Reed, R. M., Thien, S. J., and Withee, L. V., 1971. Soils Laboratory Exercise Source Book, American Society of Agronomy, Madison, Wisconsin, U. S. A.

Murali, V. and Aylmore, L. A. G., 1983. Competitive adsorption during solute transport in soils I. Mathematical models, *Soil Sci.*, 135(3), 143-150.

Sheindorf, C., Rebhun, M., and Sheintuch, M., 1981. A Freundlich type multicomponent isotherm, *J. Colloid. Interface Sci.*, 79, 136.

Sheintuch, M. and Rebhun, M., 1988. Adsorption isotherms for multisolute systems with known and unknown composition, *Water Res.*, 22(4), 421-430.

FAST PREDICTION OF RUNOFF REMOVAL OF CHEMICAL POLLUTANTS UNIFORMLY DISPERSED IN A DRAINAGE BASIN

S.R. JOSHI
Lakes Research Branch
National Water Research Institute
Canada Centre for Inland Waters
Environment Canada
P.O. Box 5050
Burlington, Ontario L7R 4A6
Canada

ABSTRACT. A simple model is described for the prediction of the solid→solution transfer of chemical pollutants uniformly dispersed in a drainage basin. The model is based on our recent finding that the mean residence time of such a contaminant is simply proportional to its dimensionless solid/solution distribution coefficient. The proportionality constant is independent of the nature of the chemical and can be assigned a value using a constant flux, naturally occurring radionuclide. Since the model assumes that the removal primarily occurs in the solution phase, it is important to include runoff removal constant in model expressions. A parameter describing downward leaching of chemicals is also included. This is especially important for chemicals with low distribution coefficients. The model is evaluated for the Ottawa River drainage basin using a conservative (^{90}Sr) and a particle-reactive (^{137}Cs) fallout radionuclide the measured long-term (~ 20 yr) terrestrial fluxes of which compare favourably with those inferred from the model. The observed variance between the two values for few years is noted and discussed.

1. Introduction

It is a well-known fact that chemical pollutants uniformly dispersed in a drainage basin are slowly released to the receiving waters. Such releases have been inferred by trial and error calculations (*e.g.* Lerman, 1972; Korhonen, 1990) and are often predicted by the calibration and simulation techniques (*e.g.* Salo *et al.*, 1984). This approach requires monitoring of contaminant levels in the receiving waters over long periods of time, usually several years, and thus is of limited use in cases (such as the Chernobyl nuclear reactor accident) where information is needed immediately to assess the impact on populations consuming such waters.

The present article describes a simple approach whereby the transfer of contaminants from drainage basin soils to the aquatic compartment can be predicted using much fewer field

V. P. Singh and B. Kumar (eds.), Water-Quality Hydrology, 27–40.

measurements in conjunction with historical waterflow data. This approach largely derives from our recent demonstration (Joshi and Shukla, 1991) that the mean residence time of such a contaminant is simply proportional to its dimensionless solid/solution distribution coefficient. A naturally occurring radionuclide, such as ^{210}Pb (half-life, 22.3 yr), is used to estimate the value of the proportionality constant.

2. Materials and Methods

2.1 STUDY AREA

For the purpose of the present evaluation, Ottawa River water and sediment samples were collected at the Environment Canada sampling station located at Lemieux Island on the outskirts of Ottawa (Fig. 1). The Ottawa River basin covers an area of about 146,000 km^2; the area up to the sampling site is about 90,900 km^2. Approximately 9.9% of the drainage basin is comprised of waterbodies. The average annual precipitation in the watershed is nearly 870 mm, about half of which is released in the form of runoff (OWRC-QWB, 1971). Discharge is regulated through a series of dams and other control structures and excellent historical streamflow summary data are available for various stations (Environment Canada, 1987). Over the 1150-km course of the river, the mean annual discharge increases from 450 m^3 s^{-1} at the head of Lake Temiskaming to about 2060 m^3 s^{-1} near its confluence with the St. Lawrence River. The river spans about 950 km up to Ottawa, and its average width is assumed to be 1 km.

Besides fallout from the atmospheric testing of nuclear weapons, the Ottawa River also receives controlled releases of nuclear effluents from the Chalk River Nuclear Laboratories located, on its Ontario bank, about 190 km upstream from Ottawa. A small nuclear power plant also operated for about 25 years (1962-1987) at Rolphton, nearly 25 km upstream of Chalk River. The releases from this plant are estimated (Joshi and McCrea, 1992) to be <0.1% of those from the CRNL facilities, and were, therefore, deemed negligible for the present evaluation.

2.2 ANALYTICAL METHODS

The analytical aspects of the present study have been detailed earlier (Joshi *et al.*, 1991; Joshi and McCrea, 1992).

2.3 CONTAMINANT TRANSFER MODEL

In a recent communication (Joshi and Shukla, 1991), we have shown that the mean residence time (τ_i, yr) of the *i*th chemical contaminant, homogeneously distributed in the drainage basin soils, is proportional to its dimensionless solid/solution partitioning coefficient, K^*_{di}, *i.e.*

$$\tau_i = \beta K^*_{di} , \qquad (1)$$

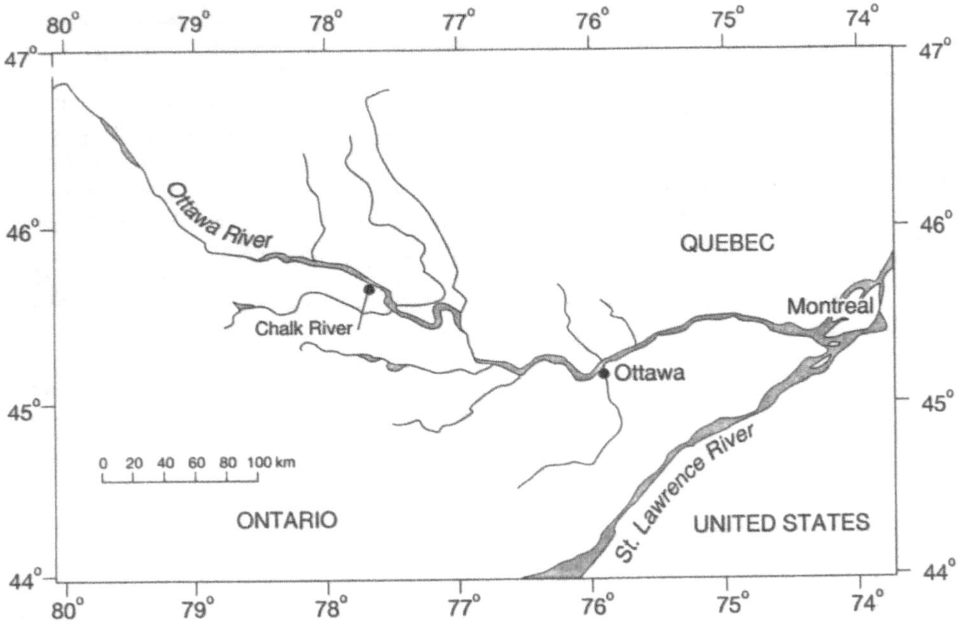

1. Map showing sampling area. The Ottawa River basin is shown in part only.

where $K^*_{di} = \rho K_{di}$, with ρ being the density of soil particles (g cm^{-3}) and K_{di} the calculated ratio (expressed in cm^3 g^{-1}) of contaminant concentration in the solid phase, C_{si} in g g^{-1}, to that in the aqueous phase, C_{wi} in g cm^{-3}; β (yr) is deemed to be constant.

Referring to Fig. 2, a known annual volume (cm^3 yr^{-1}) of precipitation impacts a unit area (1 cm^2) of the landscape receiving atmospheric contaminant flux, F_{ia} (g cm^{-2} yr^{-1}), at a rate given as V_{wi} in cm yr^{-1} (= cm^3 yr^{-1}/cm^2). This precipitation then impacts a certain volume (cm^3) of soil particles per unit area beneath the soil/water interface given as V_{si} (cm = cm^3/cm^2). Thus V_{wi} essentially represents the annual thickness of water above the interface while V_{si} represents its penetration depth. The ratio V_{si}/V_{wi} is given as β.

If F_{ia} (g cm^{-2} yr^{-1}) and F_{ir} (g cm^{-2} yr^{-1}) represent the atmospheric input and removal fluxes, respectively, the rate of change (increase) of contaminant inventory in the soils, I_i (= $C_{si}V_{si}$; g cm^{-2}), is given as

$$\frac{dI_i}{dt} = F_{ia} - F_{ir} . \tag{2}$$

If removal is proportional to I_i, *i.e.* $F_{ir} = k_i I_i$ (= $C_{wi} V_{wi}$), where k_i (=$1/\tau_i$; yr^{-1}) is the removal constant, then

$$\frac{dI_i}{dt} = F_{ia} - k_i I_i . \tag{3}$$

For time-dependent flux,

$$\frac{dI_i(t)}{dt} = F_{ia}(t) - k_i I_i(t) . \tag{4}$$

Since water phase is assumed to play the dominant role in removal, the transfer of contaminants from soils to the receiving surface waters will be delayed depending upon the removal charcateristics of the surface runoff (Fig. 3). This transfer is inversely proportional to the mean residence time of runoff in the drainage basin. The flux from soils to the receiving waters is then described by

$$\frac{dI_{wi}(t)}{dt} = F_{wi}(t) - k_w I_{wi}(t) , \tag{5}$$

where k_w (yr^{-1}) is the rate constant for the removal of surface runoff, $I_{wi}(t)$ is the contaminant inventory
in the runoff waters, and $F_{wi}(t)$ is the contaminant flux from soil to runoff [= $k_i I_i(t)$].

Equations (4) and (5) adequately describe the removal of particle-reactive contaminants, but conservative contaminants, *i.e.* those with low solid/solution distribution coefficients, will also migrate deeper into the soils. In such cases, eqns. (4) and (5) can be modified by taking into

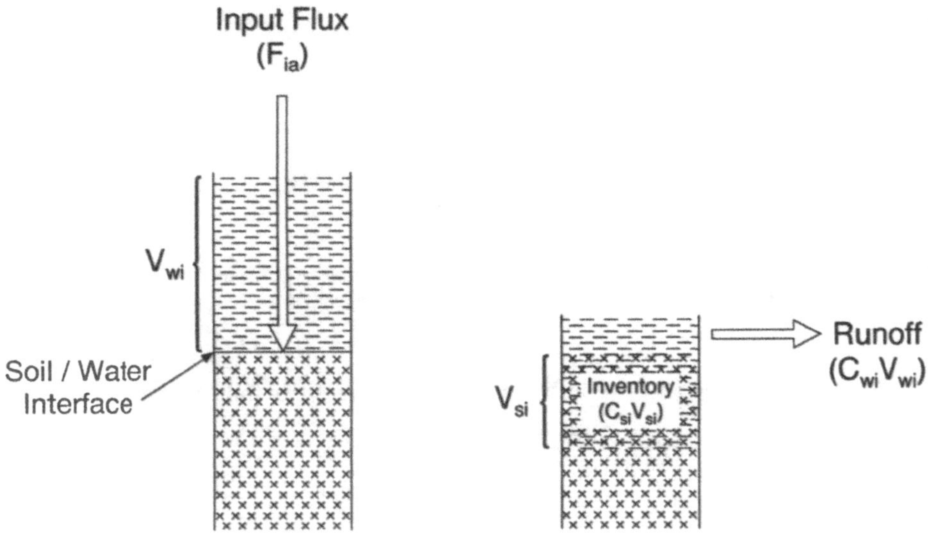

2. **Depiction of model parameters used in deriving equation (1).**

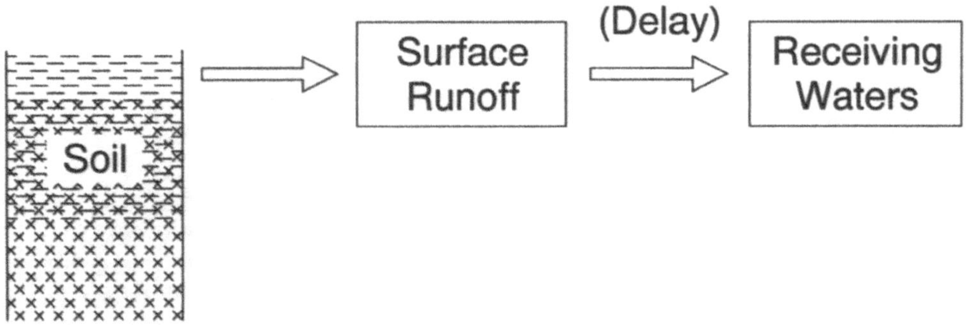

3. The role of surface runoff in contaminant transfer from soils to the receiving surface waters.

account such downward leaching represented by the removal constant k_l (yr^{-1}). Following Fig. 4, if f is the fraction of precipitation V_w leaving as runoff v_w, it can be shown (Joshi, 1994) that

$$k_l = \frac{(1-f)}{\beta f(\phi + K_{di}^{+})}, \qquad (6)$$

where ϕ (dimensionless) represents soil porosity and K_{di}^{+} the dimensionless bulk distribution coefficient of contaminant(= bulk density x K_{di}). The time-dependent inventory of contaminant in the soils is then described as

$$\frac{dI_i(t)}{dt} = F_{ia}(t) - I_i(t)[k_i + k_l]. \qquad (7)$$

The contaminant flux to the receiving surface waters would still be given by eqn. (5) using values for $F_{wi}(t)$ and $I_w(t)$ derived from eqn. (7).

3. Results

The time-dependent contaminant flux to the receiving surface waters can be readily evaluated provided K_{di}, k_i, $F_{ia}(t)$, soil density, soil porosity and hydrological parameters of the watershed are known. The soil density, porosity and K_{di} are readily measured, but a direct, facile estimation of k_i requires a knowledge of β. A direct measurement of β is much complicated, if not impossible. Since β is independent of the nature of chemical contaminant under consideration, its value can be derived (Joshi, 1994) by using a constant-flux, naturally-occurring radionuclide such as ^{210}Pb (half-life, 22.3 yr).

Thus, if k_r (yr^{-1}) is the first-order removal constant of this radionuclide and K_d^{*} its dimensionless field distribution coefficient, then

$$\beta = \frac{1}{k_r K_d''}. \qquad (8)$$

The value of k_r is established from the equation

$$k_r = \frac{\lambda_1 F_r}{(F_c - F_r)}, \qquad (9)$$

where λ_1 (yr^{-1}) is the radioactive decay constant of ^{210}Pb and F_c ($Bq\ cm^{-2}\ yr^{-1}$; 1 Bq = 1 disintegration per second) and F_r ($Bq\ cm^{-2}\ yr^{-1}$) are its atmospheric input (constant) and hydrologic removal fluxes, respectively.

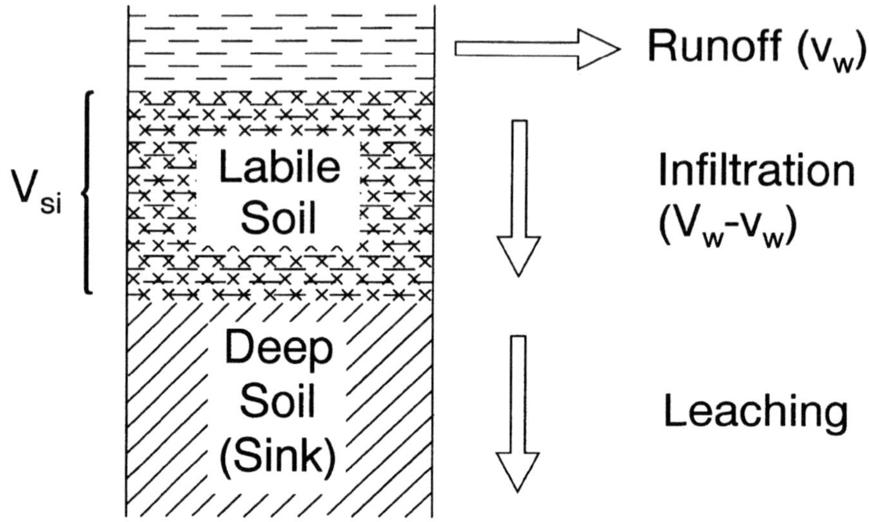

4. Depiction of model parameters considered in deriving equations (6) and (7).

The described model was evaluated using two well-known nuclear weapons testing fallout radionuclides, ^{90}Sr and ^{137}Cs. Both radionuclides are produced in similar yields of ~6% as nuclear fission products and have similar half-lives of 29 and 30.1 yr, respectively. Both, however, have very dissimilar chemistries: ^{90}Sr is a conservative radionuclide displaying very low K_d values of the order of ~500; ^{137}Cs, on the other hand, is highly particle-reactive with K_d values of ~10^5. The historical monitoring data (from 1962 onwards) for the two radionuclides in the Ottawa River waters are also available (Meyerhof, 1984; Health and Welfare Canada, 1984-1986). Thus, for the purpose of the present comparison, the two radionuclides comprise an ideal set of chemicals with common and well-documented production and time release histories.

The $F_{ia}(t)$ and K_{di} values for the two radionuclides in the study area have been computed earlier (Joshi and McCrea, 1992). The dry solid and bulk densities were estimated to be 2.5 and 1.2 g cm^{-3}, respectively, while ϕ was set (A.G. Bobba, National Water Research Institute, Personal Communication, 1991) at 0.3. A value of 0.013 yr was inferred for β from the partitioning studies of ^{210}Pb in this system (Joshi *et al.*, 1991). The value of k_w was set (Brown, 1961) at 0.27 yr^{-1}.

The measured annual fluxes of the two radionuclides from the drainage basin were inferred from their concentrations in the river waters, multiplied by the annual water discharge (Environment Canada, 1987) and divided by the applicable drainage area excluding the water surface area. The concentrations were also corrected, except for some instances in the case of ^{137}Cs (see later in the text), for contributions arising from direct fallout over the river's surface and for inputs from nuclear facilities in the Chalk River area. Figures 5 and 6 show the measured and model-predicted release profiles for ^{90}Sr and ^{137}Cs, respectively, in the Ottawa River waters.

4. Discussion

The measured and predicted fluxes of the drainage basin ^{90}Sr to the receiving waters (Fig. 5) clearly suggest that the model gives satisfactory results though for some years the two values may vary by up to a factor of three. This situation is even more pronounced in the case of ^{137}Cs (Fig. 6) where the two values may differ by up to a factor of 12 for some years. Nonetheless, it is clear that the long-term release predictions show much better overall agreement.

It is interesting to note that the observed and predicted values tend to disagree around the time period of large-scale fallout (1956-1965) of the two radionuclides over the study area (Joshi and McCrea, 1992). Figure 7 shows the radioanalytical data used in the present study along with concentrations expected from direct fallout over the river water surface. It is observed that the ^{137}Cs concentrations in the Ottawa River waters attributable to the direct fallout alone exceed those measured. No such discrepancy is discernible for ^{90}Sr. It is possible that a fraction of the direct fallout ^{137}Cs is removed with the settling particles to river bottom as the Ottawa River does have numerous (water) holding areas. We have obtained several sediment cores from the river and a further evaluation of this possibility must await completion of analytical work currently in progress.

The ^{90}Sr data shown in Fig. 7 clearly shows that the measured concentrations peaked a couple of years later than expected on the basis of the well-known 1963 fallout activity maximum alone. The model predictions shown in Fig. 5 definitely account for this observation,

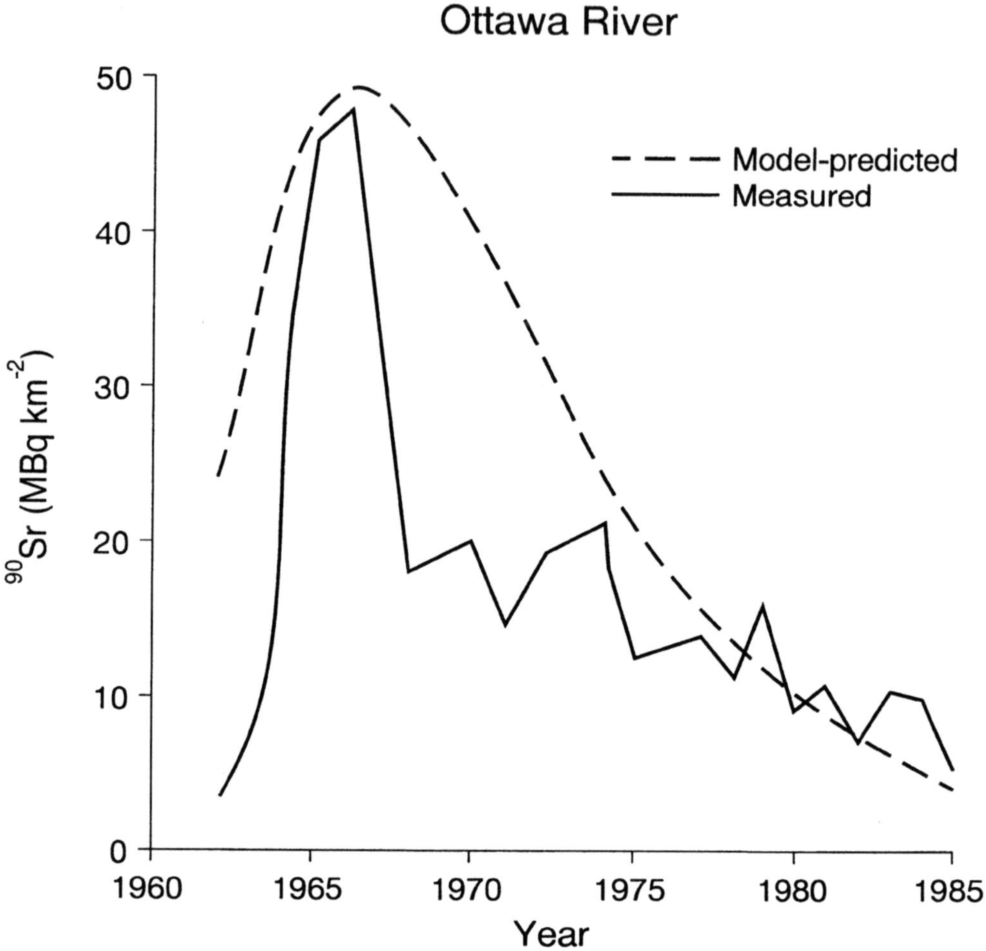

5. Measured and model-predicted drainage basin flux of fallout ^{90}Sr into the Ottawa River.

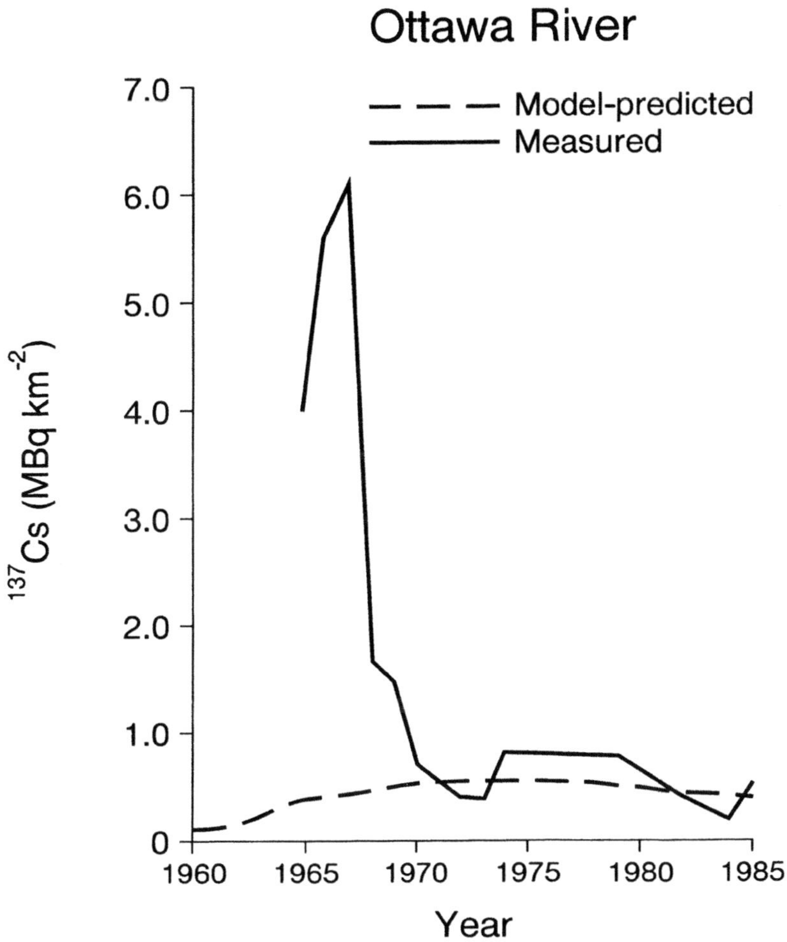

6. Measured and model-predicted drainage basin flux of fallout ^{137}Cs into the Ottawa River.

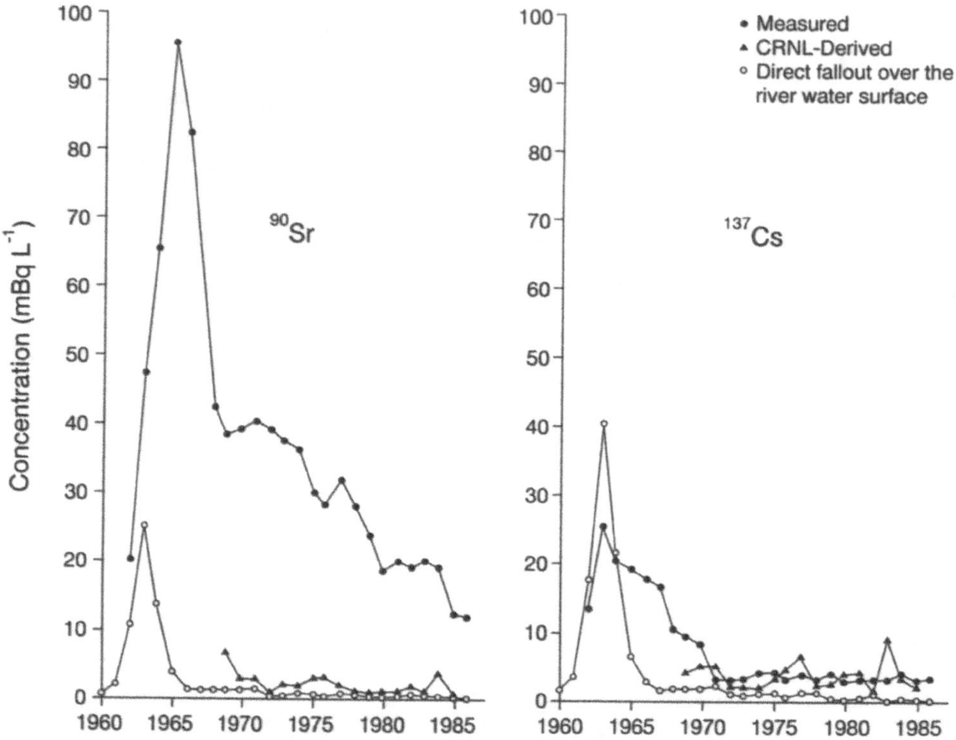

7. Measured concentrations of ^{90}Sr and ^{137}Cs in the Ottawa River waters. Estimated contributions of fallout and CRNL emissions are also given.

thus validating consideration of runoff removal time as a model parameter since the low K_d value of ^{90}Sr ensures that this radionuclide is much more preferentially removed in the water phase (~99%; Joshi and McCrea, 1992) than the particle-reactive ^{137}Cs.

It should be noted that the model does not take into account the fraction of radioactivity retained by plants and wildlife, *etc.* Undoubtedly, consideration of such parameters would lower the predicted runoff removal estimates by at least a few percent. Also, the data used for the present evaluation do not necessarily all correspond to time-series measurements as is implied in the model expressions. Removal of such constraints would perhaps yield even better agreement between predicted and measured values. Nonetheless, even at this stage of development, the concept offers a simple tool for assessment purposes.

5. Conclusion

The results from the present investigation show that the described approach can satisfactorily account for the long-term release of chemicals from soils to the receiving waters. The major advantage of the present approach is that the long-term release patterns can be quickly established from chemical parameters that can be measured in a matters of weeks. This may be contrasted to the commonly used calibration and simulation techniques requiring collection of such data over a number of years.

Such a predictive tool is especially useful in assessing the impacts of accidental releases of hazardous compounds where the long-lived contamination is spread far and wide. The 1986 Chernobyl nuclear power plant accident is a prime example of such a situation. The long-lived radioactivity released during this accident was transported and spread over the European landscape in radiologically significant amounts. However, data are only now beginning to emerge on the release of this radioactivity from soils to the receiving waters (which may also comprise public drinking water supplies in some cases thus providing additional radiation dose).

The observation that predictions for some years vary from measured values by up to several factors may be viewed in the light of discrepancies among predictions by different models themselves. For example, predictions by several leading radiological dose assessment models (where runoff contributions have usually not been included) can vary by factors of up to 30 (Hoffman *et al.*, 1984).

References

Brown, R.M.: 1961, 'Hydrology of tritium in the Ottawa Valley', *Geochim. Cosmochim. Acta* **21**, 199.

Environment Canada: 1987, *Historical Streamflow Summary to 1986 - Ontario*, Ottawa.

Health and Welfare Canada: 1984-1986, *Environmental Radioactivity in Canada*, Environmental Health Directorate, Ottawa, Reports 86-EHD-125, 87-EHD-132 and 87-EHD-136.

Hoffman, F.O., Bergstrom, U., Gyllander, C. and Wilkens, A.-B.: 1984, 'Comparison of predictions from internationally recognized assessment models for the transfer of selected radionuclides through terrestrial food chains', *Nucl. Saf.* **25**, 533.

Joshi, S.R.: 1994, 'Prediction of runoff transport of fallout ^{90}Sr', *Health Phys.* (In Press).

Joshi, S.R. and McCrea, R.C.: 1992, 'Sources and behavior of anthropogenic radionuclides in the Ottawa River waters', *Water, Air,and Soil Pollut.* **62**, 167.

Joshi, S.R., McCrea, R.C., Shukla, B.S. and Roy, J.-C.: 1991, 'Partitioning and transport of lead-210 in the Ottawa River watershed', *Water, Air, and Soil Pollut.* **59**, 311.

Joshi, S.R. and Shukla, B.S.: 1991, 'The role of the water/soil distribution coefficient in the watershed transport of environmental radionuclides', *Earth Planet. Sci. Lett.* **105**, 314.

Korhonen, R.: 1990, 'Modeling transfer of ^{137}Cs in a large Finnish watercourse', *Health Phys.* **59**, 443.

Lerman, A.: 1972, 'Strontium-90 in the Great Lakes: Concentration-time model', *J. Geophys. Res.* **77**, 3256.

Meyerhof, D.P.: 1984, 'Revisions to the guidelines for Canadian drinking water quality - Radiological characteristics', Paper presented at the Fifth Annual Conference of the Canadian Radiation Protection Association, Banff, Alberta, April 30 - May 3.

Ontario Water Resources Commission - Québec Water Board: 1971, *Ottawa River Basin: Water Quality and its Control in the Ottawa River*, Vol. 1.

Salo, A., Saxen, R., and Puhakainen, M.: 1984, 'Transport of airborne ^{90}Sr and ^{137}Cs deposited in the basins of the five largest rivers in Finland', *Aqua Fenn.* **14**, 21.

SECTION 2:

SOLUTE TRANSPORT BY WATERSHED RUNOFF

An Integrated Approach to Observe the Evolution of Pollutants in Reservoirs

V.A. Copertino, B. de Bernardinis, B. Molino, and V. Telesca
Department of Engineering and Environmental Physics
University of Basilicata
85100 Potenza, Italy

V.P. Singh
Department of Civil Engineering
Lousiana State University
Baton Rouge, LA, 70803-6405, USA

Abstract

A method for combining monitoring, mathematical modeling, and the use of a geographical information system (GIS) is proposed to investigate the transport, dispersion and diffusion of pollutants in reservoirs. Preliminary and partial results obtained from an extensive and continous observation and measurement activity on the Camastra reservoir have confirmed the validity of the proposed approach.

Introduction

The transport, dispersion and diffusion of pollutants in receiving water bodies, such as rivers, lakes, estuaries and reservoirs, is controlled by many factors. The most important are the pollutant amount and type, the location of the discharges, the biochemical processes involved, along with the hydraulic beaving of the water body itself. The knowledge of the three-dimensional evolution of the previous factors in reservoirs is essential to predict the quality of the stored water resources and then to program its management . An efficient monitoring and modeling system is needed to investigate the basic physical processes, and, moreover, to understand their hydraulic and biochemical aspects. This is to be done, either on the global scale, including the hydrological and anthropological evolutions of the territories surrounding, as well as on a smaller scale, where the mathematical approach is still meaningful to predict the spatial evolution of the dominant factors.

In the present work, an integrated approach to study the evolution of pollutants in recervoirs is proposed. The method, based on the combined and iterative use of a monitoring system, a mathematical model, and a geographical information system for territorial data is firstly used to design an optimal monitoring network to fully and continously describe the space distribution and time variation of the physical and chemical variables in the reservoir water body. Then, once that a satisfactory and safe configuration is reached, the network is used to extensively observe and investigate the basic behaviour of the transport,

V. P. Singh and B. Kumar (eds.), Water-Quality Hydrology, 43–55.

dispersion and diffusion processes of pollutants and to bead the management of the water resources itself.

A preliminary use of the monitoring system concerned with an extensive and continous observation and measurement activity on the Camastra reservoir, in Basilicata region, Italy, and the obtained results are reported.

Monitoring System

The monitoring system is composed of a mobile station, a mobile laboratory, and a base station, and it is able to transfer information and data to the mathematical models and the geographical information system.

The mobile station is organized in a small boat and equipped with the following devices:

- an ultrasound echo sounder: the ultrasonic echo sounder measures reservoir depth to 1-cm accuracy and transmits measured values in real time to the PC station;
- a multiparameter probe: the Bobcat Seametrix minicomputerized probe measures the temperature and dissolved oxygen. The galvanic measuring device for dissolved oxygen outputs a millivolt signal proportional to the amount of oxygen in the water. Measured values are transmitted in real time to the PC station;
- a PC to store the measured data and to drive the on-line transmission system;
- a transmission data system;
- an optical prism system which continously alouds to state the planimetric position of the mobile station;
- a sampler for off-line monitoring and testing at the mobile laboratory.

The mobile laboratory is equipped with several analyzers including photometers, titer measuring devices, conductivity meters, and carbon analyzers for onsite chemical analysis of samples (BOD, COD, TOC, etc.).

The base station, stationary on the shores, is able to measure continously the geographic coordinates of the mobile station, related to the reference *datum* on the dam. The station contains an MDL engineering laser track, composed of a 3A class infrared beam laser, with a 10-km activity range and an accuracy of 50 ppm, and readout photodiodes with a resolution of 0.01 m. Therefore the optical system error is:

$E_d = \pm 0.01 + 50\, X$ [ppm m], for the measuring point collimation error,

$E_a = \pm 0.00017\, X$ [radiant m], for the angular approximation,

where X is the measured distance expressed in meters.

The monitoring system presents:

- an enhanced accuracy, since the probe-acquired parameters are measured directly on site.
- the capability of measuring a large number of samples in a short time on different sites: to make an almost "istantaneous map" of the distributed variables, logistically if all the samples had to be transported to an offsite laboratory.
- measurement repeatability, since the exact planimetric position of the mobile station is known, as well as the vertical siting of the sampling.

The optimization mathematical model

The optimization model, used to optimize the monitoring activity, analyzes the information losses associated with several sets of measurements in space and time.

The inital information content is defined (Shannon and Weaver, 1949) by entropy $H(X)$ of a discrete random variable X with N elementary events of probability $P_n=p(x_n)$ $(n=1,...,N)$

$$H(X) = -\sum_{i=1}^{N} p(x_i)\log[p(x_i)] \qquad \textbf{(1)}$$

where $p(x_i)$ is the probability of the i-th value of variable X.

If the random variable is continuous, the summation sign in (1) is replaced by an integral and the density function $f(x_i)$, associated with the variable, appears in place of probability $p(x_i)$. If we consider two random processes X and Y associated with two measuring sets M_1 and M_2 that take place simultaneously and that are stochastically independent, the total information content is given by the sum of the marginal entropies of the two processes (Harmancioglu, 1981)

$$H(X,Y) = H(X) + H(Y) \qquad \textbf{(2)}$$

If X and Y are not stochastically independent, the total entropy can be written (Schultze, 1969)

$$H(X,Y) = H(X) + H(Y/X) = H(Y) + H(X/Y) \qquad \textbf{(3)}$$

in which $H(Y/X)$ and $H(X/Y)$ represent the conditional entropies, defined as the information content of variable Y, with known value of X, and vice versa.

The difference between the total entropy defined by (2) and that indicated by (3) defines the concept of transinformation $T(X,Y)$

$$T(X,Y) = [H(X) + H(Y)] - H(X,Y) = H(Y) - H(Y/X) = H(X) - H(X/Y) \qquad \textbf{(4)}$$

which represents the degree of dependence of the two sets or, in other terms, the content of the information redundancy.

Relating the transinformation to the marginal entropy, the difference

$$K = 1 - \left(\frac{T(X,Y)}{H(X)} \right) \qquad \textbf{(5)}$$

represents the loss of information introduced in the observation of the stochastically dependent process (Hermancioglu et al., 1992).

Applying equation (5) to all the sets of time and space measurements, we can thus derive the amount of information content depending on the introduction of new measuring stations (spatial resolution) and different data acquisition rate.

Monitoring: applications and preliminary results

Monitoring activities have been conducted on the reservoir located along the Camastra River, a right-hand affluent of the Basento river in southern Italy.

The Camastra Reservoir has a store capacity of approximately $36 Mm^3$ with a maximum extension of the water surface of $1.5 km^2$ and a watershed area of 350 sq km. It represents one of the most important source of water for the city of Potenza as well as for the Basilicata Region.
In Figures 1a, 1b and 1c are shown the dissolved oxygen, organic substances and the turbidity trends, recorded by the EAAP, Ente Autonomo per l'Acquedotto Pugliese since 1987, furthermore in Figure 1d the daily rainfalls, recorded by the ESIPB, Ente per lo Sviluppo dell'Irrigazione in Puglia e Basilicata at climatic parameters station control since 1987, are reported.

In Figure 2,3 and 4 the results of the optimization model, applied to dissolved oxygen data are presented: the original data, information content, and the loss of information for seven different sampling time ranging from two to thirty days have been analyzed obtaining the conditional entropies reduction, the redundant information content and the loss of information. From Figure 4 it is possible to state that, for case 7, the loss of information is equal to -6.58% and shows that the sampling is redundant and the information content is equal to 100%; whereas for case 6, the loss of information is equal to +6.30% leading to an information content of 93.70%. These percentages have to be related to the relative reduction of recording global costs to choose the most convenient sampling time.

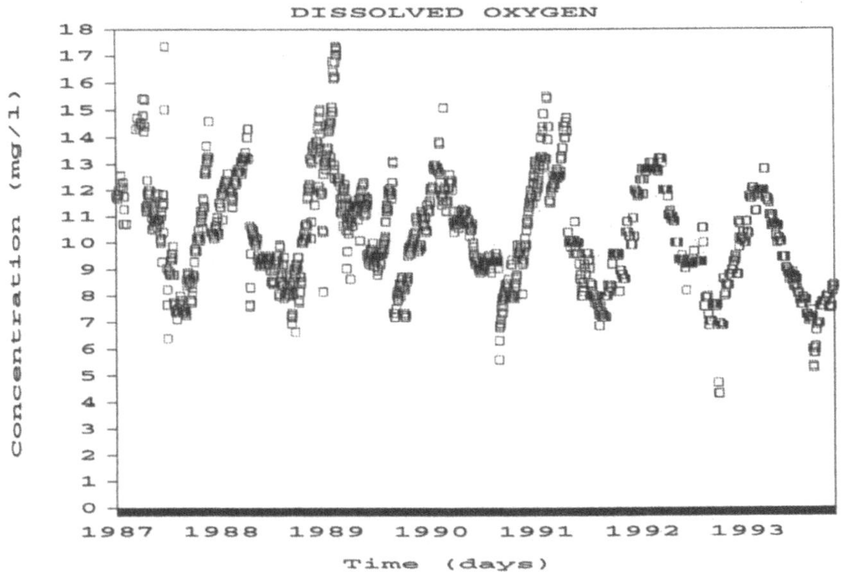

Figure 1a. Dissolved oxygen trend

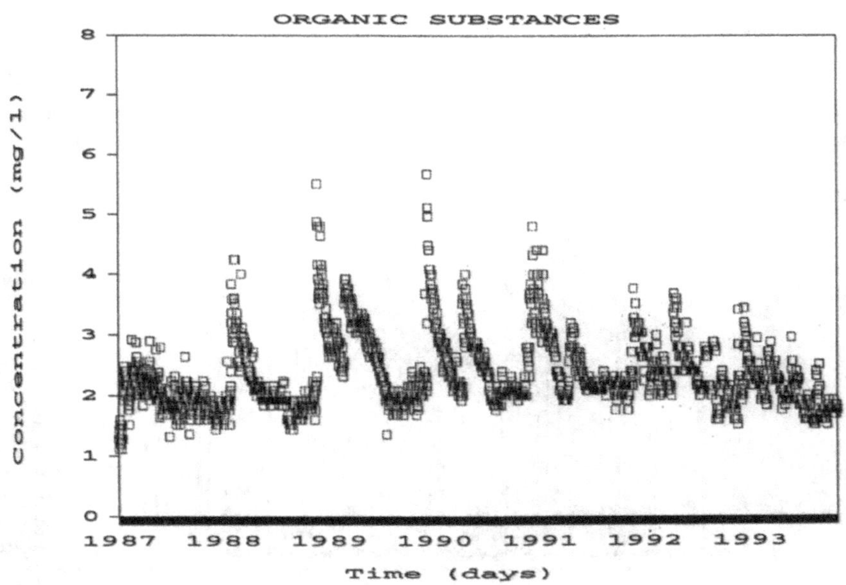

Figure 1b. Organic substances trend

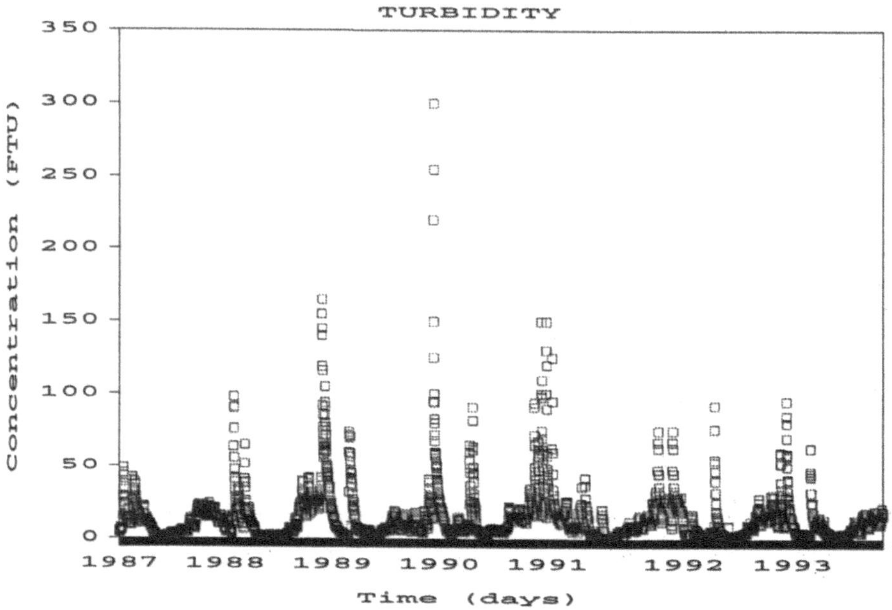

Figure 1c. Turbidity trend

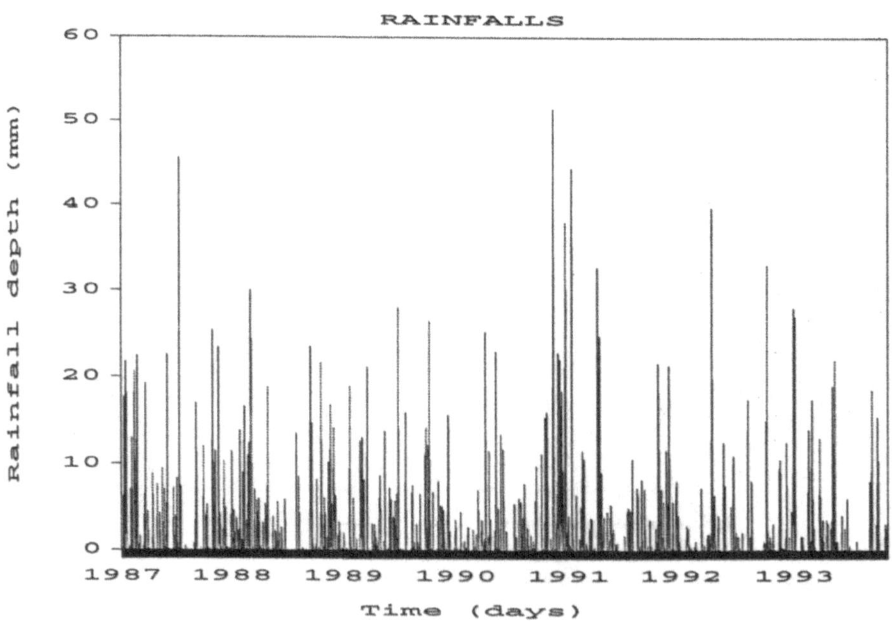

Figure 1d. Daily rainfalls trend

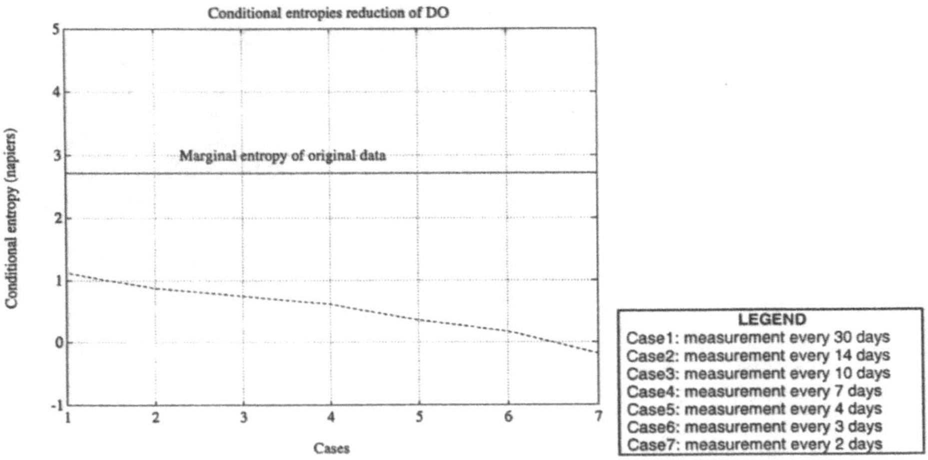

Figure 2. Conditional entropies reduction of DO

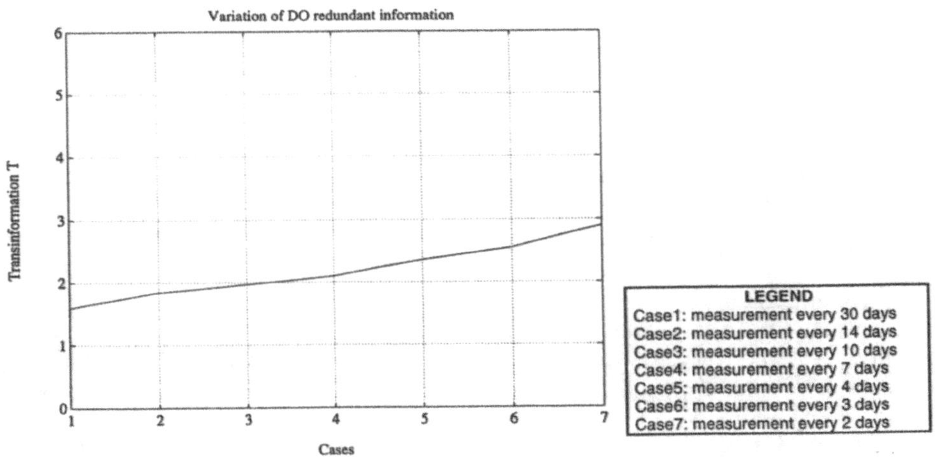

Figure 3. Redundant of DO information content

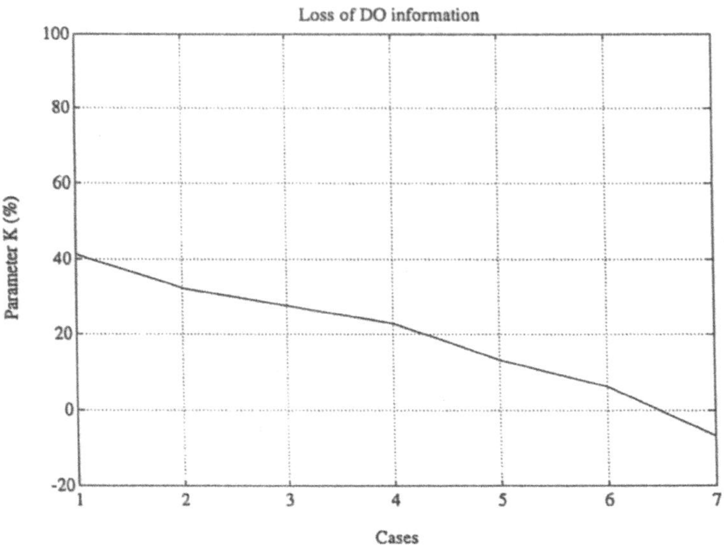

LEGEND
Case1: measurement every 30 days
Case2: measurement every 14 days
Case3: measurement every 10 days
Case4: measurement every 7 days
Case5: measurement every 4 days
Case6: measurement every 3 days
Case7: measurement every 2 days

Figure 4. The loss of DO information

Furthermore an accurate set of bathymetry and temperature measurements in the reservoir has been carried out using part of the designed monitoring system. The procedure previously described was used to design a planimetric grid organized in square meshes of 10m x 10 m average size (Boari et al., 1993), as shown in Figure 5 and 6.

The planimetric position on the relative depth were measured by the laser track and echo sounder. The temperature profile have been obtained by measurement spaced 2m along the vertical direction but a different mesh of 50m x 50m (Figure 3). The resulting map and profiles are shown in Figure 5 and 7.

With this procedure, we attained a high level of accuracy (around 1cm for depth measurements) and were able to monitor a high number of points (more then 10000, for bathymetric measurements and more then 100 for temperature measurements) in a relatively brief time (around 30 hours for batymetric measurements and around 2 hours for temperature measurements).

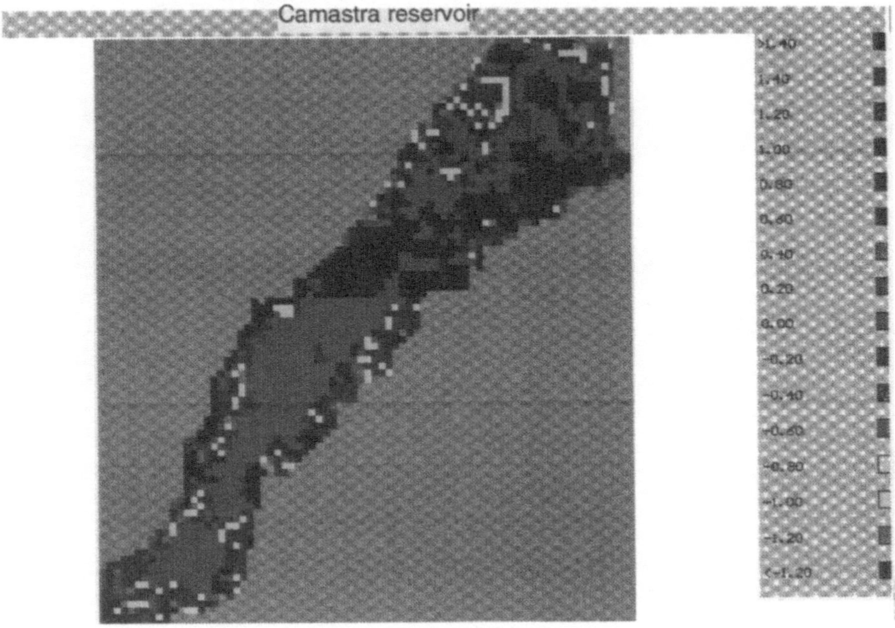

Figure 5. Bathymetry

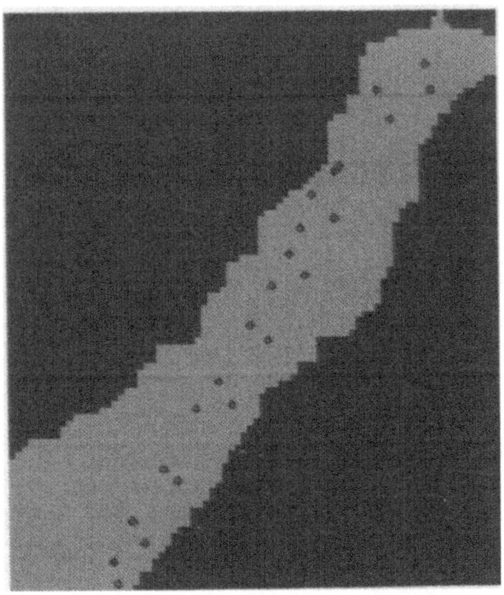

Figure 6. Measurement points of temperature profiles

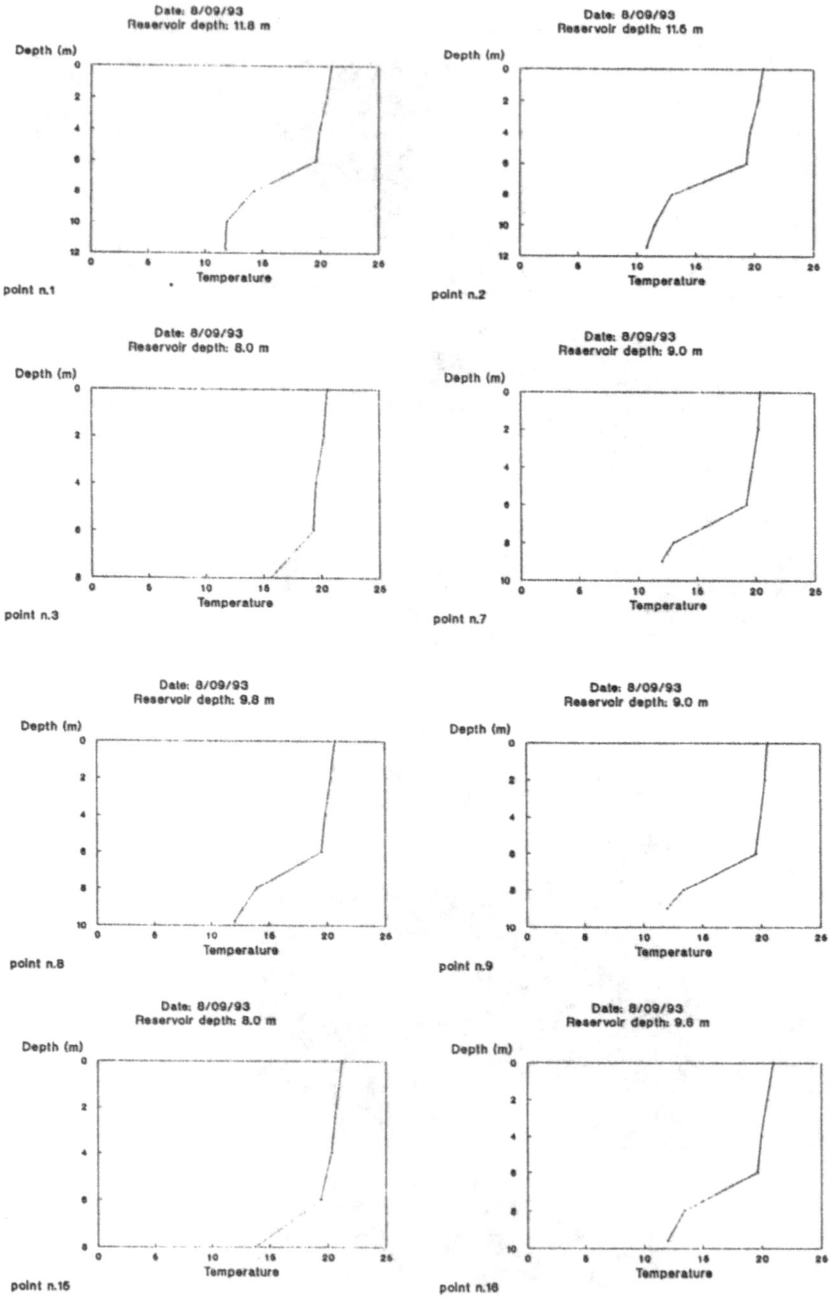

Figure 7. Temperature profiles

Interactive use of the Mathematical Models and Monitoring System

The monitoring system and the optimization model along with the experience, obtained by the preliminary experimental results previously presented, will interact with the mathematical model of the dispersion processes to determine a consistent part of the integrated approach described in figure 8.

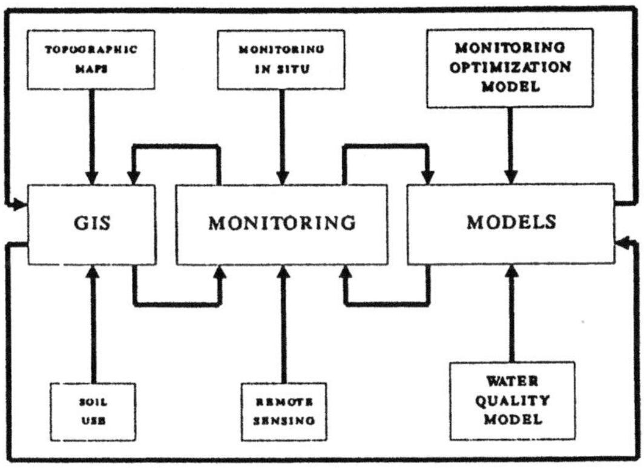

Figure 8. General scheme of integrated approach

Defining **D**, as the dispersion-diffusion tensor $[L^2T^{-1}]$, whose components due to the dispersive motion of the pollutants in the water, the diffusive motion of the water in the spaces between the pores of the sediment particles, and the sediment transport (Ambrose et al., 1986), $\bar{v}$ the advective velocity $[LT^{-1}]$, obtained by a 3-D hydrodynamic modeling of the flow motion in the reservoir, R_i the kinetic transformation due to hydrolysis, photolysis, oxidation, phosphorus/nitrogen cycle, dissolved oxygen balance, decay, growth, etc., S_i the global discharge of the spatial distribution of point loads $[ML^{-3}T^{-1}]$ obtained with the application of GIS.

The conservation of mass leads to the following equation for the concentration of the generic costituent C_i $[ML^{-3}]$

$$\frac{\partial C_i}{\partial t} + div(\bar{v} C_i) - div\mathbf{D} grad C_i = S_i + R_i \qquad \textbf{(6)}$$

Such an equation, integrated by the finite differences using a calculation mesh formed by n elementary cells lying on the horizontal and/or vertical planes, where the tensor **D** is determined by monitoring, permits the generation of synthetic data series. The integration mesh, is then verified by the optimization model, and the amount of loss information, K, obtained from equation (5).

The iterative process (Copertino et al.,1993), described in figure 9, permits then the convergence to a minimum value of parameter K, getting the best information content with the least numbers of measurements.

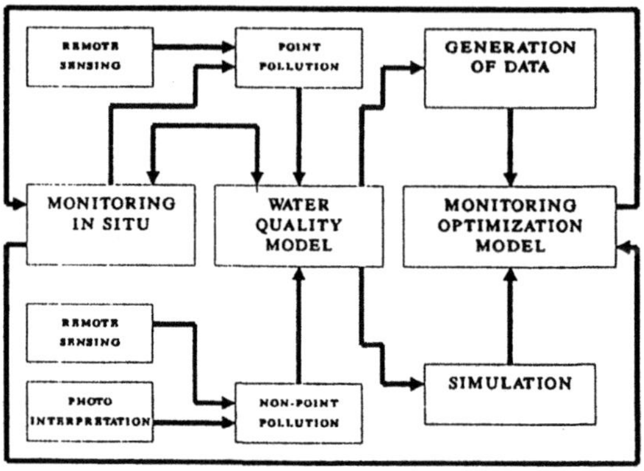

Figure 9. Scheme of interactions between monitoring and mathematical models

The described results and optimization procedure, already available, lead to a quick definition of the mathematical model partially realized and to further and necessary monitoring activities as well as to the characterization of territorial data.

Conclusions

This monitoring system, used for bathymetric characterization of the reservoir on a planimetric optimal mesh and for measurements of the temperature on a three dimensional optimal mesh, has shown its own advantages and properties allowing the observation of large number of data in a very short time.

The system, readily transportable to another water body, represents a sort of real time acquisition data system on very large spatial scale avoiding the logistic difficulty of the traditional approach and allowing for an almost "instantaneous" water quality and quantity management at the lowest possible cost. Furthermore it represent an experimental tool to investigate basic hydraulic and biochemical processes on time a spatial scales, impossible in the past.

References

AMBROSE, R.B., VANDERGRIFT S. and T. WOOL (1986):"WASP3, A hydrodynamic and Water Quality Model- Model Theory, User's Manual and Programmer's Guide", U.S. Environmental Protection Agency, Athens, G.A., EPA/600/3-86-034.

BOARI, G., MOLINO, B., VALENTI, G. and C.D'ANISI (1993): "Reservoir sedimentation: Modelling and control", in: Robin N.Chowdhury & (SIVA) M.Sivakumar, ed., Proceedings of the International Conference on Environmental Management, Geo-Water and Engineering aspects, Wollongong - New South Wales (Australia), february 8-11, 1993, (Rotterdam, A.A. Balkema), pp533-541.

COPERTINO, V.A.,MOLINO, B., TELESCA, V. and V.P.SINGH (1993): "Sviluppo di una metodologia integrata per lo studio del trasporto e della diffusione di inquinanti in corpi idrici", presented in the ANDIS Conference, Palermo, september 21-23, 1993.

HARMANCIOGLU, N. (1981): "Measuring and Information Content of Hydrological Processes by the Entropy Concept", Centennial of Ataturk's Birth, Journal of Civil Engineering, Ege University, Faculty of Engineering, pp.13-38.

HARMANCIOGLU, N., ALPASAN, N. and V.P. SINGH (1992): "Application of the Entropy Concept in Design of Water Quality Monitoring Networks", V.P. Singh and M. Fiorentino, eds., in: "Entropy and Energy Dissipation in Water Resources", (Kluwer Academic Publishers, Netherlands), pp.283-302.

SCHULTZE, E. (1969): "Einfuhrung in die Mathematischen Grundlagen der Informationstheorie", Berlin, Springer-Verlag, Lecture Notes in Operations Research and Mathematical Economics, p.116.

SHANNON, C.E. and W. WEAVER (1949): "The mathematical Theory of Communication", the University of Illinois Press, Urbana, Illinois.

IMPACT OF AGRICULTURE ON NUTRIENT CONTAMINATION OF WATER RESOURCES

M. L. Sharma
CSIRO Division of Water Resources
Private Bag PO
Wembley, Western Australia 6014

ABSTRACT: In the pursuit of meeting increasing demands for food and fibre, modernisation of agriculture has involved a rapid increase in the use of agro-chemicals (fertilisers and pesticides) and mechanisation of farm operations. This has occurred without any serious regard to the adverse environmental consequences. The major impact of modern agriculture is that over 70% of the surface waters and a large proportion of groundwaters around the world are now polluted by agro-chemicals. Many Western countries now consider non-point source pollution from agriculture as a very high priority problem. However, in developing countries such a priority has not yet been recognised, partly because of other pressing demands on resources.

This paper provides an overview of the pollution of water resources through the discharge of two major nutrients, nitrogen and phosphorus. Contamination is discussed from the view points of the quality of drinking water for humans as well as eutrophication of surface water resources. Excessive nitrates pose health risks as nitrate is the most ubiquitous agricultural pollutant in shallow groundwaters around the world. From the eutrophication view point, phosphorus is the prime element, but nitrogen can also promote algal and aquatic growth under certain circumstances.

The magnitude of the pollution is a complex function of many physical, chemical and biological interactions in relation to soil type, management practices and climatic conditions. The mechanisms of nutrient discharge to surface waters and to groundwaters are considered. The factors which affect the retention and transport of various forms of nitrogen and phosphorus in the landscape are discussed. Included is discussion on experimental and modelling techniques being developed and used for quantifying and predicting nutrient contamination. Finally, a brief discussion on the remediation/prevention methods to minimise the impact of pollution is included with some suggestions for future research.

1. Introduction

With ever increasing global population, there has been an increased pressure to maximise farm production to meet food and fibre requirements. Modern agriculture has relied, among other things, on heavy use of fertilisers and pesticides. This has largely occurred without any serious regard to the adverse environmental consequences, especially pollution of water resources. Impairment of water resources through agriculture is a world wide problem. While the problem is being recognised and systematically studied in the western countries (Strebel et al., 1989; Fried, 1991; Frink, 1991; Goodrich et al., 1991; USEPA, 1992; Cooper, 1993; Spalding and Exner; 1993), this is less so in developing countries where agricultural pollution is usually complicated by urban and industrial pollution (Duda, 1993).

V. P. Singh and B. Kumar (eds.), Water-Quality Hydrology, 57–79.

Chemicals (fertilisers and pesticides) applied in agro-ecosystems invariably enter the hydrological cycle, causing contamination of surface water and groundwater resources. Agrochemicals of special interest as contaminants of groundwaters are nitrates and pesticides; while those of surface waters are sediments, nitrogen, phosphorus, pesticides and heavy metals. When assessing the concerns of water pollution, a distinction needs to be made as to whether the water is intended for human consumption, irrigation, industrial or recreational use. The magnitude of the pollution is a complex function of many physical, chemical and biological interactions in relation to soil type, management practices and climatic conditions.

The objective of this paper is to provide an overview of the extent of pollution, mechanisms involved, techniques used for estimation/prediction of pollution, and possible solutions. The pollution of water resources is considered in relation to quality of water for human consumption and in relation to the problem of eutrophication. Hence the primary focus will be on the discharge of nitrogen and phosphorus, with some discussion on pesticides.

2. Water Quality Criteria

Both nitrogen and phosphorus are considered pollutants of water resources. From human health considerations, nitrate is of primary concern, and the suggested limit by World Health Organisation (WHO) and other international agencies for NO_3-N in drinking water is about 10 mg/l (Chapman, 1992).

Enrichment of surface waters with nitrogen and phosphorus is responsible for the widespread problem of eutrophication; as a result, excessive algal and aquatic growth occurs which can affect biological balance in lakes, rivers and estuaries. Algal blooms are known to choke lakes/rivers and aquatic life. Availability of either N or P can limit algal growth (Vollenweider, 1968), but usually P is the key element and its environmental limit is 0.025-0.050 mg/l (Ryding and Rast, 1989). The rate of algal growth is roughly proportional to the square root of the product of the concentrations of dissolved N and P, provided that light and temperature are not limiting. Generally N is considered limiting when the ratio N/P is < 5 and P is considered limiting when N/P ratio is > 16. Nitrogen has been found to be the limiting element in near-shore environments and some lakes (Viner and White, 1987; Ryding and Rast, 1989). The environmental limit for N is of the order of 0.05 to 1 mg/l (Chapman, 1992).

Ingestion of nitrate is reported to cause methemoglobinemia among infants, particularly within the first ninety days. Other adverse effects of high nitrate intake include hypertension, birth defects in the nervous system and certain cancers, e.g. stomach carcinoma (Spalding and Exner, 1993). Acute toxicity has been documented at concentrations greater than 50 mg/l NO_3-N in drinking water. However, it must be realised that water is only the second most important source of nitrates, the first being vegetables which contribute 70% of the total nitrate uptake (Vogtmann and Biedermann, 1985).

Ingestion of some pesticides by humans has been reported to be associated with cancer (e.g. leukemia), central nervous system disorders, male sterility and birth defects (Bouwer,1989; Clancy,1990). However, internationally there is considerable variation and uncertainty over the safe limits of pesticides in drinking water (Goodrich et al., 1991; Chapman, 1992). The recommended limits are at the microgram level and usually approach the limits of analytical detection. In Europe the maximum allowable EEC limit in drinking water has been set at 0.5

μg/*l* for total pesticides whereby each pesticide must not exceed 0.1 μg/*l*; while the WHO and other countries' limits are considerably different (Chapman, 1992; Sharma et al., 1993).

3. Extensiveness of Agricultural Pollution

The problem of non-point source (NPS) pollution, caused primarily by agricultural land use, is spread across the world, and its seriousness has been aptly summarised by Duda (1993) as: "Because of NPS pollution, hundreds of millions of people suffer from disease, billions of dollars of economic development investment are lost, and trillions of dollars of environmental remediation are being accumulated for future generations to address." Using specific examples from around the world, Duda (1993) argues convincingly that 'NPS pollution' must become an international priority. He submits that a coordinated effort is required by the international community to control/abate NPS pollution through a comprehensive ecosystem-based approach for a sustainable water environment.

For many, NPS pollution (=diffuse pollution) is equated to agricultural pollution, as there is no identifiable point source. However, pollution such as urban discharges (storm runoff, groundwater) containing pollutants from distributed sources (e.g. septic systems, private/public lawns) must also be considered as non-point. Compared to point source pollution, NPS pollution control is much more difficult as control/abatement measures need to be applied to a large number of polluters.

In less developed countries the difference between point and non-point sources is rather ill defined. This is because municipal sewage is usually discharged as non-point either because of lack of or inadequate treatment plants. As a result, waterborne diseases such as cholera, dysentery, typhoid and diarrhoea are of common occurrence in poor nations.

We briefly summarise the enormity and extent of the problem under two headings: groundwater pollution and surface water pollution. This artificial distinction is made for convenience, although it is realised that there is an overlap of surface and groundwater problems as the two are components of the same hydrological continuum. The problem of groundwater pollution from nitrates and pesticides is viewed primarily for drinking water by humans, while the problem of surface water pollution is viewed as a cause of eutrophication.

3.1 CONTAMINATION OF GROUNDWATERS

Groundwater contamination through nitrates and more recently through pesticides has been extensively reported (e.g. Fairchild, 1987; Addiscott et al., 1991; Follett et al., 1991; Goodrich et al.; 1991; Spalding and Exner, 1993).

Currently, groundwater represents more than 50% of the world's drinking water supply (Marsily, 1991). In the USA for example, some 90% of rural households and 75% of cities rely on groundwater (Goodrich et al., 1991). In Australia, about 20% of the population relies on groundwater, but in many rural areas groundwater is the only source of water (Dillon et al., 1989). In developing countries, because of heavy contamination of surface waters, reliance on groundwater is even greater. In many rural areas around the world, groundwater is the only source.

The contamination of groundwaters through agro-chemicals has been increasing steadily. Nitrate is the most ubiquitous groundwater pollutant around the world. Nitrate contamination of aquifers has occurred to such an extent that in many European countries (e.g. Belgium, FRG, France, Netherlands, England) up to 20% of the population consumes water which exceed EEC limit of 11.4 mg/l NO_3-N (Fried, 1991; Addiscott et al., 1991). Similarly nitrate-N concentration in groundwaters of other developed countries such as Canada, Australia and New Zealand exceed WHO's drinking limit of 10 mg/l (Burden, 1982; Lawrence, 1983; Dillon et al., 1991; Sharma, 1991). Analysis of comprehensive data sets in the USA (Canter, 1987; Spalding and Exner, 1993) showed widespread nitrate contamination of groundwater especially beneath irrigated raw cropped areas on well drained soils in the states of Minnesota, Washington, Arizona, California and Nebraska. Far less nitrate contamination was found in humid regions despite high fertiliser usage.

Nitrate contamination is expected to be widespread in groundwaters of Eastern Europe, the former Soviet Union, Asia, Africa and South America, but there is an overall lack of quantitative documentation. In India elevated levels of nitrates have been reported (e.g. Singh and Sekhon, 1976; Arora et al; 1980., Singh and Biswas, 1982; Handa, 1983, 1987). Handa (1983) observed NO_3-N concentrations of up to 290 mg/l in some wells, nitrate concentrations being much higher in wells in arid and semi-arid regions than in regions with humid climate. A further broad scale survey shows that while only less than 5% wells from confined aquifers had NO_3-N concentration > 5 mg/l, a considerably larger percentage of shallow unconfined wells had NO_3-N concentration > 45 mg/l (Handa, 1987). Even in a high rainfall humid region in Malaysia, it was found that about 25% of the drinking wells in a tobacco agrosystem contained NO_3-N in excess of 10 mg/l (Sharma et al., 1993). In many developing countries the exact source of N is not specified, as there is usually distributed input of nutrients through animal and human wastes, which is not easily quantifiable.

Although the documentation of contamination of groundwaters by pesticides has formed more recently, it is mostly in the more developed countries. This is primarily because of the very high cost of pesticide analysis and rapidly developing technology. The most documented pesticides found in groundwaters are: aldicarb, atrazine, alachlor (Canter, 1987; Goodrich et al.,1991). Goodrich et al. (1991) provide details of USEPA list of high priority pesticides and suggest that pesticides that are more water soluble and less strongly adsorbed are likely to reach groundwater. These include alachlor, atrazine, butylate and metachlor; and they account for over 50% of the herbicides used in the USA. An extensive survey in New York state showed that 13.5% of the 8300 wells surveyed exceeded the State's recommended limit for aldicarb of 7 μg/l (Canter, 1987). A country-wide survey suggested that over 50 states in the USA may be experiencing some problems with pesticides (Canter, 1987).

In a recent survey of the rural drinking wells in two agro-systems (paddy and tobacco) of northern Malaysia, > 35% samples were detected with pesticides (aldrin/dieldrin, lindane, endosulphan and DDT). In addition, carbofuran was detected in tobacco, while 2,4-D and paraquat were detected in paddy (Sharma et al., 1993).

3.2 CONTAMINATION OF SURFACE WATERS

It has been estimated that on the global scale up to 70% of impaired surface waters are as a result of diffuse pollution from agricultural practices (Olem, 1993). The USEPA (1992) estimates that agricultural activities are responsible for polluting four times more waters than municipal sources, and therefore NPS pollution remains the most widespread problem in the USA. Some 60% of the impairment in the country's surface waters is caused by agriculture. Cooper (1993) quoted USEPA reporting that some 25% of more than 12 million lakes in 34 states were impaired by pollution from nutrients and sediments.

Agriculture is recognised as the primary cause of NPS pollution in most of the developed world, including Western Europe. A large number of rivers and water bodies are found to be heavily polluted and eutrophic. Examples of these, with well documented data, include: the Seine, Rhine, Colorado and Mississippi rivers; the Adriatic, North, Baltic and Mediterranean seas; and the Great Lakes, Chesapeake Bay and Gulf of Mexico.

Although not as much quantitative information is available in less developed countries, the extent and magnitude of the problem is expected to be more serious in surface waters of Eastern Europe, Asia, Africa and South America. Some of the major Asian rivers such as the Indus, Ganges, Yellow and Chao Phraya, are highly polluted. The environmental damage through NPS pollution in Asia is especially severe due to pollutant discharge from vast areas of irrigated lands. China and India irrigate areas twice the size of the United States (Duda, 1993).

It is estimated that over 70% of the available water in India is polluted; about 42% of the population has access to safe water supply and only 20% to sanitation (Ahmed, 1990). Saify and Ali (1987) elaborate on the excessive eutrophication problem in numerous and widely spread water bodies in India and propose to utilise much of this pollution to enhance aquatic food production. Caution must, however, be exercised to ensure that food so produced is not unduly contaminated with organic toxins and/or with heavy metals.

4. Mechanisms of Nutrient Discharge

Figure 1 depicts the various processes involved on a catchment scale in the sourcing, utilisation and discharge of nitrogen and phosphorus to groundwater and streams. This shows that quantitative estimation of nutrient discharge is very complex as it will depend on: soil factors (geomorphological, physical, chemical and biological properties), input factors (source, rate, type, timing of nutrient input), agronomic factors (crop type, tillage/weed control, pesticide control, crop harvest/residue management), irrigation factors (type/source/method, drainage method) and climatic factors.Thus it should be appreciated that nutrient discharge will largely be site specific.

Both N and P are vital nutrients for plant growth. Agricultural production largely depends on the availability of these nutrients for plant uptake in the soil media. Over the last two decades, the rate of fertiliser use has increased very rapidly. Although there is a wide range in rates of fertiliser use in various countries around the world, average fertiliser use in India is much lower than the world average and about an order of magnitude lower than those of most developed countries (Singh and Biswas, 1982).

Addition of nutrients through fertilisers, manures or any other sources is done to supplement the deficiency of soil media to maximise crop production. Ideally, nutrients should be supplied at the right time and at a right amount to meet the crop needs, while minimising pollution risk

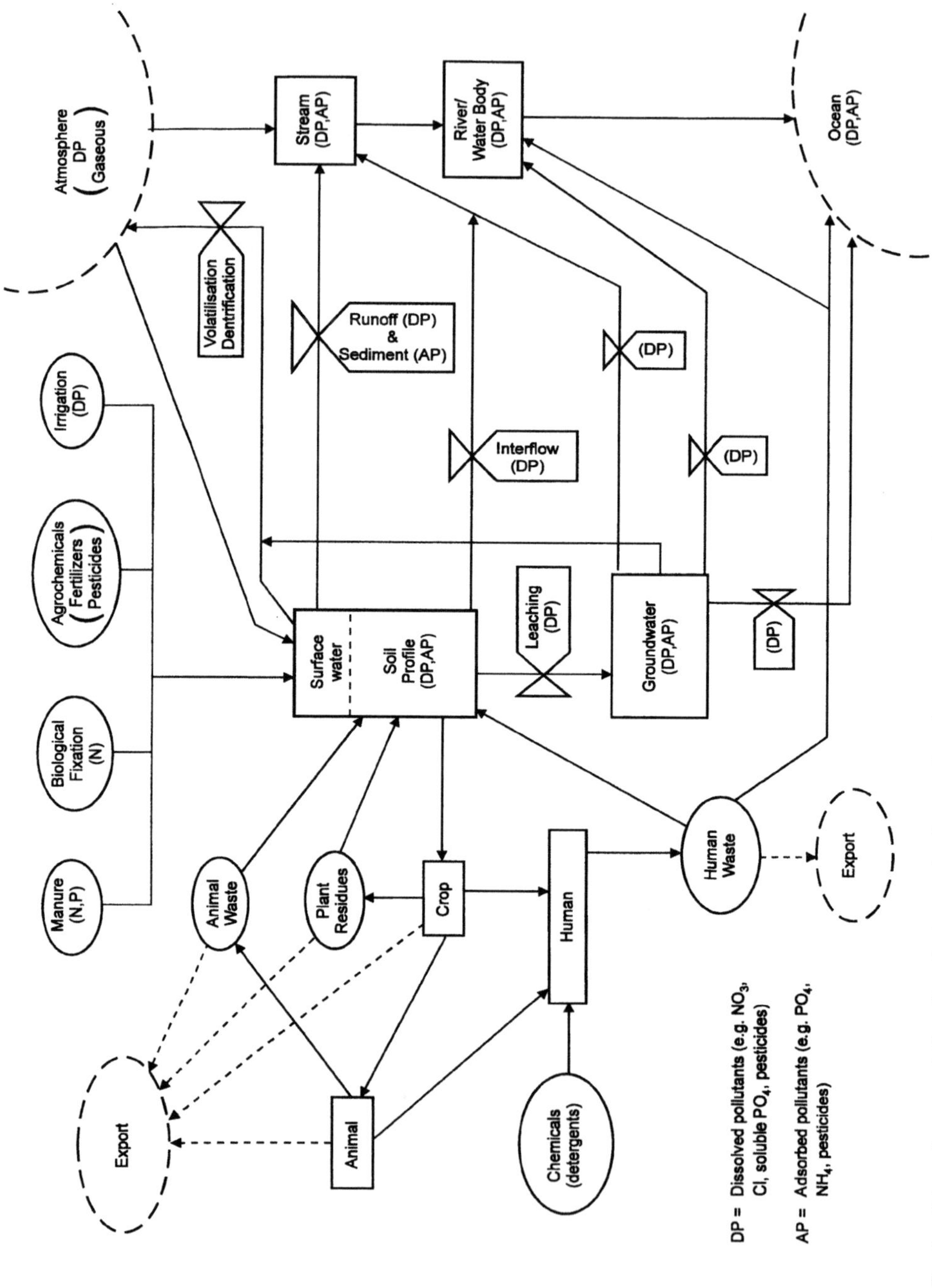

Fig. 1. Schematic representation of mechanisms and pathways of nutrient transport on a catchment scale.

to the environment. However, because of practical, economic and other factors excessive quantities of fertilisers are usually applied resulting in some discharge of nutrients in the water resources.

4.1 NITROGEN

Two primary sources of N applied in agriculture are chemical fertilisers (e.g. anhydrous ammonia, ammonium nitrate, urea, ammonium sulphate and ammonium phosphate) and organic products (e.g. manures, animal wastes, human wastes). Additional sources of N input include: symbiotic fixation by plants, irrigation and atmospheric input.

In warm a climate and under aerobic conditions, urea hydrolyses to ammonia and this in turn is oxidised quickly to nitrate. Ammonium ions are also oxidised quickly to NO_3, which is very stable under aerobic conditions. Various forms of N added in the organic form are mineralised through microbial and chemical reactions first to NH_4 and then to NO_3. Plants take up both NO_3 and NH_4 forms. NH_4 ions can be adsorbed on the soil particles, and later may be exchanged and released. Both NO_3 and NH_4 can be immobilised to organic form and accumulated in the soil. Atmospheric nitrogen can be fixed by leguminous plants and stored. N stored in the soil organic pool can be mineralised to NH_4 and then oxidised to NO_3.

Under anaerobic conditions (e.g. flooding) inorganic nitrogen is present mainly as NH_4, which is taken up by plants e.g. paddy. NO_3 can be denitrified under anaerobic conditions to N_2O and N_2 and lost to the atmosphere. Usually denitrification losses are rather small in the vadose zone below the rooting depth. However, considerable nitrate removal can occur from the saturated zone (aquifer) provided there is a supply of organic carbon for biological denitrification.

Nitrogen input to the groundwater occurs through the leaching process in the form of NO_3 and NH_4; while to the surface water it occurs via surface runoff in the form of NO_3, NH_4, organic matter, and soil particles (adsorbed NH_4). Among other things, the leaching losses of NO_3 are governed primarily by the amount of NO_3 stored in the soil and the amount of water passing through the soil profile. Losses of N through surface runoff occur mostly when fertilisers/manures are surface-applied, and/or through soil erosion. Under adequately moist conditions, N from surface-applied fertilisers/manures can be lost to the atmosphere in the form of ammonia (NH_3) gas, particularly when soil pH is more than 7.

4.2 PHOSPHORUS

Phosphorus is applied either in the chemical form (superphosphate) or in the organic form (manures, organic wastes). Depending on the physico-chemical properties of the soil, part of the applied P becomes water soluble (H_2PO_4, HPO_4) and is either taken up by the plant, adsorbed on soil, or is lost through surface runoff or through leaching. Most soils have strong affinity to adsorb P (the capacity depends on type/amount of clay fraction, contents of Fe, Al, Ca), with sandy soils having the least P adsorption capacity. Plants are able to utilise P from different soil P fractions to varying degrees. Soluble and adsorbed P remain in dynamic equilibrium. Thus a soil with large amounts of adsorbed P may be able to supply P to a crop for a considerable time.

Unlike N, phosphorus is conservative, as it cannot be lost to the atmosphere. Calcareous soils (alkali) usually have large amounts of calcium (di- and tri-) phosphate, while acid soils have Fe and Al phosphates. Organic P remains in dynamic equilibrium with soluble and not so soluble forms.

Usually losses of P through leaching are relatively small, except in heavily fertilised sandy soils with shallow water tables. Most of the P transport occurs through surface runoff, either as organic P, soluble P or as adsorbed on soil particles. P loadings in surface runoff are much higher with transport of soil particles.

In agricultural soils inorganic phosphate is the main form, and its adsorption on soil is described by an isotherm. The adsorption isotherm relates the amount of phosphate adsorbed by soil in a solution of known phosphate concentration at a given time under known pH and temperature. These isotherms are dependent on the length of time phosphate in solution reacts with the soil.

4.3 NUTRIENT BALANCES

It is very rare that all the input and output components of a nutrient are ever measured for a specified land area of an agro-system. Usually specific components of interest are measured and an attempt is made to estimate other components to check the mass balance. In the context of vertical fluxes of nutrients to groundwater, measurements can best be made by using appropriately designed lysimeters, and for surface runoff by using standard stream measuring devices for catchments or plots. Numerous studies have been performed around the world and site specific data have been collected. Here only example results will be presented for illustration.

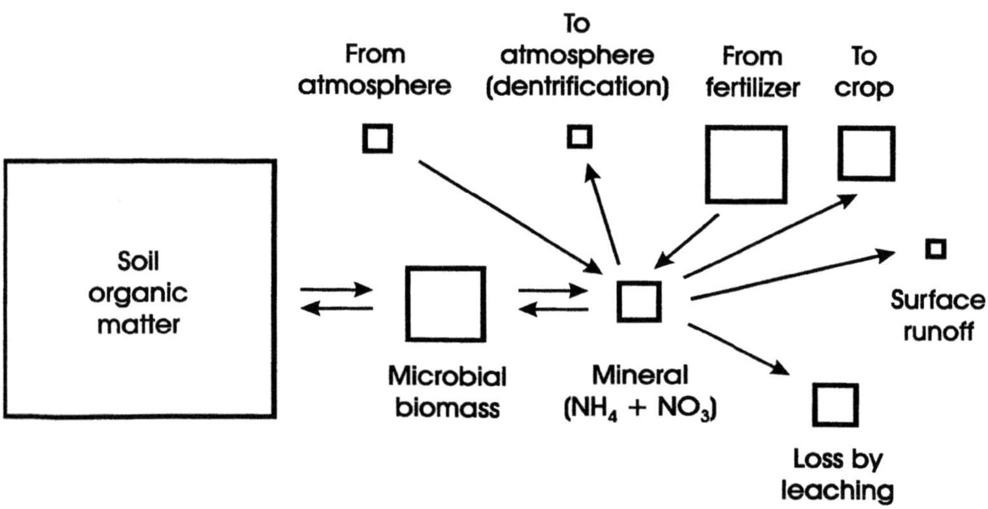

Fig. 2. Diagram showing pathways of nitrogen on a typical agricultural farm setting. The size of the square is proportional to the quantity of nitrogen (Adapted from Addiscott et al., 1991).

An approximate mass balance and pathways of flow of N in a hydro-ecosystem is presented in Fig.2. The diagram illustrates that soil organic pool can be many times larger than the annually applied N, and as a consequence NO_3 leaching from mineralisation of organic matter may continue for a very long period of time. Addiscott et al. (1991) reported that at Rothmstad NO_3-N was leaching at a rate of ~20 kg/ha/yr after 45 years. Although leaching losses depend on so many factors, NO_3-N annual loadings for a given system have been found to be linearly related to cumulative drainage water (Kolenbrander, 1982). The effects of the type of cropping/soil and of N application rate are neatly illustrated in Fig.3. This diagram was derived from a large number of European studies, in which groundwater table was > 1 m below the surface; the results were adjusted to a common drainage of 300 mm/yr (Kolenbrander, 1982).

The effect of two types of land use (horticulture- with heavy fertiliser use and urban lawn- with moderate fertiliser use) on the leaching of NO_3-N measured on two types of sand by using precise lysimeters is shown in Figs 4 and 5. These results demonstrate how intensive agriculture for high value cash crops can be a potential source of severe nutrient pollution (Sharma et al., 1991a). Cumulative yield of NO_3-N and PO_4-P, plotted as a function of cumulative recharge (Fig. 6) for horticultural land use, show almost linear relationships as suggested by Kolenbrander (1982).

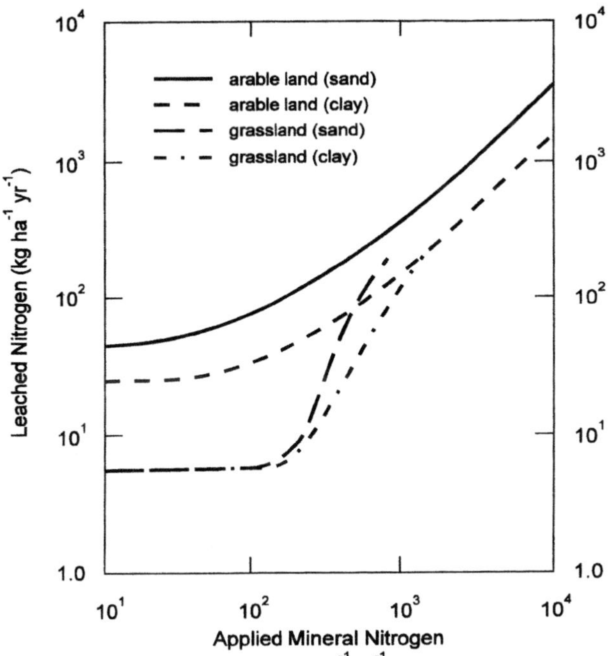

Fig. 3. A generalised diagram showing the effect of soil type, land use and amount of applied nitrogen on NO_3 leaching (Adapted from Kolenbrander, 1982).

There is vast literature on various factors affecting N balance and NO_3 leaching (e.g. Logan et al., 1987; Germon, 1990; Addiscott et al., 1991; Follett et al., 1991). A few significant points are listed below:

- Symbiotic fixation of N can be of the order of 60-80 kg/ha and an enhanced leaching of NO_3 can result under some leguminous systems (Handa, 1983; Dillon et al., 1991).

- A large proportion (up to 90%) of the surface-applied N through organic wastes and some chemical fertilisers, e.g. urea, anhydrous ammonia can be lost through volatilisation to the atmosphere (Germon, 1989; Follett at al., 1991).

- Denitrification in natural groundwaters and surface waters can result in >60% reduction in NO_3 (e.g. Kolenbrander, 1982).

Mass balance of applied P is relatively easier to assess, as major components are: soil storage, input from fertiliser/manures, crop harvest and losses through surface runoff. Usually leaching losses are small, but can be substantial in sandy soils (Sharma et al., 1991b). Phosphate

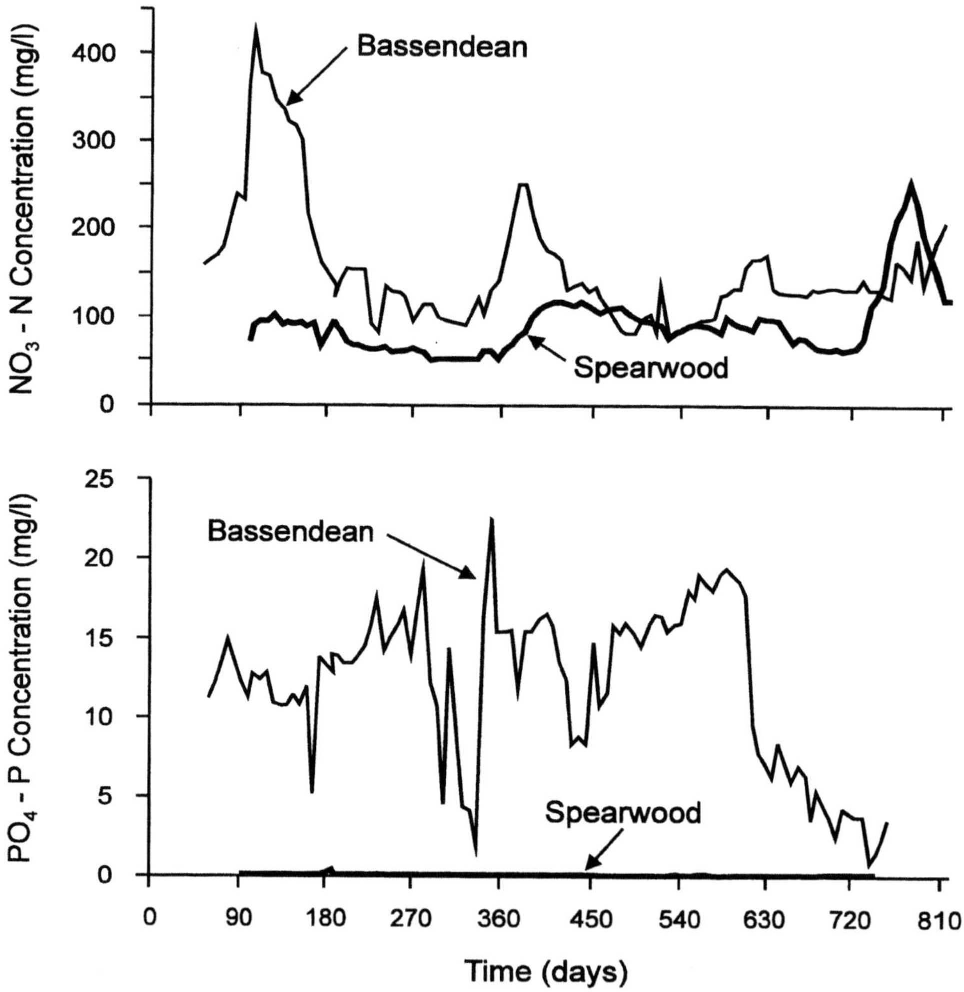

Fig. 4. Measured NO_3-N and PO_4-P concentrations in leachate beneath horticultural land use on two sandy soils.

leaching from two sandy soils under two land use conditions are compared in Figs 4 and 5. They show that the phosphate peak has broken through in one sand, while the majority of the applied P is still being adsorbed in the other sand. Serious eutrophication of lakes, estuaries and rivers has been observed due to discharge of high levels of N and P from coastal sandy soils in Western Australia.

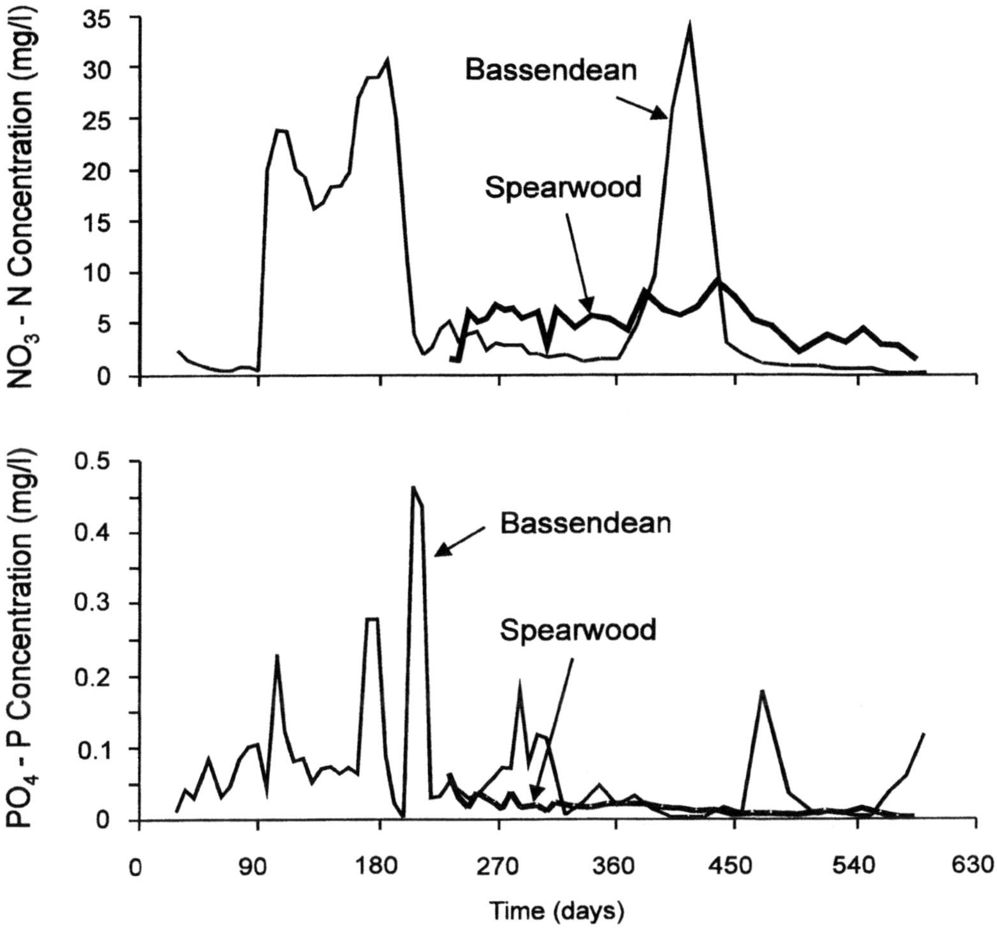

Fig. 5. Measured NO_3-N and PO_4-P concentrations in leachate beneath private lawn on two sandy soils.

Discharge of NO_3, NH_4, and PO_4 in surface waters, as well as of N and P in organic form and attached to soil particles has been extensively measured and modelled. Most studies show that sediments play a very significant role in the transport of organic N and P, and adsorbed P (Smith et al., 1983). Literature on these aspects has been reviewed (e.g. Knisel, 1980; DeCoursey, 1990; Smith et al., 1993)

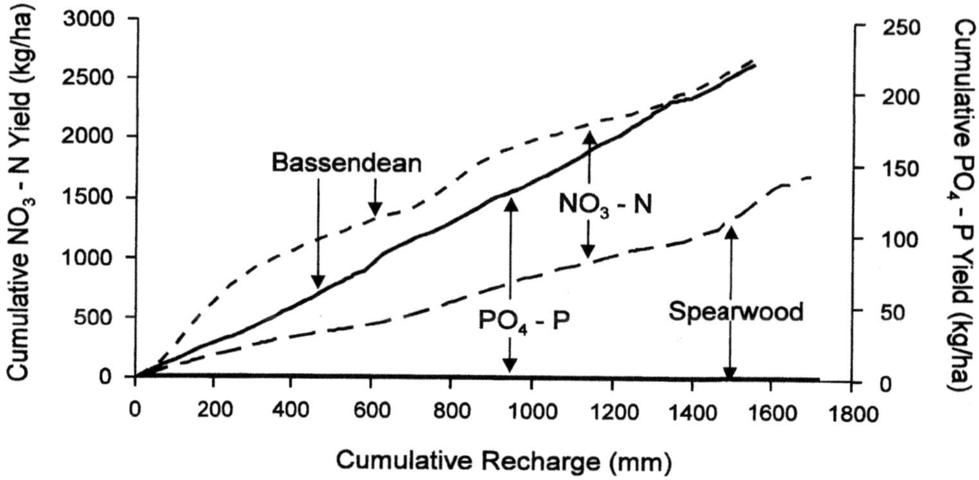

Fig. 6. Yield of NO_3-N and PO_4-P as a function of cumulative recharge beneath horticultural land use for two sandy soils.

5. Modelling Nutrient Discharge

Development efforts in modelling non-point source pollution of nutrients and pesticides have recently been reviewed (e.g. Giorgini and Zingales, 1986; DeCoursey, 1990; Moore, 1991; Ghadiri and Rose, 1992). We shall summarise mainly the types of modelling approaches being considered in relation to transport of N and P in the processes of leaching to groundwater and in surface runoff. Example model simulations of nutrient leaching through sandy soils will be presented and compared with experimentally measured data.

5.1 LEACHING MODELS

Numerous attempts have been made to model solute leaching through the unsaturated zone to groundwater. The models range from deterministic to purely empirical (e.g. Addiscott and Wagenet, 1985; Nielsen et al., 1986; Wierenga and Bachelet, 1988; Vachaud et al., 1990; de Willigen et al., 1990; Ghadiri and Rose, 1992). Many of the deterministic models which take into account various processes in detail remain largely unvalidated, while some of the semi-empirical and empirical models have found wider applications.

Since water is the primary carrier of solutes, it is essential that its movement be predicted accurately. The widely accepted equation describing the vertical distribution of a solute in a uniform soil profile under steady water flow condition is:

$$D \frac{\partial^2 c}{\partial z^2} - v \frac{\partial c}{\partial z} - R \frac{\partial c}{\partial t} = \sum \phi \tag{1}$$

Where c is solute concentration, z is soil profile depth, t is time, D is the dispersion-diffusion coefficient, v is the pore water velocity, R is the retardation factor and ϕ is the source-sink term for the solute. The water velocity is given by q/θ, where q and θ are vertical water flux and average profile water content, both being constant under steady water flow conditions. For a sorbing solute the retardation factor can be approximated by:

$$R = 1 + \rho k/\theta \tag{2}$$

where k is the equilibrium constant for a linear adsorption isotherm, and ρ is the soil bulk density.

Analytical solutions of Eqn. [1] for a limited range of initial and boundary conditions are available, but these conditions are rarely met in the field. Therefore numerical solutions using finite difference and finite element methods are usually employed. An example of such model is LEACHM (Leaching Estimation and Chemistry Model). Currently LEACHM is probably the most comprehensive process-oriented model, as it has components for movement of water, nitrogen, pesticides and organic ions, and the capability to simulate solute flow under transient flow conditions (Wagenet and Hutson, 1989). However, because of its large data requirement, this model has found very limited use.

Another approach to the modelling of solute transport is based on the use of capacity-type water inputs. An example of such approach is the semi-empirical solution of Eqn. 1 (Ghadiri and Rose, 1992). Several other models use the concept of mass balance and describe solute transport through the profile in a simplistic way (e.g. Burns, 1975; Nofziger and Hornsby, 1989; Hutson and Wagenet, 1993).

Relatively large spatial variability in water and solute transport in the field has lead to the development of stochastic approaches to predict solute transport on a land area basis (e.g. Jury et al., 1982; Sharma et al., 1988; Roth and Jury, 1993). However, to successfully apply such approaches, extensive calibration is required for specific conditions.

For a highly sorbing solute (e.g. PO_4), the terms D and ϕ can be ignored. A non-linear time dependent PO_4 isotherm can be described by (Barrow, 1980):

$$s = k\, c^{a}\, t^{b} \tag{3}$$

where s and c are concentrations of solute in the sorbed and solution phases respectively, and k, a and b are empirical constants.

Using these relationships, travel time (τ) of PO_4 for a vertical distance X can be approximated (Sharma et al., 1991a) by:

$$\tau = \frac{X\rho}{q}\, k\, a\, c^{a-1} \tau^{b} \tag{4}$$

5.2 SURFACE RUNOFF AND EROSION MODELS

The most promising models which simulate sediment and nutrient transport in surface runoff on an area (catchment) basis include: CREAMS/GLEAMS (Knisel, 1980; Leonard et al., 1987), AGNPS (Young at al., 1985), ACTMO (Frere et al., 1975) and SWAM (DeCoursey, 1982). We shall comment briefly on the first two models.

CREAMS/GLEAMS are the group of most comprehensive models which have received popular application within and outside USA. CREAMS (Chemical Runoff and Erosion from Agricultural Management Systems), developed by USDA, is an event based model with major components of hydrology, sedimentation and chemistry. N and P attached to the soil particles are lost with the sediments, soluble N/P are lost with surface runoff and soil N is lost by leaching, denitrification or plant uptake. GLEAMS (Groundwater Loading Effects of Agricultural Management Systems) attempts simulation of vertical flux and groundwater loading effect, but this aspect has not been adequately developed. As empirical relationships used in these models have been derived from US data base, local relationships need to be established for their successful application elsewhere (Moore, 1991).

A simpler alternative model is AGNPS (Agricultural Non-Point Source), which is process-oriented, distributed and simulates water quality of runoff on a catchment scale. AGNPS has been linked to Geographical Information System (GIS) (Panuska et al., 1991), enabling more efficient estimation of spatial parameters (topography, soil type and plant cover), and production and display of model outputs. However, it needs to be tested and evaluated under a wide range of conditions.

Smith et al. (1993) present and advocate use of fairly simple concepts/relationships, developed from their studies on 24 watersheds, for calculating agro-chemical discharge through surface runoff. Discharge of soluble chemicals is by empirical equations considering kinetic desorption and mixing approaches, while particulate discharge is described by the relationship between discharge enrichment ratio and soil loss. Comparisons between measured and simulated results have been very encouraging. Such approaches are attractive, but their applicability needs to be tested on a wide range of conditions, and local calibration data must be obtained.

5.3 MODELLING NO_3 AND PO_4 LEACHING

Here we present the application of process-oriented models to simulate leaching of NO_3 and PO_4 through coastal sandy soils under different land uses, and compare these with the experimentally measured data.

When considering NO_3 leaching, two important components are: (i). estimation of chemical/biological transformation and plant uptake of N, and (ii). leaching of NO_3 through the unsaturated zone. Adequate accounting to enable computation of NO_3 within the active soil zone is the primary weakness of most models. We computed NO_3 production in the active surface soil from the lysimeter data on recharge and NO_3 concentration of leachate. Assuming that plant uptake of NO_3 is directly proportional to growth, total NO_3 produced in the active soil can be calculated. Thus the NO_3 available for leaching just below active soil zone will precede the NO_3 collected in the lysimeter. Based on the movement of Cl in the same soil/crop system of a horticultural crop, Sharma et al. (1991a) estimated that about 70 mm of leachate will be required to move a parcel of water through a 2 m soil profile. A comparison of total applied N, potential leachable N (total applied N minus crop uptake N) and modelled

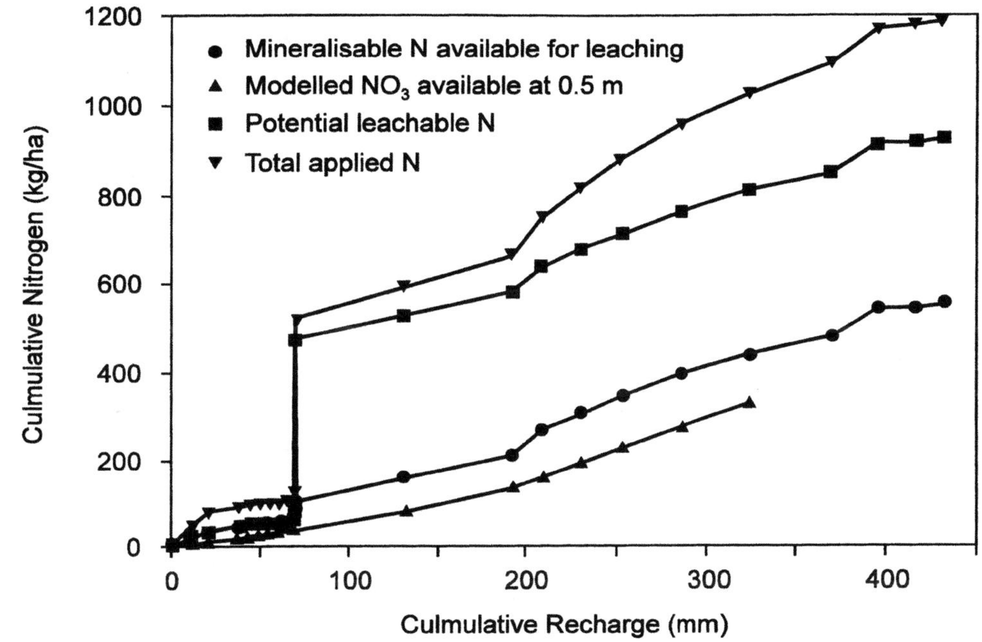

Fig. 7. Correspondence of estimated mineralisable N available for leaching (mineralisable N - crop uptake N), modelled NO_3 available at 0.5 m (estimated from lysimeter output of NO_3 but displaced by 70 mm of recharge) plotted as a function of cumulative recharge at 2 m depth at a horticultural farm. Also shown are: total N, potential leachable N (from Sharma et al., 1991a).

NO_3-N at a depth of 0.5 m as a function of cumulative leachate is shown in Fig. 7. It is clear that potential leachable N provides a very poor estimate of the actually leached N, as about 45% of applied N is lost through volatilisation/immobilisation.

Our leaching simulations of NO_3 are based on the numerical solution of the differential Eqn [1]. An inverse parameter estimation technique was used to calibrate the model. Using measured data on Cl leaching, a dispersion coefficient was estimated which was subsequently used to simulate NO_3 leaching. Using these dispersion coefficients, estimated input concentrations of mineralisable N, the source/sink term in Eqn [1] was estimated through inverse optimisation using measured NO_3 concentration data for a horticultural crop obtained by field lysimeters. A comparison of time variant NO_3-N concentrations obtained via model simulations with the measured values shows a very good agreement between the two (Fig. 8). Application of the model to simulate NO_3 leaching beneath an urban lawn also shows (Fig. 9) a reasonable prediction.

To compute PO_4 leaching through the profile, Eqn [4] was solved using adsorption isotherm

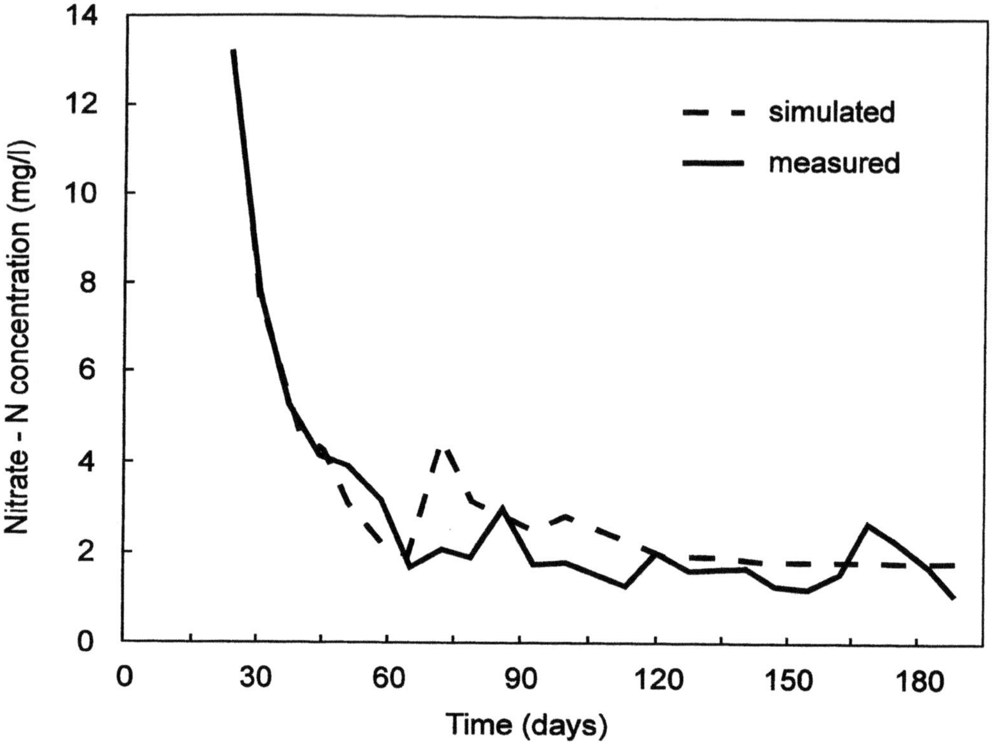

Fig. 8. Comparison of measured and modelled NO_3-N concentration in the leachate at 2 m depth at a horticultural farm. The model included zero-order production rate term for NO_3. Dispersion/diffusion coefficients were those obtained through the inverse parameter estimation technique for a conservative tracer (from Sharma et al., 1991b).

computed travel times (years) for Bassendean and Spearwood sands in Western Australia parameters obtained by Barrow (1980) and Sharma (1993-unpublished data). Table 1 presents at a recharge rate of 100 mm/yr. As can be expected the travel time for PO_4 peak is highly dependent both on recharge rate as well as the PO_4 concentration in the soil solution. These calculations show that at a PO_4 soil solution of 2 to 5 mg/l, and at an annual recharge of

C (mg/l)	Spearwood Sand	Bassendean Sand
2	255	20
5	120	10
10	68	5
50	18	1
100	10	<1

Table 1. Estimated travel times (τ-yrs) for movement of phosphate path through 1 m depth of Bassendean and Spearwood sands with an assumed recharge rate (q = 100 mm/yr) and for a range PO_4-P concentrations (C)

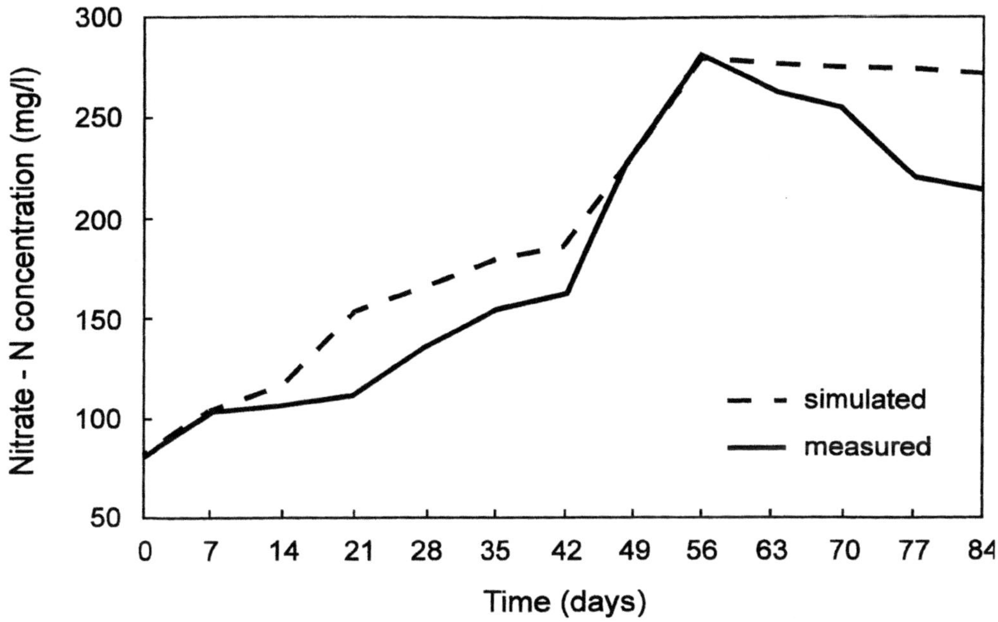

Fig. 9. A comparison of modelled and measured NO_3-N concentration in the leachate at 2 m depth beneath an urban lawn (For details please see Sharma et al., 1992).

100 mm, it will take 10 to 20 yrs for the PO_4 peak to move 1 m depth through the Bassendean sand compared with 120 to 255 yrs through the Spearwood sand. It is expected that considerably longer times will be needed to leach the adsorbed phosphate from a soil by the desorption process. Phosphate soil profiles sampled at two horticultural farms with two soils showed a sharp PO_4 peak at about 3.5 m in the Bassendean farm and about 1.7 m at the Spearwood farm. As these farms have been in horticultural operation for some 20 years, our approximate calculations appear reasonable.

6. Methods to Minimise Nutrient Discharge

To combat the problem of 'NPS pollution from agriculture' will not only require sound scientific knowledge but will also require integrated consideration of social, economic, institutional and government policies. In addition to developing appropriate local strategies, an management approach has to be developed on the scale of river and/or groundwater 'basin'.Prevention of water pollution is a preferred option as it is far more economical than remediation or abatement. Nevertheless, there are many situations where prevention is too late and treatment is the only option. Here we shall concentrate on methods which are based on modification of the management of a watershed so that it can become environmentally sustainable. Only brief comments will be made on treatment methods.

6.1 MANAGEMENT THROUGH SUSTAINABLE AGRICULTURE

It is inevitable that there will be some nutrient discharge to water resources if agriculture is to be practised. To reduce and maintain nutrient discharge at an acceptable level, it is essential that water quality is recognised as an important parameter in developing sustainable agriculture. Thus sustainable agriculture should compromise between demand for food versus cost to the environment. Various practices which can be used to reduce N and P losses through surface runoff and leaching have been recently considered (Germon, 1990; Hatfield, 1993; Olson, 1993). The possible options which may reduce nutrient discharge are discussed under three headings: input of nutrients; land use and crop management; and soil/water management.

Input of Nutrients: Nutrient losses can be potentially reduced by modifying the amount of applied fertiliser/manure and time of their application. For example surface runoff losses can be reduced if fertilisers/manures are ploughed in after surface application. The application rate and timing should be based on the plant need rather than convention. For example the amount of fertiliser to be applied should take into account the amount stored in soil, amount to be applied as part of irrigation and that which can be fixed from the atmosphere by the leguminous plants, as the case may be. Even an approximate knowledge of the mass balance of a nutrient for the system may assist in conserving fertiliser losses.

Improved management of live stock effluent can certainly reduce nutrient discharge. Such practices should include: even and intermittent distribution of slurries/manures, improved method of storing manures/fertilisers on the farm.

Following are some of the application methods which are likely to reduce N leaching :

- split application of fertilisers
- use of slow-release fertilisers
- use of organic-based inputs
- replacing chemical fertilisers by use of leguminous plants

Land Use and Crop Management: Land use/cropping systems are important factors affecting nutrient discharge. For example, intensive agriculture (e.g. horticultural/cash crops) has the potential to discharge much higher quantities of nutrients. On a regional basis it is quite feasible to modify land use and influence the hydrological as well as the nutrient balance as type of crop, cropping pattern/rotation can modify nutrient discharge. For example, arable and grassland agriculture have different management and different rates of nutrient discharge.

Several management practices are known to affect both the hydrological regime and nutrient discharge. The most commonly recognised management practice to modify non-point source pollution is reduction in tillage, sometimes referred as 'minimum tillage'. The term, 'conservation tillage' (CT), is defined by the US Soil Conservation Service as providing 30% or more cover of the land surface at the time of planting. Overall CT has shown to reduce erosion by about 50%, accompanied by reduction in sediment bound pollutants (P, organic N, NH_4). Although beneficial effects of CT in lowering total N and total P via surface runoff are widely accepted, it is not conclusive whether CT reduces/increases NO_3 leaching losses (Logan et al., 1987). Conservation tillage, accepted as 'Best Management Practice' by the USEPA, is being tried and evaluated in reducing P loadings to an important polluted US water body, the Great Lakes (Logan et al., 1987).

To minimise N losses it is important that surface-applied manures and straws are ploughed in and that the soil surface is not left bare for any length of time. This could sometimes be achieved through the use of a catch crop.

Nitrogen losses can also be reduced by use of nitrification inhibitors.

Soil Water Management: Any management practice which reduces surface runoff and recharge below the root zone will minimise nutrient losses. Improved irrigation practices which control the rate and timing of irrigation to meet the crop demand can reduce recharge and therefore N leaching (Sharma et al., 1991). Adjustment of fertiliser application with irrigation can reduce N leaching. Modification in water management through tillage (e.g. no tillage, conservation tillage), and/or through modification of soil structure through other means, can be utilised to change nutrient losses via runoff or leaching.

One of the effective methods of reducing sediment and nutrient discharge to streams is the use of riparian zones. These zones can trap sediments (and sorbed N/P), can take up N/P and can also reduce NO_3 through denitrification. Nutrients from surface waters as well as from inflowing groundwaters can be reduced (Cooper, 1993). Wooded riparian zones are the most effective, providing larger N/P storages. However, their quantitative effectiveness and long term viability and management are some of the issues that still need to be addressed. Natural and artificial wetlands can also contribute to the management of water quality, nevertheless they need to be protected from degradation from non-point source pollution (e.g. Olson, 1993).

6.2 TREATMENT OF POLLUTED WATERS

Besides the methods discussed above, treatment of instream waters for removal of N and P is likely to be very expensive. However, from drinking considerations, it may be economically acceptable to treat NO_3 contaminated waters through processes such as: biological denitrification, anion exchange, reverse osmosis, electrodialysis, distillation and chemical reaction. Currently ion exchange and reverse osmosis are considered the best technologies by the USEPA (Goodrich et al., 1991).

7. Summary and Research Needs

Non-point source pollution from agro-chemicals is a world wide problem, as it has been responsible for contamination of over 70% surface waters and a large proportion of groundwaters. This paper provides an overview of the water pollution problem caused by the discharge of nitrogen and phosphorus. Mostly the source of pollutants is anthropogenic as in many situations increased NO_3 concentrations of many shallow groundwaters, as well as increased N and P concentrations of numerous water bodies have been found to be closely associated with the increased use of fertilisers.

This paper summarises the extent of the problem, and discusses the various mechanisms involved in the cycling of N and P and how pollution is a complex function of many physical, chemical, biological and management factors. A brief overview is presented of various modelling approaches for estimating the transport of pollutants to groundwaters and water bodies. Example model simulations for the transport of NO_3 and PO_4 through the unsaturated soil profile are presented and compared with the measured field data.

Prevention/reduction of nutrient water pollution through sustainable agriculture is the only practical solution on a large scale basis. For the development and adoption of this, a trade off between agricultural production and cost of the environmental degradation of water will have to be seriously considered. A range of options which could be adopted to minimise nutrient discharge are discussed under the headings of: nutrient inputs; land/crop management; and soil/water management.

To combat the problem of nutrient pollution, besides sound scientific knowledge, it will require integrated consideration of social, economic, institutional and government policies. For a successful solution, not only the appropriate local strategies must be developed, but also a wider approach has to be considered on the scale of river and/or groundwater basins.

Among the scientific needs requiring attention are:

- Development of remote monitoring techniques to measure water quality parameters at the required frequency/cost/accuracy.
- Conduct of integrated field studies incorporating the most relevant factors affecting nutrient discharge and evaluating their complex interrelationships.
- Detailed studies of key processes (retention, transformation and transport) on the field scale to quantify their controlling effects.
- Evaluation of field-oriented areal models in predicting nutrient discharge to groundwaters and water bodies.
- Quantitative evaluation of the effectiveness of key management practices on nutrient discharge.
- Quantitative evaluation of the effectiveness of riparian zones and wetlands in the removal of nutrients.
- Feasibility and cost effectiveness of *in situ* remediation (denitrification) techniques in the removal of nitrates from aquifers.

8. Acknowledgment

I wish to thank Bernadette Waugh and her staff for assisting with the extensive literature search; David Herne and Piotr Kin for assisting with the preparation of this paper.

9. References

Ahmed, S. (1990). Cleaning the river Ganges: rhetoric and reality. *Ambio* (Oslo) **19**: 42-45.

Arora, R.P., Sachdev, M.S., Sud, Y.K., Luthra, V.K. and Subbiah, B.V. (1980). *Soil Nitrogen as Fertiliser Pollution.* International Atomic Energy Agency. pp 3.

Addiscott, T.M. and Wagenet, R.J. (1985). Concepts of solute leaching in soils: a review of modelling approaches. *J. Soil Sci.* **36**: 411-424.

Addiscott, T.M., Whitmore, A.P. and Powlson, D.S. (1991). *Farming, Fertiliser and the Nitrate Problem. C.A.B.* International, Wallingford, U.K.

Barrow, M.J. (1980). Differences amongst a wide-ranging collection of soils in the rate of reaction with phosphate. *Aust. J. Soil Res.* **18**: 215-224.

Bouwer, H. (1989). Agriculture and groundwater quality. *Civil Eng.* **59**: 60-63.

Burden, R.J. (1982). Nitrate contamination of New Zealand aquifers: a review. *New Zealand J. Sci.* **25**: 205-220.

Burns, I.G. (1975). An equation to predict the leaching of surface-applied nitrate. *J. Agric. Sci. Cambridge*. **85**: 443-454.

Canter, L.W. (1987). Nitrates and pesticides in groundwater: an analysis of a computer-based literature search. *Groundwater Quality and Agricultural Practices*. Ed. (D.M. Fairchild) Lewis Publishers. Inc. Chelsea, Michigan, USA. 153-174.

Chapman, D. (1992). *Water Quality Assessments*. Chapman & Hall, London.

Clancy, K.L. (1990). Agriculture and human health. *Sustainable Agricultural Systems*. Ed. (C.A. Edwards et al.) Soil and Water Conservation Society, Ankeny, Iowa, USA.: 653-665.

Cooper, C.M. (1993). Biological effects of surface water pollutants on aquatic systems - a review. *J. Environ. Qual.* **22**: 402-408.

DeCoursey, D.G. (1982). ARS's small watershed model. *ASAE Paper No.* 82-2094. Am. Soc. Ar. Eng. St. Joseph, Michigan, USA.

Decoursey, D.G. (1990). Proceedings of the International Symposium on *Water Quality Modelling of Agricultural Non-point Sources*. ARS-81. USDA - ARS, pp 881.

Dillon, P.J., Close, M.E. and Scott, R.I. (1989). Diffuse source nitrate contamination of groundwater in New Zealand and Australia. Hydrol. and Water Resources Symp. Christchurch, N.Z.. IE Aust. National Conf. Publ. 89/19: 351-355.

Dillon, P.J., Ragusa, S.R. and Richardson, S.B. (1991). Biochemistry of a plume of nitrate-contaminated groundwater. *Nitrate Contamination: Exposure, Consequence, and Control*. Ed. (I. Bogardi and R.D. Kuzelka) NATO ASI Ser. G: Ecological Sciences 30. Springer-Verlag, Berlin.

Duda, A.M. (1993). Addressing non-point sources of water pollution must become an international priority. *Water Sci. Tech.* **28** No 3-5: 1-11.

Fairchild, D.M. (1987). *Groundwater Quality and Agricultural Practices*. Lewis Publishers, Chelsea, Michigan, USA.

Follett, R.F., Keeney, D.R. and Cruse, R.M. (1991). *Managing Nitrogen for Groundwater Quality and Farm Profitability*. Soil Sci. of Am. Inc. Madison, Wisconsin, USA.

Frere, M.H., Onstad, C.A. and Holtan, H.N. (1975). ACTMO, an agricultural chemical transport model. *ARS-H-3*. Agricultural Research Service, US Department of Agriculture, Washington, D.C., pp 54.

Fried, J.J. (1991). Nitrates and their control in the EEC aquatic environment. *Nitrate Contamination: Exposure, Consequence, and Control*. Ed. (I. Bogardi and R.D. Kuzelka) NATO ASI Ser. G: Ecological Sciences 30. Springer-Verlag.

Frink, C.R. (1991). Estimating nutrient exports to estuaries. *J. Env. Qual.* **20**: 717-724.

Germon, J.C. (1989). *Management Systems to Reduce Impact of Nitrates*. Elsevier Applied Science, New York.

Ghadiri, H. and Rose, C.W. (1982). *Modelling Chemical Transport in Soils*. Lewis Publishers, London.

Giorgini, A. and Zingales, F. (1986). *Agricultural Non-Point Source Pollution: Model Selection and Application*. Elsevier, Amsterdam.

Goodrich, J.A., Lykims, Jr, B.W. and Clark, R.M. (1991). Drinking water from agriculturally contaminated groundwater. *J. Environ. Qual.* **20**: 707-717.

Handa, B.K. (1983). Effect of fertiliser use on groundwater quality in India. IAHS Publ No 142 Proceedings of a Symposium. Koblenz, West Germany: 1105-1119.

Handa, B.K. (1987). Nitrate content of groundwater in India. *Fertiliser News*. **32**: 11-29.

Hatfield, J.L. (1993). Sustainable agriculture: impacts on non-point pollution. *Water Sci. Tech.* **28**: 415-424.

Hutson, J.L. and Wagenet, R.J. (1993). A pragmatic, field-scale approach for modelling pesticides. *J. Environ. Qual.* **22**: 494-499.

Jury, W.A., Stolzy, L.H. and Shouse, P. (1982). A field test of the transfer function model for predicting solute transport. *Water Resour. Res.* **18**: 369-375.

Knisel, W.G. (1980). CREAMS: A field scale model for chemicals, runoff and erosion from agricultural management system. Report No 26 USDA.

Kolenbrander, G.J. (1982). Fertiliser and pollution: *Whither Soil Research.* 12th International Congress of Soil Sci. New Delhi, India: 248-266.

Lawrence, C. R. (1983). Nitrate-rich groundwaters of Australia. Australian Water Resouces Council, Tech. Rep. No. 79. Aust. Govt. Publ. Service, Canberra., pp 110.

Leonard, R.A., Knisel, W.G. and Sull, D.A. (1987). GLEAMS: Groundwater Loading Effects of Agricultural Management Systems. *Trans. ASAE,* **30**: 1403-1418.

Logan, T.J., Davidson, J.M., Baker, J.L. and Overcash, M.R. (1987). *Effects of Conservation Tillage on Groundwater Quality, Nitrates and Pesticides.* Lewis Publishers, Chelsea, Michigan, USA.

Marsily, G.D. (1991). Can we still hope to preserve uncontaminated groundwater resources for the next generation? International Hydrol. & Water Resources Symp. IE Aust National Conf. Publ No 91/2: 306-310.

Moore, I.D. (1991). *Modelling the Fate of Chemicals in the Environment.* The Australian National University Canberra, Australia.

Nielsen, D.R., Van Genuchten, T. and Bigger, J. W. (1986). Water flow and solute transport processes in the unsaturated zone. *Water Resour. Res.* **22**: 89S-108S.

Nofziger, D.L. and Hornsby, A.G. (1989). Interactive simulation of chemical movement in soil. *Ground Water Quality and Agricultural Practices.* Ed. (D.M. Fairchild) Lewis Publishers, Chelsea, Michigan: 319-328.

Olem, H. (1993). *Diffuse Pollution.* Pergamon Press, New York.

Olson, R.K. (1993). *Created and Natural Wetlands for Controlling Non-Point Source Pollution.* C.K. Smolley, CRC Press Inc. N.Y.

Panuska, J.C., Moore, I.D. and Kramer, L.A. (1991). Terrain analysis: integration into the agricultural non-point source (AGNPS) pollution model. *J. Soil and Water Conserv.* **46**: 59-63.

Roth, K. and Jury, A. (1993). Modelling the transport of solutes to groundwater using transfer functions. *J. Env. Qual.* **22**: 487-493.

Ryding, S.O. and Rast, W. (1989). *The Control of Eutrophication of Lakes and Reservoirs.* Parthenon Publishing Group, Paris.

Saify, T. and Ali, S.N. (1987). Management of surface water quality in India. *Wat. Sci. Tech.* **19**: 87-96.

Sharma, M.L., Watson, J.D. and Cresswell, I.C. (1988). Estimation of areal recharge using a non-reactive chemical tracer. *Infiltration Principles and Practices.* Ed. (Yu-Si Fok). Water Resour. Res. Center, Univ. Hawaii: 316-325.

Sharma, M.L. (1991). Contamination of groundwater beneath horticultural land use, Swan Coastal Plain, WA. International Hydrol & Water Resour Symp. The Institute of Engineers, Australian National Confer. Publ No 91/22: 330-334.

Sharma, M.L., Kin, P.G., Mok, J.Y.N. and Gerritse, R.G. (1991a). Leaching of nitrate and phosphate beneath irrigated horticultural systems. *Modelling the Fate of Chemicals in the Environment.* Ed. (I. Moore). Australian National Univ, Canberra: 109-114.

Sharma, M.L., Byrne, J., Herne, D.E. and Kin, P.G. (1991b). Impact of horticulture on water and nutrient fluxes to a sandy aquifer. CSIRO Division of Water Resources Report No 91/33, pp. 105.

Sharma, M.L., Herne, D.E., Kin, P.G. and Byrne, J. (1992). Impact of urban lawns on nutrient contamination of an unconfined aquifer. Urban Water Research Association of Australia, Research Report No. 49, pp. 86.

Sharma, M.L., Aminuddin, B.Y. et al (1993). Impact of agricultural practices on groundwater quality in Kelantan Plain, Malaysia: survey results. CSIRO Division of Water Resources Report No 93/9, pp. 77.

Singh, B. and Sekhon, G.S. (1976). Nitrate pollution of groundwater from nitrogen fertilisers and animal wastes in the Punjab, India. *Agric. and Environment* **3**: 57-67.

Singh, R. and Biswas, B.C. (1982). Fertiliser use and environmental quality in the tropics and subtropics. *Whither Soil Research,* 12th International Congress of Soil Sci., New Delhi, India: 226-247.

Spalding, R.F. and Exner, M.E. (1993). Occurrence of nitrate in groundwater: a review. *J. Environ. Qual.* **22**: 392-402.

Smith, S.J., Sharpley, A.N. and Ahuja, L.R. (1993). Agricultural chemical discharge in surface water runoff. *J. Environ. Qual.* **22**: 474-480.

Strebel, D., Duynisveld, W.H.M. and Bottcher, J. (1989). Nitrate pollution of groundwater in western europe. *Agri. Ecosystem and Environment* **26**: 189-214.

US Environmental Protection Agency (1992). *1990 National Water Quality Inventory.* 305(b) Report to Congress. Washington.

Vachaud, G., Vauclin, M. and Addiscott, T.M. (1990). Solute transport in the vadose zone: A review of models. Proc. of the Int. Symp on *Water Quality Modelling of Agr. Non-Point Sources. Part 1.* USDA-ARS 81, Washington, D.C., pp 81-104.

Viner, A.B. and White, E. (1987). Phytoplankton growth. *Inland Waters of New Zealand.* Ed. (A.B. Viner) DSIR Sci. Inf. Publ. Centre, Wellington, New Zealand.

Vollenweider, R.A. (1968). Scientific fundamentals of the eutrophication of lakes and flowing waters with particular reference to phosphorus and nitrogen as factors in eutrophication. Organisation for Econ. Coop. & Development, Paris.

Vogtmann, H. and Biedermann, R. (1985). The nitrate story--no end in sight. *Nutrition and Health* **3**: 203-216.

Wagenet, R.J. and Hutson, J.L. (1989) LEACHM: Leaching Estimation and Chemistry Model. Continuum June 1989, Water Resour. Institute. Ithaca, New York.

Wierenga, P.J. and Bachelet, D. (1988). Proc. of International Conference and Workshop on *Validation of Flow and Transport Models for the Unsaturated Zone.* New Mexico University, Las Cruces, N.N. USA.

De Willigen, P., Bergstrom, L. and Gerritse, R.G. (1990). Leaching models of the unsaturated zone: Their potential use for management and planning. Proc of the Int. Symp on *Water Quality Modelling of Agr. Non-Point Sources Part 1.* USDA-ARS 81, Washington, D.C: 105-128.

Young, R.A., Onstad, C.A., Bosh, D.D. and Anderson, W.P. (1985). Agricultural Non-point Source Pollution Model (AGNES) I and II. Model Documentation. Magnesite Pollution Control Agency, St Paul, MI, and Agricultural Research Service, U.S.D.A., Washington, D.C.

Nutrient runoff modelling of an Australian urban catchment

M. Sivakumar and S. Boroumand-Nasab

Water Engineering and Geomechnics Research Program
Department of Civil and Mining Engineering
University of Wollongong, Wollongong, NSW 2522, Australia

ABSTRACT. A new semi empirical washoff method has been developed (Sivakumar, M. and Boroumand-Nasab, S. 1993) to predict pollutant washoff load from urban catchments based on the assumption that transport rate is proportional to the distribution and availability of pollutants. In this paper, the new washoff algorithm is tested against two existing washoff algorithms (SWMM4 and Nakamura) using available nutrient field data for an Australian urban catchment for total oxidised nitrogen (NOx-N), ammonia (NH_3-N) and total phosphorous (TP). The calibration results are shown to be better for the new developed washoff algorithm compared to the two existing washoff methods. Implication of the results for future modelling and catchment management strategies are also highlighted.

1. INTRODUCTION

The process of urbanisation not only changes the rainfall-runoff process but also have significant effects on the quality of runoff. Nonpoint source of pollution in the form of runoff from urban areas can contribute greatly to degradation of receiving waters. Stormwater runoff from urban areas carry pollutants in the form of sediment loads, nutrients, bacteria and toxic substances. The nutrients in particular can lead to eutrophication of water bodies which is detrimental to the various users of the water source.

The transport of pollutants from their source to the outlet of urban catchments is the least understood aspects of urban nonpoint source pollution. The newly proposed method is derived to develop process oriented semi-empirical algorithms for modelling urban runoff quality. Using this method total washoff pollutant load for each storm event can be predicted. The proposed washoff method has been calibrated and verified for two Australian urban catchments (Sivakumar, M. and Boroumand-Nasab, S. 1993). The results for total phosphorous (TP) and total oxidised nitrogen (NOx-N) showed good agreement between measured and predicted washoff loads. In this paper, the proposed washoff algorithm is tested with Huber and Dickinson (1989) and Nakamura's (1984) methods for NH_3-N, NOx-N and TP.

V. P. Singh and B. Kumar (eds.), Water-Quality Hydrology, 81–90.

2. POLLUTANT WASHOFF MODELS

2.1 Sivakumar and Boroumand-Nasab model (1993)

Based on the assumption that the washoff pollutant rate in an urban catchment is proportional to the distribution and availability of pollutants, a washoff algorithm was derived and a summary is given below.

If the pollutant washoff rate $\frac{dP}{dt}$ is assumed to be proportional to shear stress (τ_o) acting on a surface, hence:

$$\frac{dP}{dt} \propto \tau_o \tag{1}$$

and shear stress for steady, uniform flow in wide open surface is given by:

$$\tau_o = \gamma_s \, y \, S_o \tag{2}$$

where
- P = pollutant load remaining, kg
- τ_o = bed shear stress, Nm^{-2}
- γ_s = specific weight of storm runoff, Nm^{-3}
- y = depth of water, m
- S_o = slope of the surface

If the pollutant available for transport in an urban catchment can be assumed to be proportional to the specific weight of storm runoff (Nakamura 1984) then,

$$\gamma_s \propto P \tag{3}$$

By combining equations (1), (2) and (3), it can be shown that, the pollutant load remaining at the surface is,

$$P = P_o \, e^{-K v^a} \tag{4}$$

where
- P_o = initial pollutant load, kg
- v = cumulative depth of runoff, mm
- K = washoff coefficient
- a = exponent

Hence the washoff equation is:

$$P_o - P = P_o \, (1 - e^{-Kv^a}) \tag{5}$$

where $P_o - P$ = pollutant washoff, kg

Equation (4) gives a functional relationship between the amount of pollutant available for transport with the parameters Po, K, v and a in an urban catchment.

2.2 SWMM4 Washoff algorithm (Huber and Dickinson, 1989)

The washoff equation, used within SWMM4 model, describes washoff at each time step to be proportional to the runoff rate 'r' raised to a power 'a' and to the pollutant remaining at any time t, Hence,

$$\frac{dP}{dt} = -K_h P r^a \qquad (6)$$

in which K_h is a washoff coefficient.

The pollutant washoff load was derived from the area beneath the mass flux curve by numerical integration.

2.3 Nakamura's washoff model

Nakamura (1984) used laboratory and field experiments to determine factors affecting the rate of pollutant washoff in urban areas. Nakamura showed that the rate of pollutant removal is a function of cumulative runoff volume 'v', runoff rate 'r', slope of the surface 'S_o' and Manning roughness 'n'. After simplification, Nakamura's washoff algorithm can be expressed as,

$$\frac{dP}{dt} = -K_n P v^b r^a \qquad (7)$$

where a and b are exponents and K_n is washoff coefficient.

To obtain the pollutant washoff by Nakamura's method, a similar procedure shown in SWMM4 algorithm was adopted.

3. THE STUDY CATCHMENT AND DATA AVAILABILITY

A data set from Giralang catchment (Figure 1) in Australia was selected to test the above method. The Giralang catchment (National Capital Development Commission, 1980) in the Australian Capital territory (ACT) is an urban catchment which is located approximately 10 km from Canberra city centre and has an area of 94 hectares with 27% impervious surfaces. The land use includes 69.4 hectares for medium density residential and 24.6 hectares for rural and pasture. Common deficiencies in the data set for example non availability of data for complete events, timing errors between hydrograph and time-step problems have been identified.

The nutrient transport concentration (NH_3-N, NOx-N and TP) for this catchment and their dynamic variations with runoff are shown in Figure 2. From the pollutographs it is seen that the nutrient parameters behaved in different ways. In the most cases flow peak is different to the concentration peaks for TP and NOx-N. On the other hand NH_3-N does not show a peak concentration. The instantaneous mass flux of the various nutrient loads have been calculated and shown in Figure 3. It shows the variations of mass flux rate (loadograph) and hydrograph during storm event. From the observed hydrographs and loadographs (mass flux rate), it is seen, as expected, that the three pollutant parameters (NH_3-N, NOx-N and TP) tend to behave in the same way as the hydrograph, however the NOx-N loadograph peaks earlier. Peak mass flux rates for other two parameters almost coincide with peak flow.

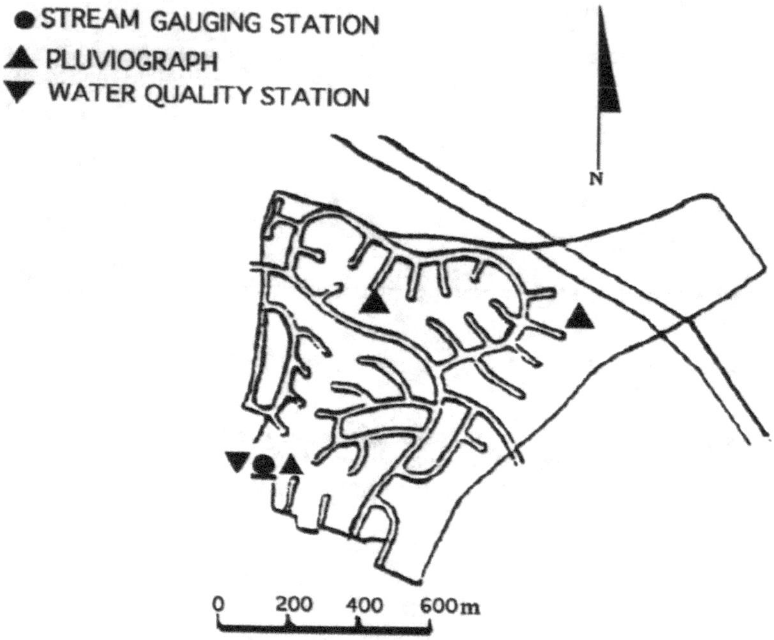

Figure 1 Giralang Catchment, ACT, Australia

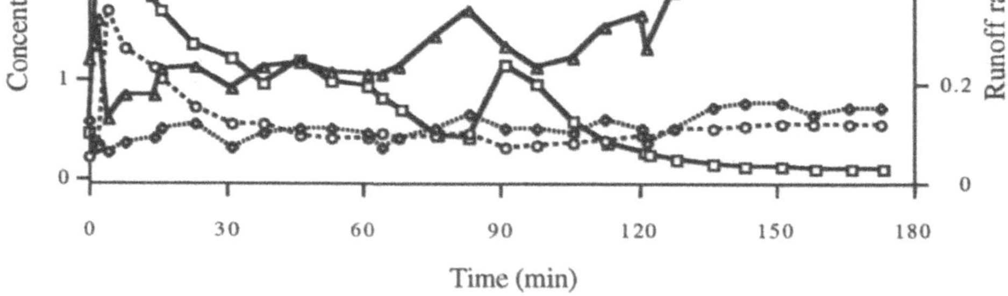

Figure 2 Hydrograph and Pollutographs for NOx-N, NH_3-N and TP for Event 770812.

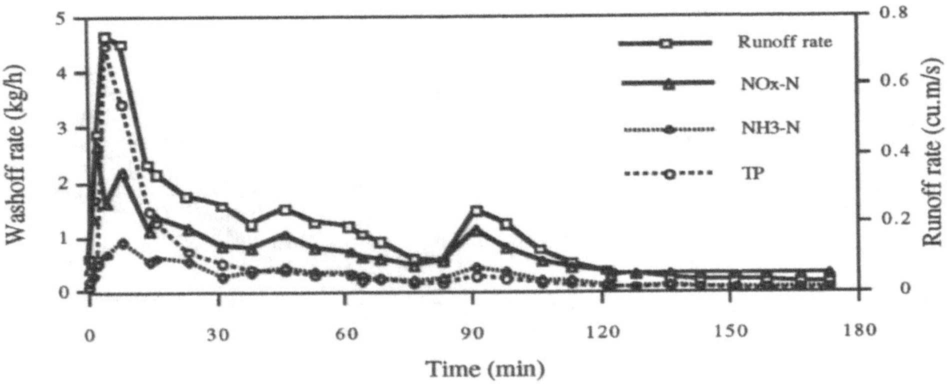

Figure 3 Hydrograph and Loadographs for NOx-N, NH_3-N and TP for Event 770812.

4. POLLUTANT BUILDUP

The pollutant accumulation is derived by the dust and dirt method (Huber and Dickinson, 1989). An exponential form is used to predict the dust and dirt build up. The initial NOx-N loads, ammonia (NH_3-N) loads and total phosphorous (TP) loads are determined from the quantities of dust and dirt by specifying the fraction of pollutant present per unit mass of dust and dirt. Figure 3 gives the calibrated graphs for each water quality parameter used to estimate initial nutrient loads.

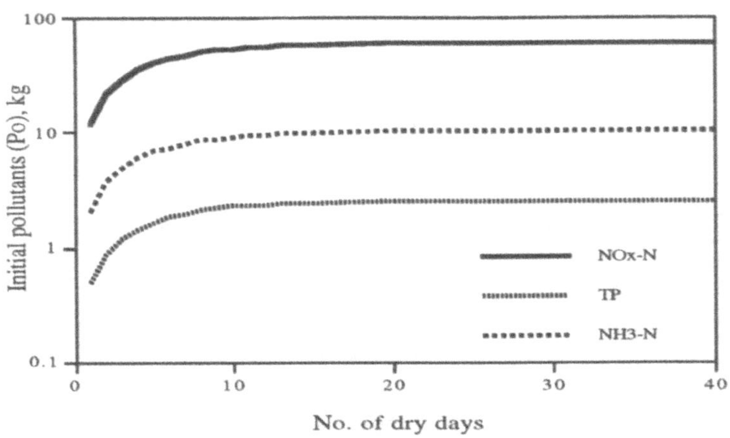

Figure 3. Variation of nutrients accumulation in the Giralang Catchment.

5. CALIBRATION RESULTS

The simulation of the washoff algorithms was achieved by minimising an error function using nonlinear regression option of the statistical analysis package SAS for selected four storms (as identified by year, month and date of event) in the study area. The variations in calibrated parameter values are given in Tables I, II and III for three selected methods for NH_3-N, NOx-N and TP respectively. The calibrated coefficients (K and a) varies for each method as well as for each water quality parameter. However, the relative variation of K and a for the proposed method is small in comparison to other methods for each storm event.

Table I Coefficients and exponents used by different methods for NH_3-N.

Event No.	Proposed		Nakamura			SWMM4	
	K	a	K_n	a	b	K_h	a
780510	0.08	1.07	0.09	0.84	0.066	0.08	0.84
770930	0.17	1.12	0.22	0.83	0.004	0.21	0.84
770812	0.17	1.35	0.21	0.77	0.158	0.18	0.58
780308	0.12	0.75	0.12	0.89	-0.170	0.14	0.89
Mean	0.135	1.07	0.16	0.83	0.014	0.15	0.79

Table II Coefficients and exponents used by different methods for NOx-N.

Event No .	Runoff depth (mm)	Proposed		Nakamura			SWMM4	
		K	a	K_n	a	b	K_h	a
780510	1.82	0.008	1.08	0.009	0.70	-0.14	0.010	0.70
770930	3.52	0.011	1.12	0.014	0.69	-0.019	0.014	0.69
770812	1.88	0.016	1.26	0.017	0.63	-0.062	0.019	0.67
780308	2.65	0.015	1.05	0.024	0.49	-0.191	0.027	0.51
Mean	2.47	0.0125	1.13	0.016	0.63	-0.103	0.0175	0.643

Table III Coefficients and exponents used by different methods for TP.

Event No.	Proposed		Nakamura			SWMM4	
	K	a	K_n	a	b	K_h	a
780510	0.058	0.809	0.041	1.449	-0.054	0.043	1.445
770930	0.050	0.932	0.039	1.462	0.004	0.039	1.458
770812	0.093	0.625	0.060	2.052	0.124	0.049	2.041
780308	0.130	0.530	0.105	0.966	-0.378	0.178	0.813
Mean	0.083	0.724	0.060	1.482	-0.076	0.077	1.440

The calibration of pollutant washoff load for each water quality parameter is shown in Table IV. Two parameters, relative error "RE" and absolute error "AE" are used for the purpose of assessing the goodness of validation. The parameters are defined as follows.

$$RE = \frac{\text{Meas.} - \text{Pred.}}{\text{Meas.}} \tag{8}$$

$$AE = \frac{|Meas. - Pred.|}{Meas.} \qquad (9)$$

where Meas. = Measured event total load, kg
Pred. = Predicted event total load, kg

Values of "AE" and "RE" near zero give the best prediction. However for average value of "RE", it is possible to get small "RE" values even though results may not be good, because positive values of "RE" are compensated by negative values of "RE". Hence it is important that both AE and RE values are close to zero for a reliable model. Baffaut and Delleur (1990) considered good calibration when average values of "AE" is less than 20%. The assessment of goodness of calibration for each water quality parameter is given in Table V. The average values of "AE" for the proposed method varies between zero to 0.03 for different water quality parameters. The average values of "AE" for SWMM4 and Nakamura methods vary between 0.13 to 0.28 and 0.06 to 0.14 respectively. A similar pattern can be observed for values of "RE". Hence it can be concluded that the calibrated parameters of the proposed method show better agreement with measured data than the other two methods.

Table IV Calibration results for different water quality parameters.

Event No.	Pollutant washoff (kg)											
	Measured			Proposed			Nakamura			SWMM4		
	NH_3-N	NOx-N	TP	NH_3-N	NOx-N	TP	NH_3-N	NOx-N	TP	NH_3-N	NOx-N	TP
780510	0.37	0.98	0.88	0.37	0.94	0.89	0.35	0.98	0.89	0.37	0.94	0.89
770930	1.22	2.71	1.63	1.23	2.66	1.51	1.34	2.94	1.32	1.23	2.66	1.51
770812	0.78	2.02	1.22	0.77	1.94	1.24	0.78	1.97	1.14	0.77	1.94	1.24
780308	0.56	2.56	1.91	0.56	2.47	1.98	0.65	2.93	2.45	0.56	2.47	1.98

Table V Assessment of goodness of calibration.

Event No.	Proposed			Nakamura			SWMM4		
	NH_3-N	NOx-N	TP	NH_3-N	NOx-N	TP	NH_3-N	NOx-N	TP
780510	0.00	0.04	-0.01	0.00	-0.05	-0.01	0.08	-0.08	-0.02
770930	-0.01	0.02	0.07	-0.08	-0.10	0.19	-0.15	-0.08	0.09
770812	0.01	0.04	-0.01	0.02	0.00	0.07	0.04	-0.03	0.16
780308	0.00	0.04	-0.04	-0.14	-0.16	-0.28	-0.32	-0.33	-0.87
RE	0.00	0.03	0.03	-0.05	-0.05	-0.01	-0.09	-0.13	-0.16
AE	0.00	0.03	0.00	0.06	0.08	0.14	0.15	0.13	0.28

6. VERIFICATION RESULTS

Verification is an indicator of the predictive ability of a chosen washoff method. The verification of the proposed washoff algorithm, using the parameter values specified from the calibration and without using measured washoff loads, are demonstrated for storm events which are separate from the events selected for the calibration. The measured and predicted washoff values for NH_3-N, NOx-N and TP are given in Tables VI, VII and VIII respectively. The variation of the assessment of goodness of verification results are shown in Tables IX, X and XI for the respective nutrient transport loads.

It may be considered that for satisfactory verification of nutrient washoff load with independent events a value of "AE" less than 40% is appropriate. Considering the values of both "RE" and "AE", the proposed method predicts NOx-N and TP nutrient loads significantly better than other two methods, However for NH_3-N nutrient loads all three method shows similar prediction. It should also be considered that the proposed method is much simpler than the two other washoff algorithms.

Table VI Verification results for NH_3-N.

Event No.	Runoff depth mm	Pollutant washoff (Kg)			
		Measured	Proposed	Nakamura	SWMM4
780409	2.18	1.09	0.46	o.45	0.44
770912	1.32	0.34	0.24	0.26	0.25
790303	2.89	0.54	0.85	1.02	0.99
780903	11.10	2.07	2.06	2.03	1.96
790315	11.48	1.29	1.92	1.71	1.66

Table VII Verification results for NOx-N.

Event No.	Runoff depth mm	Pollutant washoff (Kg)			
		Measured	Proposed	Nakamura	SWMM4
770113	1.94	1.55	1.52	1.88	2.08
780409	2.18	0.79	1.22	1.21	1.41
770912	1.32	1.41	0.58	0.75	0.79
790303	2.89	2.57	2.40	3.53	3.90

Table VIII Verification results for TP.

Event No.	Runoff depth mm	Pollutant washoff (Kg)			
		Measured	Proposed	Nakamura	SWMM4
780409	2.18	0.810	0.95	1.06	1.32
770912	1.32	0.663	0.56	0.55	0.65
790303	2.89	0.878	1.63	1.48	1.80
780115	2.45	2.219	1.00	1.43	1.66

Table IX Relative error (RE) and absolute error (AE) for different methods for NH_3-N (Verification).

Event No.	Proposed		Nakamura		SWMM4	
	RE	AE	RE	AE	RE	AE
780409	0.57	0.57	0.59	0.59	0.60	0.60
770912	0.29	0.29	0.23	0.23	0.26	0.26
790303	-0.57	0.57	-0.89	0.89	-0.83	0.83
780903	0.00	0.00	0.02	0.02	0.05	0.05
790315	-0.49	0.49	-0.32	0.32	-0.29	0.29
Mean	-0.04	0.39	-0.07	0.41	-0.04	0.41

Table X Relative error (RE) and absolute error (AE) for different methods for NOx-N (Verification).

Event No.	Proposed		Nakamura		SWMM4	
	RE	AE	RE	AE	RE	AE
770113	0.02	0.02	-0.21	0.21	-0.34	0.34
780409	-0.54	0.54	-0.53	0.53	-0.78	0.78
770912	0.59	0.59	0.47	0.47	0.44	0.44
790303	0.07	0.07	-0.37	0.37	-0.52	0.52
Mean	0.03	0.30	-0.16	0.39	-0.30	0.52

Table XI Relative error (RE) and absolute error (AE) for different methods for TP (Verification).

Event No.	Proposed		Nakamura		SWMM4	
	RE	AE	RE	AE	RE	AE
780409	-0.17	0.17	-0.31	0.31	-0.63	0.63
770912	0.15	0.15	0.17	0.17	0.02	0.02
790303	-0.86	0.86	-0.69	0.69	-1.05	1.05
780115	0.55	0.55	0.36	0.36	0.25	0.25
Mean	-0.08	0.43	-0.12	0.38	-0.35	0.49

7. SUMMARY AND CONCLUSION

Three different washoff methods to predict nutrient loads in an urban catchment are presented. The models involve calibration of washoff coefficients and initial pollutant load. The establishment of the correct washoff coefficients such as K and a for each water quality parameter is important. The calibrated values shown in this paper may be catchment specific and should be applied with caution. Fine tuning of K and a can be done if extensive data is available, a process which is considered to be expensive.

As catchment undergoes urbanisation the percentage of imperviousness as well as the land use pattern changes which can also affect K and a coefficients. Once the K and a values were established, Eq. 5 does provide a simple algorithm for the estimation of nutrient loads emanating from the catchment for a given storm event. This will provide a significant tool for practising engineers in the modelling of nutrient washoff loads as well as in the design of water pollution control structures which can be used in urban catchment management.

The application of the proposed equation to a limited amount of nutrient data from Giralang catchment in Australia has shown very satisfactory results. In comparison with the two existing washoff algorithms (SWMM4 and Nakamura), the new equation has indicated reasonably better predictions. The simplicity of the proposed algorithm avail itself for incorporation with existing catchment water quality models. However the coefficients associated with the methods should be tested by a large number of reliable urban water quality data.

ACKNOWLEDGMENT

The authors wish to kindly acknowledge the ACT Electricity and Water for providing nutrient and runoff data. The second author wishes to thank the Iranian Government for providing a postgraduate research scholarship.

REFERENCES

Baffaut, C. and Delleur, J. W. (1990). 'Calibration of SWMM runoff quality modelling with expert system' *Journal of Water resources Planning and Management,* **ASCE , 116, pp. 247-261.**

Huber, W.C. and Dickinson, R.E (1989). 'Stormwater Management Model-Version 4,' User's Manual, U.S. Environmental Protection Agency, Athens, Georgia 30613, U.S.A.

Nakamura, E. (1984). 'Factor affecting the removal of street surface contaminants by overland flow,' *Journal of Research,* Public Works Research Institute, Japan, **24, 1-49**

National Capital Development Commission (1980). 'Monitoring stormwater flow and water quality in paired rural and urban catchment in ACT', **Technical paper No. 29.**

Sivakumar, M. and Boroumand-Nasab, S. (1993). 'Urban storm water quality modelling of NOx and TP pollutant loads' *Proc. International Conference on Environmental Management: Geo-Water and Engineering Aspects,* Wollongong University, Wollongong, N.S.W. Australia, R.N. Chowdhury and M. Sivakumar (eds), Balkema, **1 pp. 207-212.**

SECTION 3:

SEDIMENT TRANSPORT IN UPLAND AREAS

A FRAMEWORK FOR MODELING SEDIMENT PROCESSES IN UPLAND WATERSHEDS

Vicente L. Lopes
School of Renewable Natural Resources, College of Agriculture
University of Arizona
Tucson, Arizona 85721
USA

and

Peter F. Ffolliott
School of Renewable Natural Resources, College of Agriculture
University of Arizona
Tucson, Arizona 85721
USA

ABSTRACT: A modeling framework is presented as the basis for developing numerical models of sediment processes on upland watersheds. Sediment processes are described in terms of both broad shallow flow areas and concentrated flow systems. Basic assumptions and governing equations are presented for broad shallow flow and concentrated flow sediment processes. The effects of space and time averaging on the predictive equations are described. Starting from the most general one-dimensional, unsteady model, simpler model structures are obtained and the successive simplifications made on the governing equations to obtain simpler and less detailed formulations are described. Although progress has been made in modeling sediment processes on upland watersheds, the development of improved modeling approaches must be based on a better understanding of runoff mechanisms and the mechanics of sediment generation, transport, and deposition occurring over a range of catchments scales and environmental conditions. Future research must address the development of boundary conditions, model components, the scales over which the governing equations are valid, and integration of the equations in space and time.

1. Introduction

Upland watersheds form the headwaters of streams and, therefore, river systems, and are important for surface water supplies and for recharge of many groundwater aquifers. As such, their proper management is essential to water resource developmental projects to increase water supplies, for hydropower production or flood control, or to maintain the streamflow levels needed for the dilution of pollutants, fish and wildlife, or navigation in downstream channels.

V. P. Singh and B. Kumar (eds.), Water-Quality Hydrology, 93–111.

Water-based outdoor recreation and fisheries industries also benefit, and local and regional economical growth can be stimulated from the construction of large reservoirs. However, rarely is water resource planning and management on upland watersheds based on a fundamental understanding of the hydrologic processes in effect on the watershed in question.

Of particular importance for water resources management is understanding of the processes of soil erosion, the transport of eroded sediments, and sediment deposition in stream channels and reservoir bottoms. The ability to predict soil erosion and sediment yield under current and alternative land-use conditions is an important step in identifying the underlying causes of erosion and sediment damages, determining the source of sediment causing such damages, and formulating feasible control measures. There is a current perspective that process-based mathematical modeling of erosion and sediment yield can provide an approach to achieve such understanding. Therefore, there is a need for developing a modeling framework from which different modeling approaches can be derived according to specific needs and/or conditions. It is to this topic that this paper addresses itself.

Sediment processes embodies the processes of detachment, entrainment, transportation and deposition of sediment by flowing water. For convenience, surface runoff is described as a combination of broad shallow flow areas and concentrated flow systems. Broad shallow flow areas refer to interrill areas where soil erosion is controlled by raindrop impact and sediment is transported by shallow sheet flow. This definition does not exclude the possibility of some detachment by flow and deposition. Concentrated flow systems are geomorphic features of the landscape that convey water and sediment originated from adjacent contributing broad shallow flow areas and their banks and beds to river systems. These features include rills, gullies, and small stream channels. Distinguishing among rills, gullies and small stream channels is difficult. They are all components of a system that interact with geomorphic and hydrologic processes, and, therefore, act as sediment sources or sinks, depending upon the dynamic characteristics of the water-sediment mixture that they carry (Schumm 1977).

2. Broad Shallow Flow Sediment Modeling

Sediment processes on broad shallow flow areas begin when the impact of raindrops exceeds the soil's ability to withstand the impulse at the soil surface. Detachment also can occur when shear stresses caused by flowing water exceed the soil's ability to resist these erosive forces. Variables influencing sediment processes on broad shallow flow areas are soil type, soil condition, and soil moisture content at the start of the event; slope gradient and slope length; above and below ground biomass; residual land use; tillage; rainfall amount, intensity, and duration; and rainfall excess rate. In this phase, the soil particle sizes of major interest are in the silt and clay ranges.

Once detachment occurs, soil particles are transported by raindrop splash and flowing water. Conditions that limit raindrop detachment can limit the sediment supply available for transport by splash and flow mechanisms. The rate of sediment transport by broad shallow flow is influenced by factors controlling the amount of sediment available for transport, the sediment supply, flow disturbances caused by raindrop impact, flow depth, flow velocity, and accelerations due to microtopographic flow patterns as affected by slope steepness, length, and shape.

Deposition can occur on broad shallow flow areas when sediment particles come to rest on the soil surface, which occurs when sediment load in the flow exceeds the flow's sediment transport capacity. The rate of deposition is determined by both flow characteristics affecting energy, momentum, and turbulence, and sediment characteristics, including particle interactions, affecting fall or settling velocity.

Two main approaches to modeling broad shallow flow sediment process have been attempted: one that is empirically-based typified by the Universal Soil Loss Equation, and another that is mostly process-based exemplified by a set of coupled overland flow-sediment equations based on the concepts of kinematic flow.

The Universal Soil Loss Equation (USLE) is the equation most widely used for upland erosion prediction since the late 1960s (Wischmeier and Smith 1978), particularly in the USA. Developments resulting in the USLE were summarized by Meyer (1984) and Nyhan and Lane (1986). The equation was developed as a method to predict long-term average annual soil loss from interrill (sheet erosion) and rill areas. It was designed to meet the need for a relatively simple technique for predicting soil loss rates for specific situations.

Despite its widespread use and breadth of experience that it incorporates, the USLE is based on data restricted to uniform slope test plots, and it suffers from the conceptual defect that rainfall and soil factors (among others) cannot be simply multiplied together largely because of the subtractive effect of infiltration on overland flow from a given rainfall. In addition, the equation does not incorporate the processes of detachment and deposition by flowing water.

Research on process-based sediment modeling has taken two basic approaches. For simplicity, the first approach assumes steady state conditions even though the sediment processes are unsteady. Major contributions using this approach are those of Meyer and Wischmeier (1969), Foster and Meyer (1972), Komura (1976), Meyer et al. (1983), Rose (1985), and Lopes (1991). The second approach has its focus in attempting to mathematically describe the major significant features of the flow-sediment system without steady state assumptions. The kinematic-wave approximation to the dynamic flow equations has been used largely to drive the hydraulics of the sediment processes, which has been modeled using the continuity equation for advective sediment transport and empirical relationships for detachment by raindrop impact and shear stress. The flow and (coupled) sediment equations have been solved analytically using the method of characteristics or numerically by finite difference methods. Analytical kinematic-wave models have been restricted in applications because of the simplifying assumptions (constant rainfall intensity and constant infiltration rate, for instance) required to solve the equations. Contributions to this approach are those by Bennett (1974), Hjelmfelt et al. (1975), Simons et al. (1975), Smith (1976), Li (1979), Borah et al. (1981), Smith (1981), Lane and Shirley (1982), Singh and Prasad (1982), Croley (1982), Singh (1983) and Lopes and Lane (1988). The two approaches are described in this paper in the context of their application to upland watersheds.

2.1 GOVERNING EQUATION OF SEDIMENT TRANSPORT BY BROAD SHALLOW FLOW

The sediment continuity equation describing the movement of suspended sediment particles in a one-dimensional, infinitely wide, free surface flow is:

$$\frac{\partial(ch)}{\partial t} + (1-\lambda)\gamma_s\frac{\partial z}{\partial t} + \frac{\partial(cq)}{\partial x} = \frac{\partial}{\partial x}\left[h\epsilon_p\frac{\partial c}{\partial x}\right] \tag{1}$$

where h = depth of flow (L), q = flow discharge per unit width (L^2T^{-1}), c = sediment concentration (FL^{-3}), λ = porosity of deposited sediment, γ_s = specific weight of sediment (FL^{-3}), z = local surface elevation (L), ϵ_p = sediment particle mass transfer coefficient (L^2T^{-1}), x = distance in the direction of flow (L), and t = time (T). (Note: L represents length, T time and F force or mass for all variables.)

There are two traditional approaches in solving Eq. (1). One approach assumes unsteady state (time variant) conditions (Hjelmfelt et al. 1975; Lane and Shirley 1982; and Lopes and Lane 1988). The other assumes steady state (time invariant) conditions (Meyer and Wischmeier 1969; Komura 1976; and Rose et al. 1983). These approaches are discussed briefly in the following sections.

2.1.1 *Time Variant Approach.* Equation (1) can be solved by numerical methods once the flow velocity field and boundary conditions for the sediment are known. The term on the right-hand side of Eq. (1) accounts for dispersion of the material while it is suspended in the flow. In comparison with the effects of other processes represented in Eq. (1), dispersion is normally negligible and, as a consequence, is usually not considered in the development of simplified versions of Eq. (1). A simplified version of Eq. (1) is:

$$\frac{\partial(ch)}{\partial t} + \frac{\partial g_B}{\partial x} = \Phi(x,t) \tag{2}$$

where the term $\partial(ch)/\partial t$ is the storage rate of sediment within the flow depth, $\partial g_B/\partial x$ is the buildup or loss of sediment with distance, and $\Phi(x,t)$ is the net sediment flux to the flow ($F/L^2/T$).

Equation (2) describes flow and sediment processes which are uniformly distributed across the slope. Therefore, variables such as h(x,t) are averages and q(x,t) and $\Phi(x,t)$ are expressed on a per width or per area basis, although the processes can actually occur on a limited area (Foster 1982).

2.1.1.1 Transport Capacity Approach. Soil erosion researchers have made different hypothesis on formulating the right-hand side of Eq. (2). A reasonable hypothesis seems to be that the net sediment flux through the bottom surface is given by a reaction type equation:

$$\Phi(x,t) = k(c_e - c) + e_i \tag{3}$$

where e_i is the rate of sediment entrainment by raindrop impact ($FL^{-2}T^{-1}$), k is a first-order reaction coefficient (LT^{-1}), and c_e is the sediment concentration at equilibrium transport capacity near the soil surface. The transfer rate coefficient k describes soil erodibility as constrained by cohesiveness and related factors when c_e is greater than c (net entrainment case), and it alternatively represents directly the characteristic fall velocity of the particles when c exceeds c_e (net deposition case). In the case of local net deposition, k is usually calculated from particle size and density. Particle fall velocity is usually calculated with the assumption that the particles have drag characteristics and terminal fall velocities similar to spheres. The equilibrium sediment concentration, c_e, for a local instantaneous flow condition can be computed as $c_e = T_c/q$, where T_c is the equilibrium transport capacity for the given flow condition.

By replacing the right-hand side of Eq. (2) by Eq. (3), the former equation becomes:

$$\frac{\partial(ch)}{\partial t} + \frac{\partial g_B}{\partial x} = \frac{KT_c}{q}\left(1 - \frac{g_B}{T_c}\right) + e_i \tag{4}$$

Equation (4) is similar to the non-equilibrium equation for solid transport obtained by Foster (1982). Given an initial condition of known concentration for all x and an upstream boundary condition for all t, Eq. (4) can be solved numerically at each time step.

2.1.1.2 Simultaneous Sediment Exchange Approach. Another approach to time variant sediment process modeling involves the concept of simultaneous (dynamic) sediment exchange between the slope surface and the flow. This approach describes the rates at which each separate source and sink contributes to sediment concentration in the broad shallow flow system. Sediment entrainment in broad shallow flow areas is accomplished by two processes during a runoff storm event: sediment entrainment through raindrop impact, and sediment entrainment by hydraulic shear. Entrainment and deposition can occur simultaneously at different rates. The resultant sediment concentration is controlled by the relative magnitude of these processes as explained earlier. The rate of each of these processes depends on different factors and, therefore, should be described separately.

Equation (1) can be written for a single particle size on a uniform slope (Lopes and Lane 1988) as:

$$\frac{\partial(ch)}{\partial t} + \frac{\partial(cq)}{\partial x} = e_r - d + e_i \tag{5}$$

where $c(x,t)$ = the local mean sediment concentration for the ith particle size (FL^{-3}), $e_r(x,t)$ = the rate of sediment entrainment by hydraulic shear ($FL^{-2}T^{-1}$), $d(x,t)$ = the rate of sediment deposition ($FL^{-2}T^{-1}$), and $e_i(x,t)$ = the rate of sediment entrainment by raindrop impact ($FL^{-2}T^{-1}$).

Equation (5) reflects the concept that entrainment is not the same thing as net erosion, and deposition is not the same thing as net deposition. Entrainment (upward flux) is the actual rate at which detached sediment particles lying on the soil surface are picked up and carried into the flow region. Deposition (downward flux) is the actual rate at which suspended sediment particles fall to the soil surface. In many turbulent flows that carry sediment, there is a continuous exchange of particles between the flow and the soil surface (Croley 1982; Meyer et al. 1983; Rose 1985). Consequently, entrainment and deposition can take place simultaneously at a location in those cases when net deposition or net detachment do not occur along the soil surface. However, if deposition exceeds entrainment, net deposition occurs, and, conversely, when entrainment exceeds deposition, net erosion takes place. If deposition is equal to entrainment, the transport process is in equilibrium at this point.

Sediment entrainment and transport occurs when the forces tending to entrain and transport sediment exceed those tending to resist removal. Water flowing over the soil surface exerts forces on the soil particles that tend to entrain them. The forces that resist entrainment are caused mainly by the weight of the particles for cohesionless or coarse sediment. For cohesive or fine sediment that contain significant fractions of silt, clay, or both, the forces resisting entrainment differ according to the relative effect of interparticle physico-chemical interactions.

Groups of particles (aggregates) also are entrained as units rather than moving as individual particles in cohesive sediments. Sediment entrainment by hydraulic shear is described by a relationship expressing the entrainment rate as proportional to a power of the effective shear stress acting on the soil surface (Foster 1982). For a single particle size:

$$e_r = a(\tau_e)^b \tag{6}$$

where a = a soil detachability parameter for hydraulic shear ($F^{1-b}L^{2b-2}T^{-1}$), τ_e = effective hydraulic shear (F/L^2), and b = an exponent in the range 1.0 to 2.0.

An effective shear stress incorporates the concept of a critical shear stress, which represents the minimum requirement for initiation of particle movement. Critical shear stress values are available from test results conducted in a single rill (Foster 1982). However, it is difficult to evaluate critical shear stress for upland watershed applications.

Assuming that the sediment concentration distribution is uniform with flow depth above the soil surface, the mass rate of sediment deposition (downward flux) on a broad shallow flow area is proportional to the local mean sediment concentration (Mehta 1983). For a single particle size:

$$d = \epsilon V_s c \tag{7}$$

where ϵ = a dimensionless parameter depending on the soil and fluid properties, and V_s = the particle fall velocity (LT^{-1})

Sediment entrainment by raindrop impact, e_i, has been modeled as a function of the rate of detachment by raindrop impact and rate of transport of sediment by shallow flow (Foster 1982). On steep slopes, entrainment by rainfall is limited by the rate of detachment by raindrop impact, whereas entrainment by rainfall is limited by the rate of transport of sediment particles by shallow flow on flat slopes (Foster et al. 1977). A simple expression of entrainment by raindrop impact incorporates rainfall intensity, i, as a measure of the erosivity power of raindrop impact and rainfall excess, q_*, to reflect the rate of sediment transport by shallow flow (Lane and Shirley 1985):

$$e_i = k_i i^2 (q_*/i) = k_i (i)(q_*) \tag{8}$$

where k_i = a parameter to measure soil detachability by raindrop impact ($FL^{-4}T$). The ratio of rainfall excess rate, q_*, to rainfall intensity, i, can be interpreted as a measure of normalized runoff intensity for sediment transport by shallow flow. (Notice that when $q_* = 0$ (pre- or post-rainfall excess phase) or when $i = 0$ (post-rainfall phase), there is no entrainment by raindrop impact, and when $q_* = i$, (impermeable surface), entrainment rate by raindrop impact is not limited by transport of shallow flow).

2.1.2 *Time Invariant Approach.* For simplicity, the first term of Eq. (1), that is, the storage variation of sediment within the flow depth can be neglected. If this is done, the conservation of mass for the sediment in one-dimensional, steady state and gradually varying flow on a slope surface can be written as:

$$\frac{dg_B}{dx} = \Phi(x) \tag{9}$$

where $g_B(x) = c(x)q(x)$ = the weight rate of sediment transport per unit width, $(FL^{-1}T^{-1})$ and $\Phi(x)$ = net detachment or net deposition rate $(FL^{-2}T^{-1})$.

Different approaches have been taken to formulate the right hand-side of Eq. (9). Foster and Meyer (1975), Meyer and Wischmeier (1969) and Foster and Meyer (1972) suggested the following relation based on a steady-uniform approximation and composed of the smaller of interrill detachment (supply) or interrill transport capacity:

$$\frac{dg_B}{dx} = \min \cdot (e_s, e_t) \tag{10}$$

where e_s = sediment supply rate $(FL^{-2}T^{-1})$ and e_t = transport capacity $(FL^{-2}T^{-1})$. If $e_s > e_t$ sediment transport is said to be transport-limiting; if $e_s = e_t$ sediment transport is at equilibrium conditions, and if $e_s < e_t$, sediment transport is detachment-limiting.

On steep slopes, the rate of detachment by raindrop impact can limit soil erosion, whereas transport capacity can limit the rate of sediment transport on flat slopes (Foster et al. 1977).

An approach to solving Eq. (10) is to assume the supply-limiting sediment transport case. This approach assumes that sediment supply is a function of detachment by raindrop impact and transport by broad shallow flow (detachment by flow is neglected). Following this assumption, the sediment delivery rate from interrill areas has been described (Lane and Nearing 1989) as:

$$\frac{dg_B}{dx} = C_i k_i (i_e)^2 SF \tag{11}$$

and

$$SF = 1.05 - 0.85e^{-4\sin\alpha} \tag{12}$$

where C_i = a dimensionless cover parameter; k_i = a parameter to measure soil detachability by raindrop impact $(FL^{-4}T)$ and i_e = effective rainfall rate (LT^{-1}), defined as the rainfall intensity during periods of rainfall excess, i.e., when rainfall intensity is greater than infiltration capacity, and zero otherwise, SF = slope adjustment factor, α = slope angle.

The parameter C_i in Eq. (11) is a function of canopy height and the fraction of the ground surface covered by plant canopy and other materials in direct contact with the soil surface. The erodibility parameter, k_i, is defined for a base-line condition, but its actual value depends on soil texture and other soil properties that are affected by climate, tillage operations, and other disturbances. Laflen et al. (1991) present values for k_i for a number of soil types in the United States.

An alternative approach to solving Eq. (10) is to assume that sediment transport at equilibrium conditions ($e_s = e_t$) can be expressed as a function of sediment transport only. Following this assumption, Lopes (1991) presented Eq. (10) as:

$$\frac{dg_B}{dx} = \gamma_s P_s \tag{13}$$

where γ_s = specific weight of sediment (FL^{-3}) and P_s = the volume rate of fine-sediment entrainment (pickup) per unit area ($L^3L^{-2}T^{-1}$).

Using this approach, soil particle detachment by raindrop impact is not considered explicitly and deposition is neglected.

To determine g_B from Eq. (13), P_s can be obtained from an equation for sediment transport. Lopes (1991) used the sediment transport formula of Einstein-Brown (Vanoni 1975) to compute P_s. That is:

$$P_s = 40(1/\psi)^3F_1[(S-1)gd_s^3]^{1/2} \tag{14}$$

where

$$1/\psi = \frac{\tau}{\gamma(S-1)d_s} \tag{15}$$

$$F_1 = [(2/3)+36v^2/(S-1)gd_s^3]^{1/2} - [36v^2/(S-1)gd_s^3]^{1/2} \tag{16}$$

where d_s = the representative particle size usually taken as the median size d_{50} (L). The quantity, F_1, appears in the Rubey's formula for fall velocity (Vanoni 1975). (Note that $1/\psi$ is the same as the dimensionless shear stress, τ_*, shown by the Shields' diagram.)

Li et al. (1973) presented an expression for boundary hydraulic shear, τ, for turbulent flow (Reynolds number between 2000 and 12600) as:

$$\tau = \gamma\left[\frac{(k+0.012)q^{7/4}v^{1/4}}{8g}\right]^{1/3}S_o^{2/3} \tag{17}$$

where k = a constant describing the Darcy-Weisbach friction factor without rainfall and v = the fluid kinematic viscosity (L^2T^{-1}). (Li et al. (1973) obtained the values of k = 0.24 for flow over smooth surfaces. Komura (1976) suggested the value of k = 0.60 for flow over rough surfaces).

The continuity equation for flowing water at equilibrium on a slope surface is:

$$q = q_o + q_* x \tag{18}$$

where q_o = the inflow rate per unit width in the upper end of slope (L^2T^{-1}).

A steady state erosion equation for broad shallow flow can be defined by substituting Eqs. (17) and (18) in Eq. (15), substituting the resulting expression for P_s into Eq. (13), and integrating with respect to x. For q_o = 0, the resulting equation is:

$$g_B = \frac{CF_{*}q_{*}^{1.75}L^{0.75}S_o^{\ 2}}{(d_s)^{1.5}} \tag{19}$$

where

$$F_{*} = [(2/3)+(0.0016/d_s^{\ 3})]^{1/2} - (0.0016/d_s^{\ 3})^{1/2} \tag{20}$$

in the derivation of Eq. (19), the following values were used: $k = 0.6$, $\gamma = 1000$ kgf/m^3, $\gamma_s = 1300$ kgf/m^3, $g = 9.8$ m/s^2, $S = 2.65$ and $\nu = 8.5\text{x}10^{-7}$ m^2/s, for which $C = 3.52\text{x}10^5$, g_B in Kgf/m^2.s, L in m, d_s in mm and q_{*} in m/s.

Equation (19) is valid for bare soil conditions with high contents of silt and sand. However, this equation must be used with caution for other types of soil and surface cover conditions. Equation (19) should be multiplied by C_v (ratio of bare-soil area to total area) and C_e (soil erodibility parameter) to include the effects of surface cover and soil resistance to soil particle detachment by raindrop impact and hydraulic shear, respectively.

3. **Concentrated Flow Sediment Modeling**

Sediment processes by concentrated flow on upland watersheds are difficult to model due to complex interactions between independent variables. The interrelationships involve not only the available sediment supply from adjacent contributing broad shallow flow areas, but also flow velocity, flow depth, channel widths, bed slope and energy gradients, water temperature, particle size distribution, shape and density, bank roughness and bed forms, transmission losses into the streambeds, and (at times) chemical composition of the water-sediment mixture. Most factors affecting sediment processes by concentrated flow change not only with time and distance along the stream channel, but also with flow depth and lateral distance at an individual cross-section. These complex interrelationships have not been well understood.

Recently, sediment processes by concentrated flow on upland watersheds have been modeled by using the equations of continuity and momentum for unsteady free surface flow, and the continuity equation for advective sediment transport. Both, unsteady (time variant) and steady (time invariant) state conditions have been assumed. Contributions to time-variant sediment process modeling are those of Bennett (1974), Simons et al. (1975), Smith (1976), Li (1979), Borah et al. (1981), Smith (1981), Alonso and Decoursey (1983), and Lopes and Lane (1988).

A major contribution using the time invariant approach to modeling sediment processes by concentrated flow systems was presented by Foster and Meyer (1972, 1975). The time variant and time invariant approaches to modeling sediment processes by concentrated flow systems on upland watersheds are discussed in the following sections.

3.1 GOVERNING EQUATION OF SEDIMENT TRANSPORT BY CONCENTRATED FLOW

The differential equation that describes suspended sediment movement in stream channels is nearly the same as for broad shallow flow systems (Lopes and Ffolliott 1992). The major difference in the equation is the source of lateral sediment inflow into these systems. Lateral sediment inflow is controlled mainly by raindrop impact processes in broad shallow flow areas.

In concentrated flow systems, the lateral sediment inflow is controlled by sediment processes occurring on adjacent broad shallow flow areas, which are usually conceptualized as soil particle detachment by raindrop impact and detachment, transport, and deposition by broad shallow flow. The differential equation used to describe sediment transport in one-dimensional, unsteady flow in a single concentrated flow system is commonly written as:

$$\frac{\partial(CA)}{\partial t} + B(1-\lambda)\gamma_s\frac{\partial z}{\partial t} + \frac{\partial(CQ)}{\partial x} = \frac{\partial}{\partial x}\left[A\epsilon_p\frac{\partial C}{\partial x}\right] \tag{21}$$

where A = cross-sectional area of flow (L^2), Q = flow discharge (L^3T^{-1}), C = sediment concentration (FL^{-3}), B = width of streambed (L), λ = porosity of deposited sediment, γ_s = specific weight of sediment (FL^{-3}), z = local surface elevation (L), ϵ_p = sediment particle mass transfer coefficient (L^2T^{-1}), x = distance in the flow direction (L), and t = time (T). (Note that L represents length, T time and F force or mass for all variables.)

Sediment discharge and concentration are usually given in weight units, although they should be given in mass units. This confusion presents no difficulty, since mass in pounds and kilograms are numerically equal to those of weight in the traditional systems of units. However, in the SI system of units, sediment discharge and concentration are given in terms of Newtons that numerically are 9.806 times as large as the same quantities expressed in kilograms of mass.

3.1.1 *Time Variant Approach.* Equation (21) can be solved by numerical methods once the flow velocity field is known and boundary conditions for the sediment are prescribed. The term on the right-hand side of Eq. (2) accounts for dispersion of the material while it is suspended in the flow. In comparison with the effects of other processes represented in Eq. (21), dispersion normally is negligible and, as a result, Eq. (21) can be rewritten as:

$$\frac{\partial(cA)}{\partial t} + \frac{\partial(cQ)}{\partial x} = \Phi(x,t) \tag{22}$$

where the term $\partial(cA)/\partial t$ represents the storage rate of sediment within the flow depth, $\partial(cQ)/\partial x$ represents the buildup or loss of sediment with distance, and $\Phi(x,t)$ represents the net sediment flux to the flow ($FL^{-1}T^{-1}$) which is the difference between the entrainment rate E (upward sediment flux) and the deposition rate D (downward sediment flux).

In terms of Eq. (21):

$$\Phi(x,t) = - B(1-\lambda)\gamma_s\frac{\partial z}{\partial t} = E - D \tag{23}$$

3.1.1.1 Transport Capacity Approach. Several authors have made hypothesis on formulating the right-hand side of Eq. (22). A reasonable hypothesis is that the net sediment flux through the bottom surface is given by a reaction type equation:

$$\Phi(x,t) = K(C_{eb} - C_b) + q_s \tag{24}$$

where K is a first-order reaction coefficient (L^2/T), C_{eb} is the "equilibrium" transport capacity sediment concentration at a reference elevation "b" near the channel bed (FL^{-3}), C_b is the actual sediment concentration at the same reference elevation near the channel bed (FL^{-3}), and q_s = lateral sediment inflow from adjacent contributing broad shallow flow areas per unit length of channel ($FL^{-2}T^{-1}$).

Letting $r_e = C_{eb}/C_e$ and $r_b = C_b/C$ to be the ratios between near channel bed sediment concentrations and the transport-averaged concentrations in equilibrium and nonequilibrium conditions, respectively, letting those ratios remain constant and equal ($r_e = r_b = r$) over the flow cross-sectional area, and replacing the first term of the right-hand side of Eq. (22) by Eq. (24), the former equation becomes:

$$\frac{\partial(AC)}{\partial t} + \frac{\partial(CQ)}{\partial x} = Kr(C_e - C) + q_s \tag{25}$$

where C_e = equilibrium transport-averaged concentration (FL^{-3}) and C = nonequilibrium transport-averaged sediment concentration (FL^{-3}).

The ratio r in Eq. (25) is usually assumed to be equal to a unity in one-dimensional flow modeling of upland watersheds, that is, in assuming a uniform vertical distribution of sediment concentration along the flow depth which is correct for sediment transport by shallow concentrated flow in rills. However, the residual terms issuing from averaging over the cross-section and represented by r_e and r_b (the vertical distribution of particle concentration) are not negligible.

The equilibrium sediment concentration, C_e, for a local instantaneous flow condition can be expressed as $C_e = T_c/Q$, where T_c is the equilibrium transport capacity for the given flow condition. Equation (25) represents the concept that there is a maximum sediment load which the flow can carry under equilibrium conditions for a given flow situation and sediment characteristics. This load is called the transport capacity, and the flow is said to be under saturation conditions. The full capacity situation is also the equilibrium situation for flow over a loose bed. Two different nonequilibrium situations can occur: when C is larger than C_e, net deposition occurs so that C decreases until the carrying capacity of the flow is reached; and when C is less than the capacity and sediment is available from the bed, net entrainment occurs so that C increases until again the carrying capacity of the flow is reached. Use of Eq. (25) requires a relation for determining transport capacity. Many bed-load formulas have been proposed in the literature which can produce different results of the order of 100 to 500%. None of these formulas has been found to be of universal validity. (Alonso et al. 1981; Julien and Simons 1985; Rijn 1984; Nakato 1990; and Woo et al. 1990).

The transfer rate coefficient K in Eq. (25) represents soil erodibility as constrained by cohesiveness and related factors when C_e is greater than C (net entrainment case). It also represents directly the characteristic fall velocity of the particles when C exceeds C_e (net deposition case). In the case of local net deposition, K usually is determined from particle size and density. Particle fall velocity is calculated assuming the particles have drag characteristics and terminal fall velocities similar to spheres, a limiting case for irregular shapes and aggregates.

3.1.1.2 Simultaneous Exchange Approach. As an alternative to the above, Lopes and Lane (1988) presented a modeling approach in terms of the actual rates of entrainment and deposition rather than using the traditional concept of sediment transport capacity. They presented Eq. (25) as:

$$\frac{\partial(CA)}{\partial t} + \frac{\partial(CQ)}{\partial x} = e_f - d + q_s \tag{26}$$

where $e_f(x,t)$ = source term for bed-material load ($FL^{-1}T^{-1}$) and d = rate of sediment deposition ($FL^{-1}T^{-1}$). The other variables are the same as defined earlier.

Equation (26) reflects an important concept that entrainment is not the same as net erosion, and settling is not the same thing as net deposition. Entrainment (upward flux) is the actual rate at which detached sediment particles lying on the soil surface are picked up and carried into the flow region. Settling (downward flux) is the actual rate at which suspended sediment particles fall to the soil surface. There is a continuous exchange of particles between the flow and the soil surface in many turbulent flows that carry sediment. Consequently, entrainment and settling can take place simultaneously in those cases when net deposition or net detachment at a location do not occur along the soil surface. However, if settling exceeds entrainment, then net deposition occurs, and, conversely, when entrainment exceeds settling, net erosion takes place. If settling is equal to entrainment, the transport process is in equilibrium at that point. Lopes and Lane (1988) used an equation initially developed for bed-load transport capacity to describe entrainment rates by concentrated flow systems. For a single particle size, the equation is:

$$e_r = \begin{cases} a(\tau_o - \tau_c)^m & \text{for} \quad \tau_o \geq \tau_c \\ 0 & \text{otherwise} \end{cases} \tag{27}$$

in which

$$\tau_o = \gamma f u^2/8g \tag{28}$$

$$\tau_c = \delta(\gamma_s - \gamma)d_s \tag{29}$$

where a = a coefficient for sediment entrainment by concentrated flow ($F^{1-b}L^{b-1}T^{2b-1}$), τ_o = average shear stress (FL^{-2}), τ_c = average critical shear stress (FL^{-2}), and m = an exponent in the range 1.0 to 2.0; f = Darcy-Weisbach roughness coefficient (dimensionless), u = local flow velocity (LT^{-1}), δ = a coefficient depending on the sediment and fluid properties (dimensionless), and d_s = particle diameter size (L).

The rate of sediment deposition (downward sediment flux) is proportional to the local mean sediment concentration and particle fall velocity (Mehta 1983). The deposition rate for a single particle size is described by Lopes and Lane (1988) as:

$$d = \epsilon B V_s C \tag{30}$$

where ϵ = a coefficient depending on the soil and fluid properties (dimensionless) and the other variables were as defined earlier.

Particle fall velocity is the primary variable defining the interaction of sediment transport with the channel bed, banks, or suspended load in the fluid. Particle fall velocity is estimated assuming standard drag force relationships for a sphere of a given diameter and density falling in still water. For large sediment particles (Re $\geq$ 0.1):

$$V_s = R_e \nu / d_s \tag{31}$$

in which

$$C_D = \lambda / R_e^2 \tag{32}$$

$$\lambda = (4/3)(S-1) g d_s^3 / \nu^2 \tag{33}$$

where Re = particle Reynolds number (dimensionless), ν = kinematic viscosity of the fluid (L^2T^{-1}), C_D = drag coefficient (dimensionless), and S = particle specific gravity (dimensionless).

In regions where Stokes law applies (small spherical particles with the Reynolds number less than 0.1), the terminal particle fall velocity is estimated using Stokes equation:

$$V_s = (S-1) g d_s^2 / 18\nu \tag{34}$$

3.1.2 *Time Invariant Approach.* If the first term of Eq. (25) is neglected, that is, the storage variation of sediment within the flow depth, Eq. (25) can be rewritten as:

$$\frac{\partial(CQ)}{\partial x} = \Phi(x) \tag{35}$$

Equation (35) has been presented by Foster and Meyer (1972) as:

$$\frac{dG}{dx} = D_r + q_s \tag{36}$$

where G = CQ = weight rate of sediment load per unit width ($FL^{-1}T^{-1}$); D_r = net detachment rate or net deposition rate by concentrated flow ($FL^{-2}T^{-1}$), and q_s = lateral sediment inflow from adjacent contributing broad shallow flow areas per unit length of channel ($FL^{-2}T^{-1}$).

In Eq. (36), G is the dependent variable of interest. To calculate this variable, the independent variables D_r and q_s will have to be obtained from other considerations. Foster and Meyer (1972) model flow detachment as a rate process proportional to the difference between the flow's ability to transport detached sediment, transport capacity, and the existing sediment load in the flow. These authors used this concept to model both detachment and deposition of sediment as follows. If $T_c > G$, where T_c = sediment transport capacity of the flow ($FL^{-1}T^{-1}$), net detachment occurs and is modeled as:

$$D_r = D_c(1 - G/T_c) \tag{37}$$

where D_c = detachment capacity of the flow ($FL^{-2}T^{-1}$).

Detachment capacity D_c is in turn computed as:

$$D_c = K_c(\tau - \tau_c) \tag{38}$$

where K_c = channel erodibility parameter (T^{-1}), τ = average shear stress acting on the flow wetted perimeter (FL^{-2}), and τ_c = critical shear stress required for detachment to occur (FL^{-2}).

Equations (37) and (38) mathematically represent the concept that net detachment occurs in concentrated flow areas when the transport capacity, T_c, is greater than the sediment load, G, and the hydraulic shear, τ, exceeds the critical shear stress. When the sediment load is greater than the transport capacity ($T_c < G$), Eq. (37) becomes negative, indicating net deposition and is rewritten to compute deposition as:

$$D_r = \alpha(Tc - G) \tag{39}$$

where D_r = net deposition rate ($FL^{-2}T^{-1}$) and α = first-order reaction coefficient (L^{-1}), which is computed for a single particle size distribution as:

$$\alpha = \beta V_s/q \tag{40}$$

where β = dimensionless turbulence parameter, and $q = Q/B$ = flow rate per unit channel width (L^2T^{-1}).

Equations (27) and (28) represent the concept that deposition rate is small when β is small (for example, in the case of shallow rill flow highly disturbed by raindrop impact), particle fall velocity is small (in the case of fine particles such as clay-sized sediment), or q is large (where either flow velocity or flow depth is large).

Foster et al. (1980) solves Eq. (36) for a stream channel represented by a set of channel reaches. Each reach is divided into a number of segments depending on the spatial variability of contributing adjacent broad shallow flow areas or stream channel characteristics such as slope, surface roughness, etc. There are four possible detachment-deposition limiting cases for each segment: deposition can occur over the entire segment; deposition can occur on the upper end and detachment in the lower end of the segment; detachment can occur all along the segment; or detachment can occur on the upper end and deposition in the lower end of the segment. Assuming that the open-channel flow hydraulics on small streams can be described using the spatially varied, steady flow equations, Foster et al. (1980) derived closed-form equations for the four detachment-deposition cases. Case 1 occurs when $T_c < G$ over the entire segment. Deposition rate within the segment is estimated as:

$$D_r = [\phi/(1+\phi)(dT_c/dx - q_s)[1-(x_u/x)^{1+\phi}] + D_u(x_u/x)^{1+\phi} \tag{41}$$

in which

$$\phi = \beta V_s B/q_L \tag{42}$$

where dT_c/dx is assumed constant over the segment, q_L = the lateral inflow rate per unit channel length ($L2T^{-1}$), x_u = the point at the top of segment, and D_u = the deposition rate at x_u.

The deposition rate D_u is estimated as:

$$D_u = \alpha(T_{cu} - G_u) \tag{43}$$

where T_{cu} and G_u are the sediment transport capacity and sediment load at x_u, respectively.

Sediment load at x is calculated from Eq. (39) as:

$$G = T_c - D_c/\alpha \tag{44}$$

Case 2 occurs when $T_{cu} < G_u$, $dT_c/dx > 0$, and T_c becomes greater than G within the segment. This case is likely to occur in a situation where the transport capacity at the upper end of the segment drops abruptly to a level below the sediment load due to the presence of a rough surface strip (a grass buffer strip, for instance). Within the upper end of the segment, the sediment load decreases due to deposition, while the transport capacity increases from the point of the abrupt decrease, x_u. At a point upstream from the lower end of the segment, x_{de}, the sediment load equals the transport capacity ($G = T_c$) and deposition ends. This point can be determined by:

$$x_{de} = x_u\{ 1-[(1+\phi)/\phi][D_u/(dT_c/dx - q_s)] \}^{1/(1+\phi)} \tag{45}$$

where x_{de} = the point within the segment where deposition ends, x_u = the point of abrupt decrease in sediment transport, D_u = the deposition at x_u (calculated from Eq. 31), and T_{cu} = the transport capacity after the abrupt decrease at x_u.

Case 3 occurs when $T_c > G$ over the entire segment; that is:

$$D_r = K_c(1.35\tau_m - \tau_c)(1 - G/T_c) \tag{46}$$

where τ_m = the average shear stress acting on the channel bottom (FL^{-2}).

Case 4 occurs when $T_{cu} > G_u$, $dT_c/dx < 0$, and T_c becomes less than G within the segment. If $dT_c/dx < 0$ for a segment where $T_{cu} > G_u$, T_c can decrease below G within the segment. Within the upper end of the segment at a point x_u, the sediment load increases due to detachment ($T_{cu} > G_u$) while the transport capacity starts decreasing with x.

The point where the sediment load equals the transport capacity ($G = T_c$) is determined as x_{db}, which is the point where deposition begins. This point becomes x_u in Eq. (43) for computing D_r. The sediment load leaving a segment is computed as:

$$G_{(i+1)} = (D_{r(i)}+q_{s(i)}+D_{r(i+1)}+q_{s(i+1)})dx/2 + G_{(i)} \tag{47}$$

where the subscripts i and i+1 indicate the upper end and lower end of the segment, respectively, and dx is the length of the segment.

4. Conclusions

It is obvious that progress has been made in modeling sediment processes on upland watersheds. However, much still needs to be done on the theory of broad shallow flow and channel flow sediment processes, and experimental research.

In spite of some progress made in modeling sediment processes on broad shallow flow areas, further developmental modeling efforts must be based on a distinction between soil properties which influence runoff generation and, therefore, determine runoff rates; soil properties which determine the resistance of the soil to detachment by raindrop impacts and hydraulic shear and, as a consequence, determine erosion rates and the amount of sediment produced and delivered; and other soil properties which determine the way in which aggregates break down to seal the surface under raindrop impact, and soil becomes available for transport. At present, little more than intelligent guesses can be made as to the relative importance of soil properties, and the processes each controls.

The development of improved techniques for modeling sediment processes by concentrated flow on upland watersheds must be based on an improved understanding of runoff generation, runoff hydraulics, sediment detachment and transport mechanisms by flowing water, and other soil and water relationships operating at the watershed space-time scales. Clearly, if surface runoff cannot be accurately predicted, prediction of sediment processes is even more suspect.

Understanding runoff mechanisms and sediment processes as functions of space and time scales over a wide range of climatic and land-use conditions on upland watersheds, and identifying and quantifying differences and similarities of watershed behavior, remain important and active areas of experimental research on process-based sediment modeling.

5. Literature Cited

Alonso, C. V. and DeCoursey, D. G. 1983. Small watershed model. Proc. of the Natural Resources Modeling Symposium, U.S. Dept. of Agric., Agric. Res. Service, Pingree Park, CO., pp. 40-46.

Alonso, C. V., Neibling, W. H. and Foster, G. R. 1981. Estimating sediment transport capacity in watershed modeling. Trans. ASAE, 24(5):1211-1220, 1226.

Bennett, J. P. 1974. Concepts of mathematical modeling of sediment yield. Water Resources Research, 10(3):485-492.

Borah, D. K., Alonso, C. V. and Prasad, S. N. 1981. Stream channel stability. U.S. Dept. of Agric., Sedimentation Laboratory, Oxford, Miss., Appendix I.

Croley, T. E., II. 1982. Unsteady overland sedimentation. Journal of Hydrology, Elsevier, 56(1982):325-346.

Foster, G. R. 1982. Modeling the erosion process. In: C.T. Haan, H.P. Johnson and D.C. Brakensiek (Editors). 1982. Hydrologic modeling of small watersheds. ASAE, Monograph No. 5, pp. 259-380.

Foster, G. R. and Meyer, L. D. 1972. A closed-form soil erosion equation for upland areas. In: H.W. Shen (Editor). 1982. Sedimentation: symposium to honor Professor H.A. Einstein, Fort Collins, CO., Chapter 12:12.1-12.19.

Foster, G. R. and Meyer, L. D. 1975. Mathematical simulation of upland erosion by fundamental erosion mechanics. In: Present and Prospective Technology for Predicting Sediment Yields and Sources Proceedings, Sediment-Yield Workshop, USDA, Sedimentation Lab, Oxford, Miss., pp. 190-207.

Foster, G. R., Meyer, L. D. and Onstad, C. A. 1977. An erosion equation derived from basic erosion principles. Trans. ASAE, 20(4):683-687.

Foster, G. R.,Lane, L. J., Nowlin, J. D., Laflen, J. M. and Young, R. A. 1980. A model to estimate sediment yield from field-sized areas: development of model. In: Knisel, W. G., Jr. (editor). 1980. CREAMS: a field scale model for chemicals, runoff, and erosion from agricultural management systems. Conservation Research Report, No. 26, U.S. Dept. of Agric., pp. 36-64.

Hjelmfelt Jr., A. T., Piest, R. F. and Saxton, K. E. 1975. Mathematical modeling of erosion on upland areas. Proc. of the 16th Congress, Int. Assoc. for Hydraulic Research, Sao Paulo, Brazil, 2:40-47.

Julien, P. Y. and Simons, D. B. 1985. Sediment transport capacity of overland flow. Trans. ASAE, 28(3):755-761.

Komura, S. 1976. Hydraulics of slope erosion by overland flow. Journal of the Hydraulics Division, ASCE, 102(HY10):1573-1586.

Laflen, J. M., Elliot, W. J., Simanton, J. R., Holzhey, C. S. and Kohl, K. D. 1991. WEPP soil erodibility experiments for rangeland and cropland soils. Journal of Soil & Water Conservation, 46(1):39-44.

Lane, L. J., and Nearing, M. A., 1989. USDA - Water Erosion Prediction Project: Hillslope profile model documentation. NSERL Report No. 2, West Lafayette, Indiana.

Lane, L. J. and Shirley, E. D. 1982. Modeling erosion in overland flow. In: Estimating Erosion and Sediment Rangelands. Proc. Workshop, Tucson, Arizona, March 7-9, 1981, USDA-ARS, Agricultural Reviews and Manuals, ARM-W-26, June 1982.

Lane, L. J. and Shirley, E. D. 1985. Erosion and sediment yield equations: solutions for overland flow. Paper presented at the Workshop on USLE Replacement. NSEL, West Lafayette, IN, 22 p.

Li, R. M. 1979. Water and sediment routing from watersheds. In: H.W. Shen (Editor). 1979. Modeling of rivers. Wiley-Interscience, New York, pp. 9.1-9.88.

Li, R. M., Shen, H. W. and Simons, D. B. 1973. Mechanics of soil erosion by overland flow.

Proc. of the 15th Congress, Int. Assoc. for Hydraulic Research, Istanbul, Turkey, Vol. (1):437-446.

Lopes, V. L. 1991. Using erosion equations to predict sediment yield from overland flow systems. In: Proceedings of the Arizona-Nevada Academy of Science 35th Annual Meeting, Northern Arizona University, Flagstaff, Arizona, April 20, 1991.

Lopes, V. L. and Ffolliott, P. F. 1992. Modeling sediment processes on small watersheds: a conceptual framework. I. Broad shallow flow processes. International Journal of Sedimentation Research, IRTCES, Beijing, China, 7(3):21-44.

Lopes, V. L. and Lane, L. J. 1988. Modelling sedimentation processes in small watersheds. In: Proc. of the IAHS Int. Symp. on Sediment Budgets in Porto Alegre, Brazil, December 11-15, 1988, Publ. No. 174 (1988), pp. 497-508.

Mehta, A. J. 1983. Characterization tests for cohesive sediments. Proc. of the Conference on Frontiers in Hydraulic Engineering. ASCE, MIT, Cambridge, Mass., pp. 79-84.

Meyer, L. D. 1984. Evolution of the universal soil loss equation. Journal of Soil and Water Conservation, 39(2):99-104.

Meyer, L. D., Alonso,C. V. and Harmon, W. C. 1983. Modeling soil losses from nearly flat fields. ASAE Paper No. 83-2091.

Meyer, L. D. and Wischmeier, W. H. 1969. Mathematical simulation of the process of soil erosion by water. Trans. ASAE, 12(6):754-758,762.

Nakato, T. 1990. Tests of selected sediment-transport formulas. Journal of Hydraulic Engineering, ASCE, 116(3):362-379.

Nyhan, J. W. and Lane, L. J. 1986. Erosion control technology: a user's guide to the use of the universal soil loss equation at waste burial facilities. Manual LA-10262-M, UC- 708, Los Alamos National Lab., Los Alamos, New Mexico.

Rijn, L. C. 1984. Sediment transport, part I: bed load transport. Journal of Hydraulic Engineering, ASCE, 110(10):1431-1456.

Rose, C. W. 1985. Developments in soil erosion and deposition models. In: Advances in Soil Science, Springer-Verlag, New York, 2:1-63.

Rose, C. W., Williams, J. R., Sanders, G. C. and Barry, D. A. 1983. A mathematical model of soil erosion and deposition processes: I. theory for a plane land element. Soil Sci. Soc. of America Journal, 47(5):991-995.

Schumm, S. A. 1977. The fluvial system. John Wiley and Sons, Inc. New York, NY.

Simons, D. B., Li, R. M. and Stevens, M. A. 1975. Development of models for predicting water and sediment routing and yield from storms on small watersheds. U.S. Dept. of Agric., Forest Service, Rocky Mountain Forest and Range Experiment Station.

Singh, V. P. 1983. Analytic solutions of kinematic equations for erosion on a plane: II rainfall of finite duration. In: Advances in Water Resources, 6:88-95.

Singh, V. P. and Prasad, S. N. 1982. Explicit solutions to kinematic equations for erosion on an infiltrating plane. In: Singh, V. P. (Editor). 1982. Modeling components of hydrologic cycle. Proc. of the Int. Symp. on Rainfall-Runoff Modeling, Water Resources Publication, Littleton, CO, 515-538.

Smith, R. E. 1976. Simulating erosion dynamics with a deterministic distributed watershed model. Proc. of the 3th Federal Inter-Agency Sedimentation Conference, Denver, CO, 1.163-1.173.

Smith, R. E. 1981. A kinematic model for surface mine sediment yield. Trans. ASAE, 24(6):1508-1514.

Vanoni, V. A. (Editor). 1975. Sedimentation Engineering. ASCE-Manuals and Reports on Engineering Practice - No. 54, New York, N.Y., 745 pp.

Wischmeier, W. H., and Smith, D. D., 1978. Predicting rainfall erosion losses - a guide to conservation planning. Agr. handbook No. 537, USDA, Washington, D.C., 58 pp.

Woo, H., Yoo, K. and Seoh, B. 1990. Performance test of some selected sediment transport formulas. Proc. of the 1990 ASCE National Conference, Vol. 1:694-699. San Diego, CA.

APPLICATION OF LINEAR SYSTEM MODELS FOR ESTIMATION OF WASH LOAD

N. S. Raghuwanshi
Graduate Student,
Department of Biological and Agricultural Engineering
University of California, Davis
Davis, CA, 95616

R. A. Rastogi
Professor,
Department of Soil and Water Conservation Engineering,
G. B. Pant University of Agriculture and Technology
Pantnagar, UP, 247667

Santosh Kumar
Assistant Engineer,
Department of Agriculture
Govt. of M. P.
Morena, MP.

ABSTRACT - The applicability and validity of linear system theory for prediction of temporal distribution of wash load from a large hilly catchment have been investigated. Unit sediment graph and series graph models were applied to Chaukhutia catchment of Ramganga river for regeneration and prediction of sediment graphs. The sediment graphs regenerated and predicted by unit sediment graph method and series graph method in all cases resulted in better agreement with their measured counterparts. The measured and predicted peak sediment flow rates were also correlated with the mobilised-sediment. The results of the study reported herein can be served as a guideline for furthering use of linear system theory in wash load prediction.

1.0 INTRODUCTION

The time distribution of sediment within a storm provides useful information to solve variety of problems, like water quality modeling, design of efficient sediment control structures, study of transport of pollutants attached to the sediment and also in the development of sediment routing procedures, besides the total sediment yield. The sediment flow rate as a function of time during a storm at a given location is known as a sediment graph. Without sediment graph, only the average sediment rate for the storm can be determined and average sediment yield is not adequate for estimating dynamic sediment load and pollutants load during the storm.

The surface runoff-sediment yield process is generally recognised as a non-linear, time-variant process, and is governed by large number of unknown climatic and physiographic factors which may vary both in space and time. In describing sediment response of a watershed, the use of linear system models has a spurred criticism. Fortunately, during the past years, there has been

V. P. Singh and B. Kumar (eds.), Water-Quality Hydrology, 113–123.

success in applying the linear and time-invariant assumptions to sedimentation engineering and hydrology, making it possible to solve certain non-linear problems. Linearity and time-invariance are the bases for the development of the unit hydrograph theory. The same bases are used here to derive the sediment response for the watershed under consideration. A few conceptual linear models (Rendon-Herrero, 1974; Renard and Laursen, 1975; Bruce et. al, 1975; Rendon-Herrero, 1978; Williams, 1978; Chen and Kuo, 1986) were developed to determine sediment graph for gauged and ungauged watersheds. The most of these developed models were applied to small experimental watersheds to check their validity. However, these studies require further testing and extension of application of linear system theory in sedimentation engineering both with regard to their underlying assumptions and their accuracy in prediction of sediment discharge from large watersheds.

For the purpose of this study, sediment yield is defined in terms of suspended sediment load. Many authors Chow(1964), Graf(1971) and Shen(1971) found that suspended sediment contribution to total sediment yield are very high, i.e., 90-95 percent. Keeping in view, the above problems and limitations in the use of linear system models for prediction of wash load from a catchment, an attempt has been made to demonstrate the applicability of linear models for prediction of wash load from a large hilly catchment. Unit sediment graph and series graph, (Rendon-Herrero; 1974 and 1978) models have been applied to Chaukhutia catchment of Ramganga river.

2. STUDY AREA AND ANALYSIS OF DATA

A Chaukhutia sub-watershed (Fig.1) comparing an area of 452.25 sq km of Ramganga reservoir catchment lies between $29^0 46'15''$ to $30^0 6'$ N latitude and $79^0 12'15''$ to $79^0 31'$ E longitude in the Shivalik range of the Himalayas was selected as a source of information for this study. The climate of the region is Himalayan sub-tropical to sub-temperate having the mean annual temperature of about 21^0C. The average annual total precipitation in the area is 1466.76 mm. which varies from 1208.50 mm to 1773.76 mm at different locations. The maximum elevation of the watershed is 3114.14 meters above the mean sea level and the minimum elevation at Chaukhutia is 929 meters.

Sixteen rainfall-runoff-sediment events for the years 1976 through 1984 and catchment physiographic data, i. e., topographic map, drainage basin area, main-stream length , slope etc. were collected from the record of the Divisional Forest Office (Soil Conservation), Ranikhet, Uttar Pradesh, India. For these events data are available on rainfall charts, total rainfall, stage hydrograph, sediment concentration and sediment discharge. In order to define the shape of sediment graphs, the sediment samples were taken frequently during the rising and falling limbs, at peak and also during the recession part of the events.

The Thiessen weight method has been used to describe the mean areal depth of precipitation measurements at points within the catchment. The base runoff and sediment flows were separated from the total runoff hydrograph and total sediment graphs respectively using the base flow separation procedure proposed by Chow (1964). The causal rainfall-excess and duration of the rainfall-excess were obtained from the rainfall hyetograph by subtraction of the absolute loss to the system, i.e., that portion of rainfall which never emerges as runoff at the outlet point. In the present study a simple ϕ-index method was utilised. The ϕ-index value was chosen such that the volume of rainfall-excess was equal to that of the runoff due to storm.

3.0 THEORETICAL BACKGROUND AND METHODOLOGY

A hydrologic system is referred to as linear if it satisfies the properties of proportionality and superimpostion. Mathematically, it can be expressed as:

$$y = H(\theta_1 x_1 + \theta_2 x_2) - (\theta_1 H(x_1) + \theta_2 H(x_2)) \quad ...(1)$$

where x_1 and x_2 are system inputs; y is the system output; H is the system operation and θ_1 and θ_2 are the positive scale ratios. The assumption of linearity allows the prediction to be made even though the type of input that one is interested in has not occurred in the past.

A system is said to be time-invariant if the form of output depends only on the form of the input and not on the time at which the input is applied. Mathematically, it can be expressed as:

$$y = (t \pm \tau) = H[x(t \pm \tau)] \quad ...(2)$$

for any input x and time t or $t \pm \tau$. However, time-invariance assumption makes it possible to predict an output for a given input if that particular input has already occurrred some time during the period of record (Dooge, 1973).

The above mentioned two assumptions are the bases for system hydrology and unit hydrograph theory. The same bases of linear theory can be extended to sedimentation engineering, to develop unit sediment graph and series graph with the help of regressional relationship between the volume of direct or effective runoff (E_R) and direct or effective sediment yield (E_S), which holds a log-transformed linear relationship for a specific watershed (Rendon-Herrero et. al, 1980; Singh and Chen, 1981).

3.1 Unit Sediment Graph Method

The unit hydrograph concept as applied to a direct runoff hydrograph is analogous to the unit sediment graph. A unit sediment graph is defined as a direct sediment graph resulting from one ton of sediment distributed uniformly over the catchment area for an effective duration. The ordinates of the individual unit sediment graph were determined using the following relationship:

$$USO_i = \frac{S_{Di}}{E_S} \quad ...(3)$$

where USO_i is the individual sediment graph ordinates in sq. km. per hour, S_{Di} is the direct sediment graph ordinates in tons per hour, E_S is the amount of mobilised-sediment in tons per sq. km. The subscript i refers to the time at which sediment discharge is measured. The mobilised-sediment for each storm is expressed as:

$$E_S = \Delta t \frac{\sum S_{Di}}{A} \quad ...(4)$$

in which A is the watershed area in sq. km., Δt is the time interval of one hour considered in this study. The representative unit sediment graph for Chaukhutia catchment was developed by averaging the ordinates of unit sediment graphs of fourteen selected storm events at one hour time interval and is shown in Fig. (2).

3.2 Series Graph Method

The series graph method proposed by Rendon-Herrero (1974) is a set of plots of series graph ordinates against associated surface runoff at a particular time interval with reference to peak time. The individual series graph ordinates were determined by utilising the following relationship:

$$SGO_i = \frac{S_{Di}}{E_R} \quad ...(5)$$

where E_R is the rainfall-excess in mm, S_{Di} is the direct sediment graph ordinates in tons per hour and SGO_i is the individual series graph ordinates in tons per hour per mm of rainfall-excess.

The series graph ordinates versus rainfall-excess coordinates were plotted in time groups referenced to peak discharge (P) for all the selected storm events. Since the points were scattered, the lines for P, P$\pm$1, P + 2, P + 3, P + 4, P + 5, P + 6, P + 7, P + 8, and P + 9 were drawn by using the method of least square (Fig. 3). P$\pm$1 refers to the direct sediment graph ordinates one hour after and before the peak.

4.0 RESULTS AND DISCUSSION

The total amount of mobilised-sediment during the storm event would have to be known or estimated, in order to generate a sediment graph for a particular storm event using the unit sediment graph. To compute the mobilised-sediment during the storm event, a relationship between mobilised-sediment and excess-rainfall was developed. For Chaukhutia watershed, 14 storm events were selected for development of the following relationship; (Fig 4):

$$E_s = 1.219 E_R^{1.9271} \quad ...(6)$$
$$(\, r = 0.7622)$$

where E_S is the mobilised-sediment in tons per sq km, E_R is the total excess-rainfall in mm during the storm and r is the coefficient of correlation. This approach entail the estimation of total mobilised-sediment on the basis of known or predicted excess-rainfall as an initial step, followed by the selection, based on duration, of an appropriate unit-sediment graph. In this study all the storm events were resulted due to the rainfall-excess of one hour.

To demonstrate the applicability of the selected methods for regeneration of sediment graph resulting from a known storm event, storm events of August 21-22, 1983 and August 18-19, 1984 were chosen for comparison between measured and regenerated sediment graphs. These chosen storm events were also considered in the analysis. Since, all the storm events considered in this study resulted due to one hour effective rainfall, catchment representative one hour unit sediment graph ordinates were convolved with corresponding mobilised-sediment. In case of series graph, series graph ordinates were convolved with the respective rainfall-excess. The sediment graphs generated by unit sediment graph and series graph methods resulted in better agreement with their measured counterparts in the region of the rising and falling limbs, time to peak except slight variation in the peak sediment flow rates particularly in case of sediment graphs generated by series graph method (Fig. 5 and 6).

The peak sediment flow rates for the storm events analysed were estimated by convolving the unit sediment graph and series graph ordinates at peak with the mobilised-sediment and rainfall-excess respectively. The values of mobilised-sediment, measured and generated peak sediment flow rates along with the percentage departure in peak sediment flow rates for the storm events analysed are given in the Table I.

Table I - Measured and estimated peak sediment flow rate

Date of storm event	Mobilised-sediment (tons/km^2)	Peak sediment flow rate (tons/hour)			Percentage departure	
		Measured	Estimated			
			Unit sediment graph	Series graph	Unit sediment graph	Series graph
8/ 23/1976	2.005	370.00	379.86	318.31	+2.66	-13.90
7/21-22/1977	4.960	890.00	939.72	1542.79	+5.58	+73.34
8/11/1977	4.150	820.00	786.25	811.94	-4.11	-0.98
8/18-19/1978	35.179	6420.00	6665.01	1682.08	+3.81	-73.76
9/2/ 1978	4.743	860.00	898.60	1418.57	+4.48	-64.95
6/21-22/1979	2.662	535.00	496.76	484.16	-7.14	-9.50
8/31/1980	6.578	1170.00	1246.26	996.44	+6.51	-14.83
8/2-3/1981	4.223	810.00	800.00	1492.31	-1.22	+84.23
7/31/1982	27.000	5100.00	5115.42	5036.76	+0.30	-1.24
8/21-22/1983	4.451	900.00	843.28	765.61	-6.30	-14.93
9/5-6/1983	3.597	640.00	681.48	785.99	+6.48	+22.81
6/25/1984	4.157	910.00	788.58	1046.40	-13.45	+14.98
8/18-19/1984	4.544	840.00	860.90	878.29	+2.48	+4.55
8/22-23/1984	29.209	5280.00	5533.93	4682.06	+4.80	-11.32

The peak sediment flow rates generated by the sediment graph method compared well with their measured counterparts and showed maximum departure of less than 15%. Whereas, the peak sediment flow rates generated by the series graph method showed more than 64% departure for the four events out of the fourteen chosen events. This may be due to the fact that series graph ordinates were convolved with the rainfall-excess by assuming linear relationship between wash load and rainfall-excess. This assumption is not always valid, since the same amount of rainfall-excess may produce different amount of wash load depending on the antecedent moisture conditions of the watershed. However, regeneration results can be improved by selecting the storm events representing similar antecedent moisture conditions over the catchment or by considering more storm events for developing the regressional relationships. The values of measured and predicted peak sediment flow rates were plotted against the mobilised-sediment on log-log paper (Fig. 7). It is evident from Fig. 7 that the value of correlation coefficient is high in case of the unit sediment graph method indicating close relationship between sediment discharge and mobilised-sediment. However, the lower value of correlation coefficient for the series graph method may be due to error in rainfall-excess or linear relationship between wash load and rainfall-excess.

The prediction performance of the methods was tested by using data, recorded at the gauging station of the watershed, which were not used either to develop the regression equation or the unit sediment graph for the watershed. Sediment graph measured at the outlet of the watershed was compared with the corresponding sediment graphs obtained from the series graph and unit sediment graph methods for the storm event of July 15,1984 and are shown in Fig. (8). It is evident from Fig. (8) that the base length, time to peak, rising, crest and recession segments of the sediment graphs predicted by the models are in closer agreement with the measured sediment graph.

5.0 CONCLUSIONS

The basic principles of linearity and time-invariance as applied to the unit hydrograph theory are also applicable for the development of unit sediment graph and series graph. The sediment graphs generated and predicted by the unit sediment graph and series graph methods showed closer agreement with the measured sediment graphs in the region of rising and falling limbs, time to peak, base time except higher departure in the peak sediment flow rates in case of sediment graphs generated by the series method. These results showed the validity of the unit sediment graph and series graph methods for predicting time-wise distribution of wash load from gauged watersheds. However, these methods also can be used to determine wash load from ungauged watersheds having similar physiographic and climatic conditions. The application of linear system models for estimation of wash load is easier, and less time consuming, therefore, the use of these methods can be advocated for prediction of wash load from the large watersheds.

ACKNOWLEDGMENTS

The authors sincere thanks are due to the Forest Department (Soil Conservation), Ranikhet, U. P. for providing hydrological data and G. B. Pant University of Agriculture and Technology, Pantnagar, U. P., India, for providing necessary facilities. The authors also would like to thank the anonymous reviewers for their valuable comments.

REFERENCES

Bruce, R. R., L. A. Harper, R. A. Leonard, W. M. Synder and A. W. Thomas (1975). model for runoff of pesticides from small upland watersheds. *J. Environ. Qual.*, 4(4): 541-548.

Chen, V. J., and C. Y. Kuo (1986). A study on synthetic sediment graphs for ungaged watersheds. *J. Hydrol.*, 84: 35-54.

Chow, V. T. (1964). *Handbook of Applied Hydrology*. Mc Graw Hill. New York, NY.

Graf, W. H. (1971). *Hydraulics of Sediment Transport*. Mc Graw Hill, New York, NY.

Dooge, J. C. I. (1973). *Linear theory of hydrologic systems.* Agric. Res. Serv., U. S. Dep. Agric., Tech. Bull. No. 1468.

Raghuwanshi, N. S. (1986). *Mathematical watershed models for runoff and sediment flow from Chaukhutia catchment of Ramganga river.* Thesis. Master of Technology in Agricultural Engineering. G. B. Pant University of Agriculture and Technology, Pantnagar, UP, India.

Renard, K. G. and E. M. Laursen (1974). Dynamic behavior model of ephemeral stream. *J. Hydraul. Div., Proc. ASCE*, 101 (HY 5): 511-528.

Rendon-Herrero, O. (1974). Estimation of washload produced on certain small watersheds. *J. Hydraul. Div. Am. Soc. Civ. Eng.*, 100 (HY 7): 835-848.

Rendon-Herrero, O. (1978). Unit sediment graph. *Water Resour. Res.*, 14 (5): 889-901.

Rendon-Herrero, O., Singh V. P. and V. J. Chen (1980). ER-ES watershed relationship. *Proc. Int. Symp. Water Resources Systems,* Univ. of Roorkee, Roorkee, 1: 11-8-41- 11-8-47.

Shen H. W. (1971). *River Mechanics*. Colorado State University, Fort Collins. Vol. 1, pp. 11-13.

Singh, V. P. and V. J. Chen (1981). On the relation between sediment yield and runoff volume. *Proc. Int. Symp. Rainfall Runoff Modeling,* Mississippi State Univ., State College, Miss., 7: 555-570.

Williams, J. R. (1978). A sediment graph model based on an instantaneous unit sediment graph. *Water Resour. Res.*, 14(4): 659 - 664.

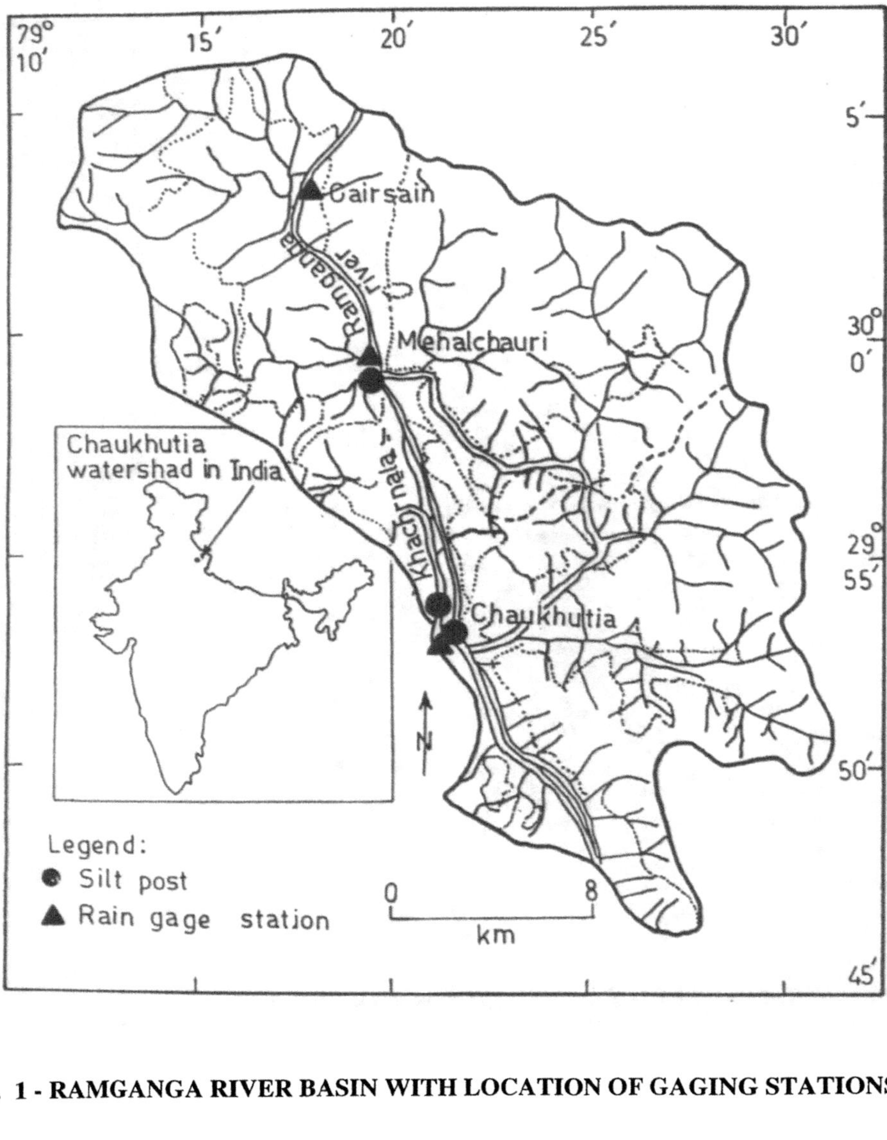

FIG. 1 - RAMGANGA RIVER BASIN WITH LOCATION OF GAGING STATIONS.

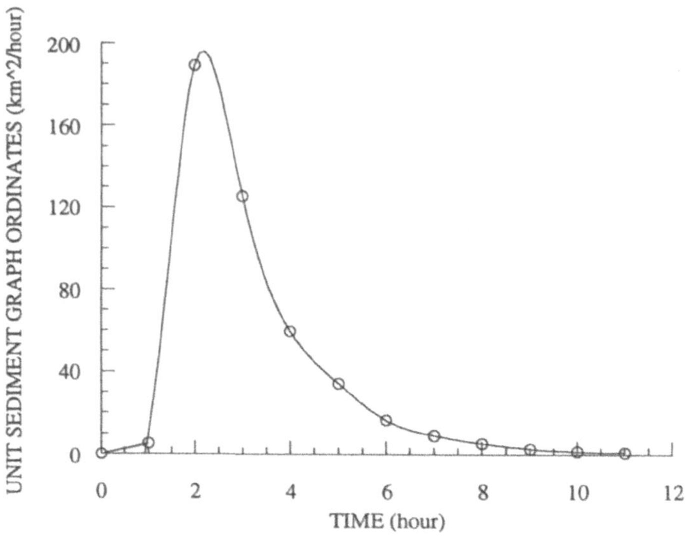

FIG. 2 - REPRESENTATIVE UNIT SEDIMENT GRAPH OF THE CATCHMENT

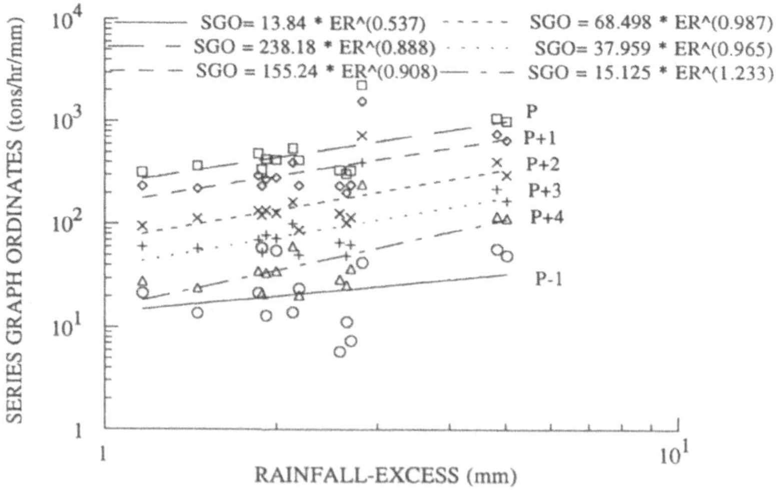

FIG. 3A - SERIES GRAPH ORDINATES VERSUS RAINFALL-EXCESS

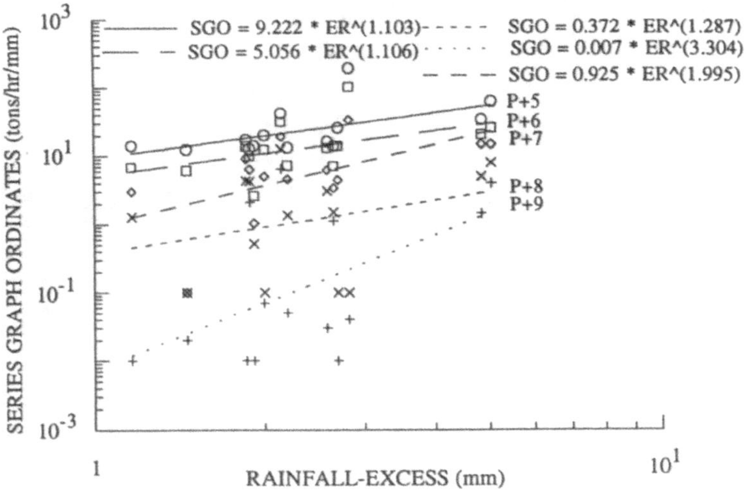

FIG. 3B - SERIES GRAPH ORDINATES VERSUS RAINFALL-EXCESS

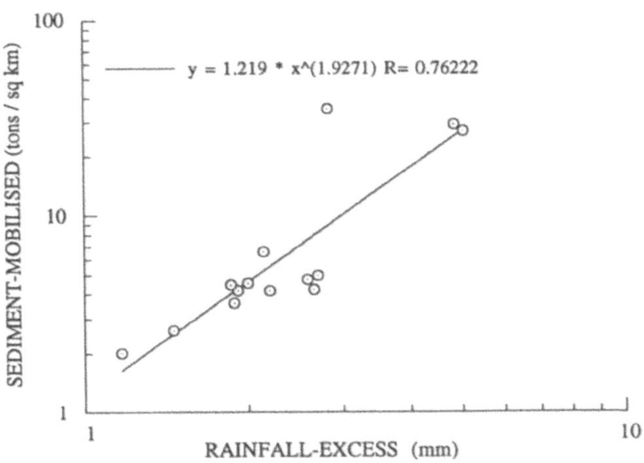

FIG. 4 - RELATIONSHIP BETWEEN SEDIMENT- MOBILISED AND RAINFALL- EXCESS

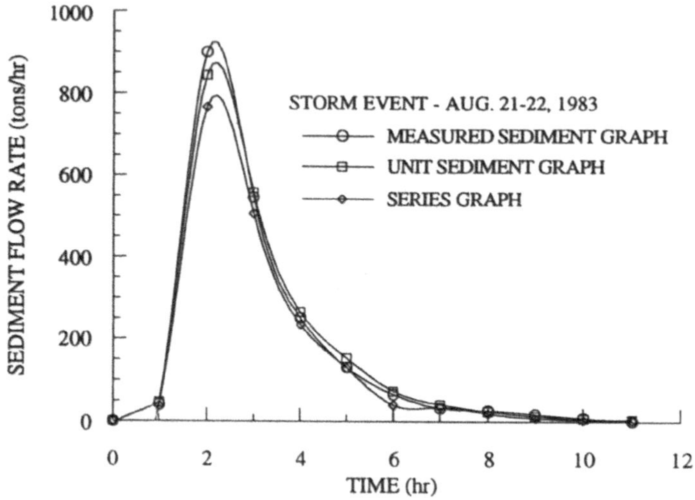

FIG. 5 - MEASURED AND REGENERATED SEDIMENT GRAPHS

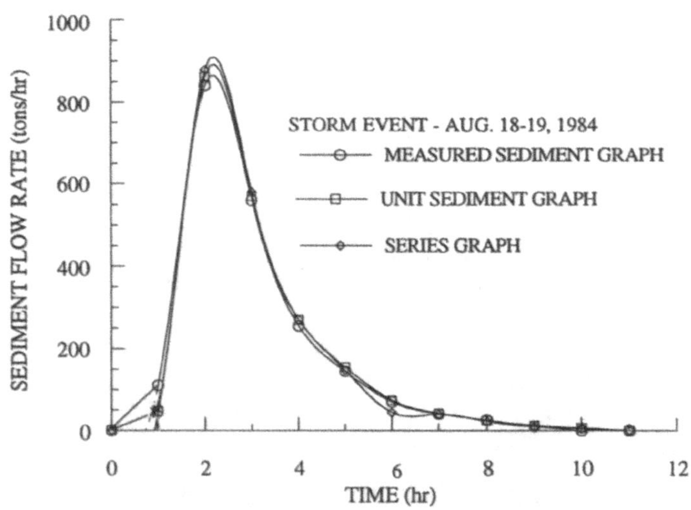

FIG.6- MEASURED AND REGENERATED SEDIMENT GRAPHS

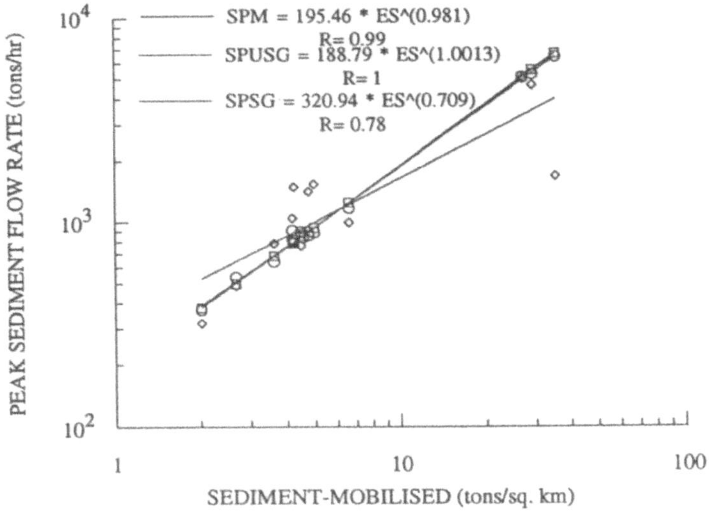

FIG.7 - PEAK SEDIMENT FLOW RATE VERSUS MOBILISED-SEDIMENT

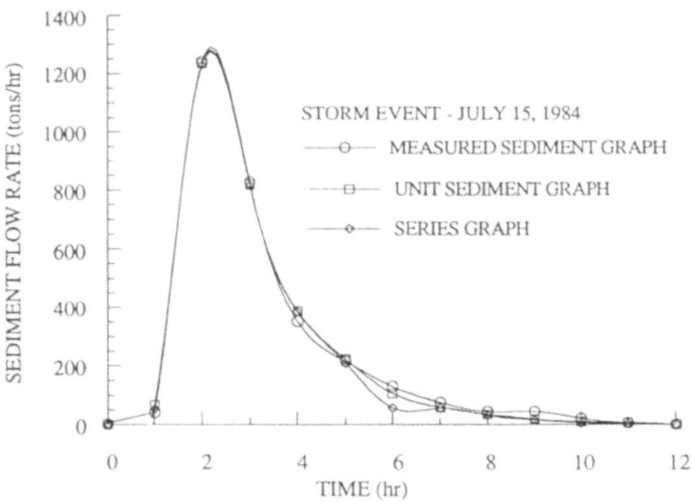

FIG. 8 - MEASURED AND PREDICTED SEDIMENT GRAPHS

SECTION 4:

SOLUTE TRANSPORT IN VADOSE ZONE

RECENT ADVANCES IN MODELLING VADOSE ZONE TRANSPORT

J.-YVES PARLANGE, TAMMO S. STEENHUIS,
Department of Agricultural and Biological Engineering,
Cornell University,
Ithaca, NY 14853, U.S.A.

FRANK STAGNITTI,
School of Computing and Mathematics,
Deakin University, P.O. Box 423,
Warrnambool, 3280, Australia.

ERIK SIMMELINK, BART NIJSSEN
Department of Agricultural and Biological Engineering
Cornell University,
Ithaca, NY 14853, U.S.A.

ABSTRACT: Modelling solute transport through the vadose zone is difficult due to the complicated networks of interconnected pathways in the soil which can transmit water and its solutes at varying velocities. Preferred pathways resulting from biological and geological activity may transmit water (and it's solutes) at very much higher rates than those predicted by Darcy's theory. Therefore, current transport models for water and solutes based on the assumption of Darcian flux, eg. Richards' equation and the Convective-Dispersive equation may significantly under-estimate the risk to groundwater supplies because they do not consider explicitly preferential flow paths but model the bulk flow in an "average" path. A preferential solute transport model developed by us simulates the effects of natural heterogeneities in the soil layer by identifying paths responsible for the transport of water and solutes. The soil-water hydraulic conductivity function is used to identify these paths and the interaction between these paths are described by mixing functions. This paper discusses the nature of preferential flow and illustrates application of the model to several field experiments.

1. Introduction

Without taking into account the water stored in icecaps and glaciers, then approximately 95% of the world's freshwater is stored in groundwater. Surface waters stored in lakes, rivers etc., have relatively short residence times of the order of minutes to years. On the other hand, groundwater has residence times typically in the order of hundreds to thousands of years (Freeze and Cherry, 1979). Therefore, contaminants reaching the groundwater may persist for many generations. In the United States alone, there are an estimated half million contaminated sites resulting from surface and subsurface impoundments of agricultural and industrial chemicals, mine waste tailings and pesticide-contaminated sites that need to be addressed under various U.S. statutes (National Research Council, 1990). Previously held belief that the soil can act as a "living filter" capable of retaining and degrading such wastes, has given way to the reality that many chemicals do not degrade (biologically, chemically or radiologically) in a sufficiently finite time and that eventual migration of such contaminants to the groundwater is highly probably (Gee, et al., 1991; Parlange, et al., 1988). This problem is further exacerbated in structured soils consisting of networks of macropores which may rapidly transmit contaminants to the underlying aquifer, bypassing much of the vadose zone.

V. P. Singh and B. Kumar (eds.), Water-Quality Hydrology, 127–151.

Much experimental research in soil heterogeneities and preferred pathways generated particularly in the last decade is forcing soil scientists and hydrologists to refine their analyses of the movement and fate of contaminants in the soil, and to consider processes which have become important because even small concentrations of highly toxic chemicals must be accurately accounted for in order to protect groundwater reserves from contamination. However, the transport of solutes and other contaminants through the vadose zone is complicated by the networks of interconnected pathways which result from biological and geological activity, such as subsurface erosion, faults and fractures, shrink-swell cracks, animal burrows, worm holes, decaying roots. Such pathways may transmit water, solutes and contaminants at very much higher velocities than those predicted by Darcy's theory (Steenhuis, et al., 1991, Stagnitti, et al. 1991). The best way to account for all possible soil types and climatic conditions would be to conduct field experiments and monitor the chemical movement on each site, but this is manifestly impractical. So, methods have been developed that simulate, measure and model the way in which water transports solutes through the soil. This paper reviews some recent developments in monitoring and modelling preferential moisture and solute transport.

2. Moisture And Solute Transport

The approach to simulate water and solute transport currently used by many soil scientists is derived from the work of Richards' who formulated a theory of water movement in unsaturated soils (Richards, 1931). Analytical solutions to Richards' equation provide important information to analyse unsaturated flow processes such as infiltration, drainage and redistribution. Recent work by Sander, et al., 1988; Sander, et al., 1991; Broadbridge and White, 1988; Broadbridge et al., 1988; Warrick et al., 1990 and White and Broadbridge, 1988; Barry and Sander, 1991 and Barry et al., 1993b) have developed a variety of analytical solutions for water infiltration into homogeneous soils for different surface fluxes and boundary conditions.

Van der Molen (1956) combined aspects of Richards' model with the theory of dispersive movement to predict desalinisation of land that had been inundated by seawater in the Netherlands. The resulting convective-dispersive equation assumes that water and solutes follow an *average* path through the soil. That is, any particular molecule of water and soluble contaminant, starting at the soil surface, is equally likely to follow any one of a multitude of available paths (Steenhuis and Parlange, 1990). Other soil scientists and hydrologists, (eg. Steenhuis et al., 1980; Parker and Van Genuchten, 1984; Sposito and Jury, 1988 and Barry and Sposito, 1988), have improved the solute transport model by incorporating a stochastic description of soil heterogeneities (see Barry, 1993a for an excellent review). However, these results are usually limited to cases of constant or uniform mean velocity as well as constant dispersion (Gee, et al., 1991). Transfer function models have received considerable field testing and have shown that scale-dependent dispersion is present in the field (Butters and Jury, 1989; White, 1989). The conclusion from these studies is that the convective-dispersion equation with a constant dispersion coefficient does not give as great a solute dispersion with depth than is observed (Gee, et al., 1991).

Models based on the convective-dispersive equation have proven quite satisfactory for predictions of the overall flow of water and the transport of nutrients in soil columns and in field experiments (eg. see Steenhuis and Van der Molen, 1986). However, predictions of pesticide movement with the convective-dispersive equation, except for a very few cases, fail to represent the observed values. The difference in predictive accuracy between the pesticides on one hand and fertilisers on the other hand result partly from the amount of applied chemicals that will cause health standards to be exceeded. For example, according to limits set by the EPA in the U.S., nitrates are tolerable in concentrations up to 10,000 ppb, which are not attained unless more than 40 percent of the amount commonly applied reaches the groundwater (Steenhuis and Parlange, 1990). In contrast, the maximum permissible pesticide concentration is usually below 10 ppb. This limit can readily be exceeded if small amounts of the pesticides enters macropores connecting the soil surface to the groundwater. Models based on the convective-dispersive equation which use a mean-flow path, even those with built-in stochastic descriptions, are too simplistic to describe heterogeneous flow patterns in soils with preferred pathways. Other models that explicitly considers preferential flow are required to accurately predict preferential solute transport. Such models must recognise that water does not percolate through the soil uniformly, but shows a preference for certain pathways.

3. Preferential Flow Mechanisms

3.1 FLOW IN WELL-STRUCTURED SOILS

Preferential flow is the rapid and non-uniform transport of soil-moisture and solutes through macropores and sub-surface channels. Flow in macropores and channels can occur with little interaction with the surrounding soil-matrix. On the other hand, matrix flow is the comparatively slower movement of water and solutes percolating though micropores of the soil. The relative importance of the two forms of percolation is dependent on the soil type and rainfall intensity (Parlange, et al., 1993). For example, well-structured soils consisting of clay and loam mixes typically experience low permeability rates. In such soils, typically less than 1% of the pore-volume consists of cracks and subsurface channels. However, often when it rains, water infiltrating from the soil surface flow through these channels in preference to the surrounding soil-matrix, whose small pores are penetrated comparatively slowly. Even though these channels consist of a relatively small percentage of the total pore volume, they may be responsible for the bulk of moisture and solute transport for certain rainfall events (Ward, 1984 and Steenhuis and Muck, 1988). Further, preferential flow in these soils can be initiated well below soil-water saturation.

The concentration of solutes in the soil matrix and in preferred paths may also be very different. The chemical composition of solutes in the macropores and subsurface channels often reflects the concentration of the water near the surface during an infiltration event. On the other hand, the concentration of solutes in the soil matrix reflects an equilibrium concentration that was established prior to the infiltration event. Thus for certain infiltration events, it is possible for high concentrations of contaminants to leach to the groundwater before dispersion has the opportunity to incorporate them with the rest of the soil matrix (Hegert, et al., 1981; Steenhuis and Muck, 1988; van Es, et al., 1991; and Richard and Steenhuis, 1988). These studies also confirm that the distribution of the solutes is highly variable even over very short distances.

3.2 FINGERED FLOW IN HOMOGENEOUS SOILS

Contrary to widely held belief, even in homogeneous soils, water does not necessarily take an average flow path. Hill and Parlange (1972) documented for the first time, preferential flow in homogeneous sand at low infiltration rates. Further research by others has led to an understanding of many of the mechanisms involved in this phenomena (Parlange and Hill, 1976; Raats, 1973; Philip, 1975; Hoa, 1981; Hillel and Baker, 1988; Glass et al., 1989a,b,c; Glass et al., 1991; Selker, et al., 1992a,b). This line of research has shown that water can find its way down through homogeneous, sandy soils in a number of channels called fingers. A poorly conducting layer of topsoil at the surface produces a wetting front instability. Gravity drives the instability and surface tension has a contrary, stabilising effect. The balancing forces of gravity, expressed as $(k_w - q)$ cm/sec, and surface tension, expressed as $s_w^2 / (\theta_w - \theta_i)$, gives the diameter, d, of the fingers. as

$$d = \lambda \frac{S_w^2}{(\theta_w - \theta_i)(k_w - q)} \qquad (1)$$

where λ is a universal number, q is the entering flux, θ_w and θ_i are the average water contents behind and ahead of the wetting front, S_w is the sorptivity, and $k_w = k(\theta_w) - k(\theta_i)$, where $k(\theta_w)$ and $k(\theta_i)$ are the soil water conductivities at their respective water contents. For two-dimensional fingers, $\lambda = \pi$ (Parlange and Hill, 1976; Glass, et al., 1989a,b) and for three dimensional fingers $\lambda = 4.8$ (Glass, et al., 1991). In the field, k_w tends to be much larger than q, and, for most sandy soils, the quantity $S_w^2 / (\theta_w - \theta_i) k_w$ is only a slightly increasing function of θ_w. Thus d is essentially independent of q and the amount of water carried by the fingers is a property of the soil. The distribution of water within a given finger is not uniform. Not only is there a sharp gradient at the fingers' surface which is responsible for lateral diffusion, there is an increase of water content with depth (Steenhuis and Parlange, 1990a). This variation with depth can be described by solving the transport equation for a constant velocity of finger growth, v, yielding (Selker, et al., 1992a,b);

$$\frac{k(\theta)}{\theta} = \frac{v}{\left[1 - \frac{\partial h}{\partial t} v^{-1}\right]} \qquad (2)$$

Until recently, it was generally thought, with the exception of Raats (1973) that instabilities would only occur when there was a distinct difference in layering (eg. Hillel and Baker, 1988). However, Selker, et al., (1991c) showed experimentally that instabilities also occur by raining at a much lower rate than the saturated hydraulic conductivity. Thus, fingering may occur widely in the field when soils are coarse or water repellent (van Dam, et al., 1990 and Glass, et al., 1988). In homogeneous sand, the fingers are nearly vertical and generally do not merge with one another. On the other hand in less homogeneous sand, the fingers can merge. In the field, where heterogeneities in the soil are far more pronounced than in laboratory experiments, most merging takes place in a shallow layer near the surface called the induction zone. Below the induction zone, merging is less common as long as the water flows through a relatively homogeneous medium. But, further down in the soil, sloping interfaces of compositionally different layers may act as funnels (Kung, 1990a,b). Such structures can concentrate the water from a large area into a single finger. Fingers may also split. This is the probable outcome when a finger reaches saturation and cannot handle the imposed flow.

4. Implications For Monitoring Groundwater Contamination

One of the greatest uncertainties in monitoring groundwater contamination is the possibility that solutes flowing in preferred paths bypass samplers (Biggar and Nielsen, 1976). Thus scaling point-source measurements to field-scale estimates may be grossly unreliable and misleading even in the case when geostatistical methods are employed (Steenhuis, et al., 1990b).

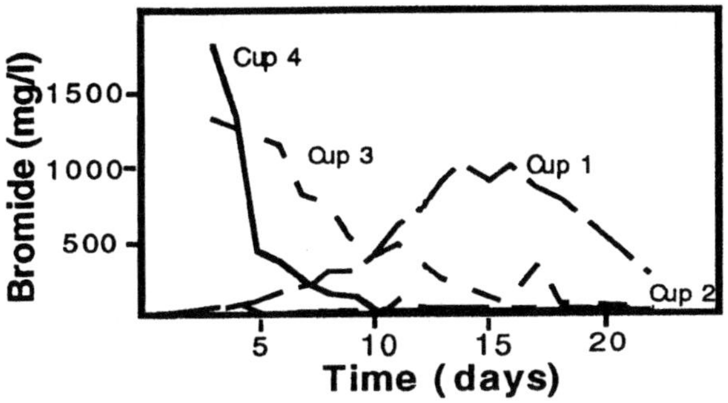

Figure 1a. Breakthrough curves for four porous cup samplers located at the 60 cm depth. (From Boll, et al., 1991)

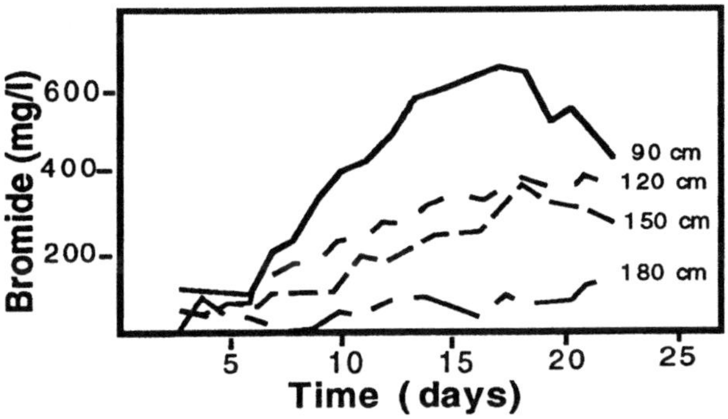

Figure 1b. Breakthrough curves for porous cups at 90, 120, 150, and 180 cm. Each line is the average of two replicates. (From Boll, et al., 1991).

How sampling is affected by preferential flow patterns was studied by Beasly (1976) for forested watersheds, by Shaffer, et al., (1979) and Shuford, et al., (1977) for water sampling with suction cups in field soils, and by Wilson (1980) in a review on sampling methods in the vadose zone. However, no solutions were proposed. Figures 1a and 1b reproduced from Boll, et al., (1991), illustrate how dramatic differences in sampled concentration of solutes can occur within a very small volume. Four suction lysimeters were installed at a depth of 60 cm, and two each at 90 cm, 120 cm, 150 cm and 180 cm depths in a sandy loam near Freeville, New York. The soil was noted as having worm holes (Boll, et al., 1991). The depth to groundwater was 2 m. A 4 cm pulse of bromide, with a concentration of 8000 mg/l, was applied on first day of the experiment and this was followed by 4 cm irrigation each day. The suction lysimeters located at 60 cm were affected by preferential flow paths directly connected to the surface (some induced along the sampling tubes) giving rise to the widely varying breakthrough curves (Figure 1a). The remaining sampling devices did not appear to have this problem. Based on the convective-dispersive model, one would expect to see the solute peaks decrease in concentration with depth. However, as shown in Figure 1b, the peak concentration observed in the capillary fringe at 180 cm, for instance, occurred prior to the peak seen at 150 cm, which, in turn, arrived prior to the maximum concentration seen at the 120 cm depth. This pattern was caused by preferential flow of water and solutes from the finer soils at the 90 cm depth through the coarse soil directly to the groundwater table, bypassing the soil suction lysimeter.

There are currently two general approaches, claimed to be independent of the sampling device, that can be used to sample solute flow in the field. The first is Bouma's (1990) morphological approach in which a Representative Elementary Volume (REV) is used. Soil volumes can be considered "representative" if they are large enough so that individual flow path differences can be averaged. According to Lauren, et al., (1988), sampling volumes of 30 peds or more result in "representative" samples. Thus, REV's characterise "an average flow path" in which the soil heterogeneities can be studied as a stochastic or statistical phenomena. However, in situations where concentrations on the order of parts per billion are considered (eg. pesticide transport), then this approach will be inadequate (Parlange, et al., 1988). Rather, individual preferential pathways that may be responsible for the transport of chemicals must be sampled. In this case, the approach by Barcelona and Morrison, (1988) to locate the sampling devices in the likely pathways of water and contaminants is better. When the groundwater is at shallow depth, then the use of artificial tile lines for sampling are useful even though the tile lines integrate samples over a field scale (Richard and Steenhuis, 1988; Everts and Kanwar, 1988).

5. Modelling Preferential Transport in the Vadose Zone

Despite many experimental findings, the preferential flow phenomena has not been adequately incorporated into simple mathematical models for simulating solute flow in the vadose zone. The major difficulty of modelling this transport phenomena is primarily due to characterising the flow-paths. Many models separate matrix flow from macroporous flow. These dual-porosity models (eg. see Gerke and van Genuchten, 1993), typically write separate transport equations for each domain of flow (ie. matrix and macropore). Mixing of water and solutes across the interface is permitted and described by transfer coefficients which are often functions of the pore-size. We follow a similar conceptual framework and propose a mathematical model that considers not two domains but any number of domains. Our solute transport model is based on the method of characteristics solution of the one-dimensional continuity equation and a simplified flux law. Unlike the usual modelling assumptions applied in the convective-dispersive equation, the concentration of solutes in the percolating water is dependent on the varying rate of applied water and the time period between rainfall and chemical application. This is achieved by relating the solute flux to the water flux. Thus transient field conditions can be simulated. The model explicitly considers preferential paths by taking various piece-wise linear approximations of the hydraulic conductivity. In this method two parameters are identified: the number of pore groups with mobile water (N) and the transport velocity of each pore group (v_p). With these parameters, any soil medium can be simply simulated. These parameters can be related to physical properties but they usually require calibration to a particular field or laboratory experiment. The model can be applied to both large-scale field experiments and small-scale laboratory experiments. We present a brief, systematic overview of the model.

5.1 MOISTURE TRANSPORT

Water and solutes may travel through a soil by any number of possible routes and at differing velocities. Thus the residence times of water-soluble contaminants in the vadose zone is highly variable. However, for the purposes of modelling, even though these paths may be highly distributed throughout the soil, they may be collected into groups of pathways in which the moisture and solutes travel with approximately the same flux. These groups are called *pore-groups*. The obvious and important question of how one determines these pores groups in addressed in the next section. The total amount of moisture, $\theta(x,t)$ in the soil at time t and point x is the sum of all individual moisture contents for each pore-group.

$$\theta(x,t) = \sum_{p=0}^{N} \theta_p(x,t) \tag{3}$$

where θ_p is the individual moisture content for the pth pore-group. The maximum amount of moisture that each group can hold and transmit is

$$\Delta M_p = M_p - M_{p-1} \tag{4}$$

where M_p are various moisture contents representing upper and lower limiting values for the pth group and are a function of the size of pores in each group. When $\theta_p = \Delta M_p$, then all the pathways for the pth group are completely saturated with water. The soil is locally saturated when all flow paths in all groups are saturated, ie. $\theta(x,t) = \theta_s$ and $\theta_p(x,t) = \Delta M_p$ for all p where θ_s is the saturated moisture content of the soil. The group's moisture content, θ_p, is a function of the downslope flux q_p and the effects of precipitation, evapotranspiration and loss or gain of moisture from interactions and exchanges with other groups. Therefore, from continuity, the transport equation may be written as

$$\frac{\partial[\theta_p(x,t)]}{\partial t} + \frac{\partial[q_p(x,t)]}{\partial x} = A_p(x,t) \tag{5}$$

$$\text{for } \ t \geq 0 \,;\, x \geq 0 \ ;\, 0 \leq \theta_p \leq \Delta M_p \,;\, p = 0 \,..\, N.$$

where $A_p(x,t)$ is a source/sink term representing the effects of precipitation, evapotranspiration and mixing between other groups, t is time and x is a distance with x = 0 being the soil surface; N is the number of mobile pore-groups. From consideration of Darcy's law.

$$\begin{aligned} q_p(x,t) &= k(\theta_p)\, H(x,t); \qquad \text{for } p = 1 \,..\, N \\ q_o(x,t) &= 0 \end{aligned} \tag{6}$$

$$0 \leq \theta_p \leq \Delta M_p; \qquad q_o \leq q_1 \leq q_2 \ldots \leq q_N$$

H(x,t) is the hydraulic gradient and $k(\theta_p)$ is the hydraulic conductivity. For moisture in the vadose zone, a unit gradient is assumed, ie. $H(x,t) = 1$. This is a useful and accurate approximation when gravity is the main driving force and is appropriate for percolation in macropores (Parlange, et al., 1993). Therefore, after substitution

$$\frac{\partial\theta_p}{\partial t} + \upsilon_p \frac{\partial\theta_p}{\partial x} = A_p \qquad \text{where} \quad \upsilon_p = \frac{dk(\theta_p)}{d\theta_p} \tag{7}$$

where v_p is the velocity of percolating moisture in the pth group. Note that for $p = 0$, $v_p = 0$. The solution of equation (7) obtained by the method of characteristics (eg. see Charbeneau, 1981) is given by

$$x = \upsilon_p t + x_o, \tag{8}$$

and

$$\theta_p(x,t) = \int_o^t A_p\, dt + \theta_p(x_o,0) \quad ; \quad x \geq \upsilon_p t \tag{9}$$

where $\theta_p(x_o,0)$ is an arbitrary function of x_o when t = 0. Equation (9) cannot be easily integrated because A_p is generally not a simple analytical expression. A simpler approach is to consider a fixed time step numerical solution for the moisture content θ_p. Consider the solution at the times (t - Δt) and t where Δt is small fixed time interval. Therefore, from equation (8), in a time Δt, moisture will travel a distance Δx_p given by

$$\theta_p(x,t) = \theta_p(x-\Delta x_p, t-\Delta t) + I_p(x,t); \qquad \Delta x_p = \upsilon_p\, \Delta t \tag{10}$$

where I_p is the integral of A_p between (t - Δt) and t. Let Δx_p be a simple integer multiple γ_p of Δx_1, the distance travelled by moisture in the slowest moving group (p = 1) ie. $\Delta x_p = \gamma_p \Delta x_1$ with $\gamma_1 = 1$. Note that in the "smaller" time step $(\Delta t / \gamma_p)$, the pth group moved a distance Δx_1. Therefore, define a grid of points in (x-t) space based on Δx_1 and $(\Delta t / \gamma_T)$ where T is the label for the pore-group with the greatest flux. (Note that in the next section we show why T does not necessarily have to equal N). Then

$$x = i\, \Delta x_1 \;;\; i = 0, 1, 2 .. \; L / \Delta x_1 \tag{11}$$

and

$$t = j\, (\Delta t / \gamma_T) \;;\; j = 0, 1, 2 .. \tag{12}$$

where L is the depth to groundwater. On this grid, only one parameter, either the spatial coordinate, Δx_1 or the time coordinate $(\Delta t / \gamma_T)$ needs to be specified. It is usually more convenient to fix the time rather than the distance. Therefore the following equation is a computational formula for moisture transport.

$$\theta_p^{(i,j)} = \theta_p^{(i-\delta_p \gamma_p, j-1)} + I_p^{(i,j)}$$

where

$$\delta_p = \begin{cases} 0; & j \bmod \gamma_p \neq 0 \\ 1; & j \bmod \gamma_p = 0 \end{cases} \tag{13}$$

Figure 2 illustrates the operation of the model defined by eq. (13). In this example, suppose N = 3 (ie. there are three mobile pore-groups, p = 1, 2, and 3), one immobile pore group (p = 0) and one macroporous pore-group, (p = 4). The transport velocities for each pore group are summarised in table 1.

Table 1. Pore velocities and transport coefficients for the example given in Figure 2.

Category	*Pore Group Number* (p)	*Transport Coefficient* (γ_p)	*Pore Velocity* (cm/hr)
Immobile	0	1	0
Mobile	1	1	19.5
Mobile	2	2	39.1
Mobile	3	4	78.1
Macropore	4	64	1250

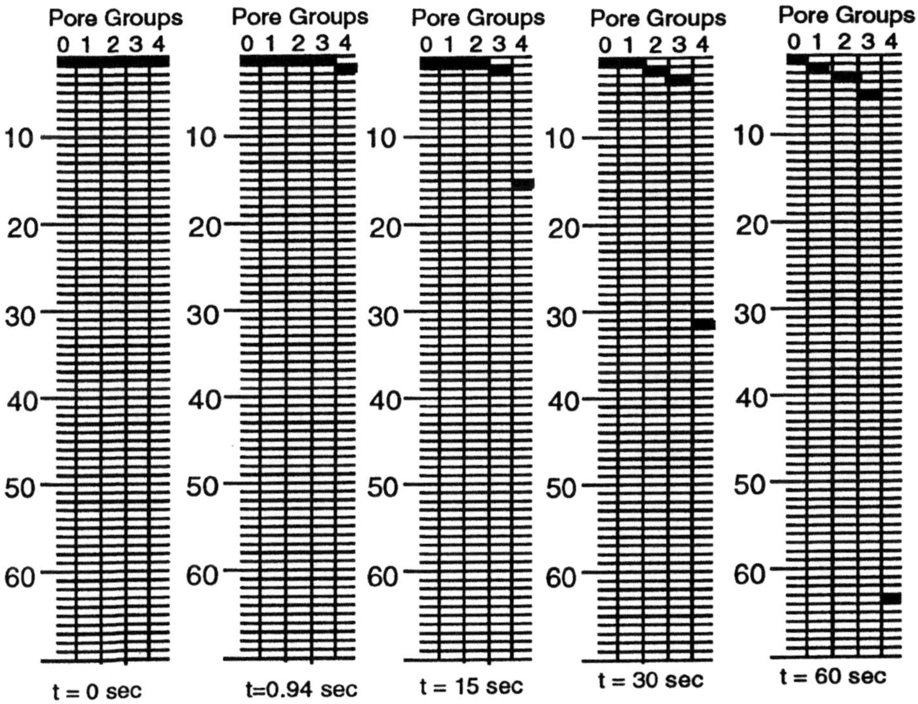

Figure 2. Transport of water and solutes in the preferential flow model (see eq. 13). The time step is defined by eq. (12). The spatial coordinate x is directed downward and the small divisions represent the space step Δx_1 (see eq. 11). (Adapted from Nijssen, et al., 1991).

For this example, $\Delta t = 1$ minute. Therefore, the "small" time step as defined by eq. (12) is $(\Delta t/\gamma_T) = 1/64$ min, a little less than one second. The reason for maintaining a second "small" time step $(\Delta t/\gamma_T)$ in the model is to provide a uniform rate of mixing of water (and solutes) during Δt. The mechanism of mixing water and solutes for each time step is described in section 5.7. The spatial step, $\Delta x_1 = 0.33$ cm. This is the distance moisture (and solutes) will move in the first pore-group in one minute. After one minute, moisture (and solutes) in the second, third and fourth pore-groups would have travelled a distance of $2\Delta x_1$ (0.66 cm), $4\Delta x_1$ (1.32 cm) and $64\Delta x_1$ (21.3 cm), respectively.

5.2 DETERMINING THE NUMBER OF GROUPS FOR A SOIL

The total number of pore-groups, the amount of water in each group and its flux cannot be individually measured by experiments and generally will have to be estimated or determined by calibration. Indeed, it is difficult to conceive of an experiment that could determine such groupings. However, we propose a simple but general method for selecting any number of pore-groups for a soil based on a piece-wise linear approximation of the hydraulic conductivity function as shown in figure 3. The derivation of the following equations is presented in the appendix. The moisture content and conductivity for the pth pore is given by,

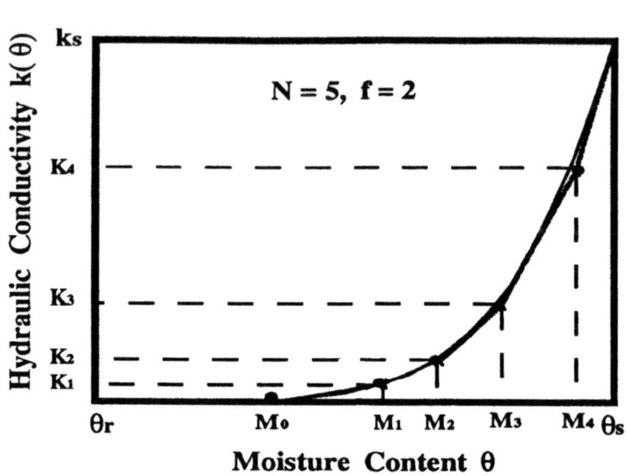

Figure 3. Piece-wise approximation to the unsaturated hydraulic conductivity function. Used to simulate pore-groups in the soil layer.

$$M_p = \theta_r + (\theta_s - \theta_r)(1-n)\left[\frac{f^{(a+1)(N-p)}(1-f^{-a})}{1-f}\right] \qquad p = 1 \,..\, N-1 \qquad (14)$$

and

$$K_p = \left[\frac{k_s(1-n)}{n}\right]\left[\frac{f^{a(N-p)}(f-f^{-a})}{1-f}\right] \qquad p = 1 \,..\, N-1 \qquad (15)$$

where, N is the number of pore groups with mobile water, $a = 1/(n-1)$ is a property of the soil and n is the reciprocal exponent in the power law of the hydraulic conductivity function which depends on soil type (Brooks and Corey, 1964) given by

$$k(\theta) = k_s\left[\frac{\theta(x,t) - \theta_r}{\theta_s - \theta_r}\right]^{1/n} \qquad (16)$$

where k_s is the saturated hydraulic conductivity, θ_s is the moisture content at saturation. Note, equations (14) and (15) are valid for $p = 1 \,..\, N-1$. For $p = N$, we have

$$M_N = \theta_s \text{ and } K_N = k_s \qquad (17)$$

and for $p = 0$, we have

$$M_o = \theta_r + (\theta_s - \theta_r)\,(1-n)\, f^{(a+1)(N-1)} \text{ and } K_o = 0 \qquad (18)$$

Therefore, the velocity for each pore-group can be written as

$$\upsilon_p = k_s f^{(p-N)} / [n (\theta_s - \theta_r)] ; \quad p = 1 .. N \tag{19}$$

and

$$\gamma_p = \upsilon_p / \upsilon_1 = f^{p-1} \quad ; \quad \gamma_p \leq L/\Delta x_1 \tag{20}$$

Note, f is an integer greater than one (eg., 2, 3 etc). If $f = 2$, for example, the velocity of the second pore group is twice as fast as the first and the velocity of the third is twice as fast as the second and so on. If $f = 3$, then the velocity of the second pore group is three times faster than the first, etc (see the appendix). Any soil may be simulated by selecting values for f and N.

5.3 MACROPORE FLOW

The number of pore-groups, their limiting moisture contents and flux, can be linked to physically measurable parameters such as soil type and hydraulic conductivity. Any number of additional pore-groups representing flow through macropores and subsurface channels can be added to the model. The transport of moisture in macroporous groups is catered for in the same manner to the other pore-groups, ie. by defining transport coefficients which are a simple integer multiple of the velocity of moisture in the slowest pore group.

$$\gamma_p = \upsilon_p / \upsilon_1 \quad \text{for} \quad p = N+1, N+2 .. T \tag{21}$$

and where in this case, γ_p represents how much faster moisture in the pth macropore group travels compared to moisture travelling in the slowest moving pore-group ($p = 1$). Although, in principle, any number of macropore groups can be modelled in this way, in practice it is hard to justify selecting any more than one macropore group without experimental evidence.

5.4 SOLUTE TRANSPORT

The solute concentration, c(x,t), in the soil is given by

$$c(x,t) = \frac{s(x,t)}{\theta(x,t)} \tag{22}$$

where s(x,t) is the total solute load given by

$$s(x,t) = \sum_{p=0}^{T} s_p^{(i,j)} \tag{23}$$

and s_p is the solute mass in pore-group p. Assuming that chemical processes like adsorption and solute precipitation are negligible, then the solute flux may be simply written as

$$s_p^{(i,j)} = s_p^{(i - \delta_p \gamma_p, j-1)} \tag{24}$$

where $\psi_p^{(i,j)}$ is the solute added to or subtracted from the pth group during $\Delta t / \gamma_p$. In each time step the concentration of solutes in each pore group is affected by three processes: evapotranspiration (moisture extracted from the soil), precipitation (moisture entering the soil after infiltrating the soil surface) and mixing or interchange between pore-groups in order to attain chemical equilibrium. These three processes determine the value of $\psi_p^{(i,j)}$. Each process is described in more detail below. We shall use the notation: E (evapotranspiration), R (precipitation) and M (mixing) in the following equations to describe how each process is modelled. Note that in our formulation at each time step, j, we first account for evapotranspiration, then precipitation and then mixing. The order in which these processes are accounted for within the time-step is arbitrary and any order can be adopted.

5.5 EVAPOTRANSPIRATION

Moisture is extracted from each pore-group in the soil in various amounts as a result of evapotranspiration. Evapotranspiration also helps to redistribute moisture and consequently solutes between groups. Let $E^{(i,j)}$ represent the evapotranspiration loss from the soil at time j and location i. Since this loss does not have to be uniform with depth nor with respect to pore groupings, let ξ_p represent the fraction of the total loss that was taken from group p during $\Delta t / \gamma_p$. If $\theta_p^{(i,j)}$ represents the moisture content just prior to accounting for evapotranspiration in the model, then the new moisture content after subtracting the evapotranspiration loss for the pth pore-group is given by

$$\theta(E)_p^{(i,j)} = \theta_p^{(i,j)} - \xi_p \frac{E^{(i,j)}}{D}; \qquad \text{for } 0 \le \xi_p \le \theta_p^{(i,j)} D / E^{(i,j)} \tag{25}$$

where D is the depth to the root zone. Although the concentration is affected, the amount of solutes remains unchanged.

5.6 PRECIPITATION

The net precipitation (precipitation less interception) is distributed in the soil in the various pore-groups. Exactly how this distribution occurs in the soil is a complex process and depends on the infiltration rate and the moisture status of the soil. Since this distribution is often non-uniform, let μ_p represent the fraction of the total precipitation that is added to the pth group following rain. Therefore,

$$\theta(ER)_p^{(i,j)} = \theta(E)_p^{(i,j)} + \mu_p \frac{P^{(i,j)}}{D}; \qquad \text{for } 0 \le \mu_p \le \Delta M_p - \theta(E)_p^{(i,j)} D / P^{(i,j)} \tag{26}$$

where $P^{(i,j)}$ is the amount of precipitation at (i,j). Let $C_R^{(i,j)}$ be the concentration of solute in the precipitation and let $s_p^{(i,j)}$ be the amount of solute in the group just prior to rain. Therefore, the amount of solute in the soil after rain is

$$s(ER)_p^{(i,j)} = s_p^{(i,j)} + \mu_p \frac{P^{(i,j)}}{D} (\Delta t / \gamma_T)\, C_R^{(i,j)} \tag{27}$$

where $s(ER)_p^{(i,j)}$ represents the amount of solutes in the pth pore-group after accounting for precipitation.

5.7 MIXING BETWEEN GROUPS

The network of flow paths in the soil are interconnected. Therefore, there is an exchange of moisture and solutes between these pathways and groups of pathways. At any time there will be some exchange between the various pore-groups ranging from complete exchange of moisture and solutes to no mixing. Mathematically, this behaviour might be described by three strategies (a) complete mixing, (b) partial mixing and (c) no exchange between pore-groups within the simulated time interval. These three strategies covering the continuum of possible behaviour should be implemented within the model. For example, under option (b), an appropriate partial mixing strategy might be one in which a substantial portion of the moisture and solutes in the slower pore groups are exchanged within the time interval whereas for the larger pore groups or macroporous groups, very little exchange occurs in the time (Skopp, et al., 1981). Various methods for implementing these mixing strategies in a computer algorithm were discussed by Stagnitti, et al., (1991).

All strategies can be modelled easily by defining two sets of coefficients, $\alpha_p(\theta_p)$ and $\eta_p(\theta_p)$. Let α_p represent the fraction of moisture extracted from group p and placed into a "common" pool ($0 \leq \alpha_p \leq 1$), and let η_p represent the fraction of the "common" pool that is moved back into the pth group after mixing. Let T be the total number of groups that are mixed. Note that T contains N micropore groups with mobile water, one immobile pore-group and any number of macropore groups (eg. see eq. (21)). After mixing, the new moisture content will be given by

$$\theta(ERM)_p^{(i,j)} = (1-\alpha_p)\,\theta(ER)_p^{(i,j)} + \eta_p \sum_{g=0}^{T} \alpha_g\, \theta(ER)_g^{(i,j)} \tag{28}$$

The values for the mixing coefficients α_p and η_p depend on the soil type and would normally have to be found by careful experimentation. In a similar fashion, after mixing is accounted for, we have

$$s(ERM)_p^{(i,j)} = (1-\alpha_p)\, s(ER)_p^{(i,j)} + \eta_p \sum_{g=0}^{T} \alpha_g\, s(ER)_g^{(i,j)} \tag{29}$$

where $s(ERM)_p^{(i,j)}$ is the amount of solutes in group p after the mixing process has been accounted for and it is equivalent to the right hand side of eq. (24) after the processes of evapotranspiration, precipitation and chemical mixing have been accounted for. Thus at the end of each time step, the concentration in each individual pore-group is

$$c_p^{(i,j)} = s(ERM)_p^{(i,j)} / \theta(ERM)_p^{(i,j)} \tag{30}$$

There are certain conditions and restrictions on the range of values that the mixing coefficients can have. These are fairly obvious and result from consideration of the conservation of mass. In particular, if a set of coefficients for α_p are adopted for a particular mixing strategy, then the values for η_p will be determined at each time step and depend on the amount of moisture in the pore group, the capacity of the pore-group to store additional moisture and the value of α_p. Equations (28) and (29) are bound by the following conditions

$$\sum_{p=0}^{T} \eta_p = 1 \; ; \quad \text{and} \quad \eta_p \leq \left[\Delta M_p - (1-\alpha_p)\, \theta(ER)_p^{(i,j)}\right] \Big/ \sum_{g=0}^{T} \alpha_g\, \theta(ER)_g^{(i,j)} \tag{31}$$

Clearly, any number of mixing strategies can be adopted depending on the values for α_p. For example, no mixing can be implemented by taking $\alpha_p = 0$ for all p. Partial mixing is simulated when $0 < \alpha_p < 1$ and full mixing occurs when $\alpha_p = 1$ for all p. If only the concentration of solutes in the pore groups is changed during the time step (ie. the volume of water remains unchanged), then

$$\eta_p = \left[\alpha_p\, \theta(ER)_p^{(i,j)}\right] \Big/ \sum_{g=0}^{T} \alpha_g\, \theta(ER)_g^{(i,j)} \tag{31}$$

On the other hand, a partial mixing strategy which tends to move moisture (and solutes) from pore-groups with small available moisture capacities to groups (ie. small ΔM_p) to groups with larger moisture capacities is given by

$$\eta_p = \left[\Delta M_p - (1-\alpha_p)\, \theta(ER)_p^{(i,j)}\right] \Big/ \sum_{g=0}^{T} \Delta M_g - (1-\alpha_g)\, \theta(ER)_g^{(i,j)} \tag{32}$$

Other strategies were suggested by Stagnitti, et al., (1991). Figures 4a and 4b illustrate the effects that different mixing strategies have on the breakthrough curves for a simulated soil column. For each simulation, the values for α_p are the same for each pore group (eg. 2% per time step for all pore groups).

5.8 COMPARISON WITH THE CONVECTIVE-DISPERSIVE EQUATION

Figures 4a and 4b show that the preferential flow model behaves like a convective-dispersive model when complete mixing is adopted and like a pore-bundle model when no mixing is adopted. We simulated the breakthrough curves obtained from a hypothetical soil column by a applying an instantaneous concentration of solute under steady-state flow conditions. He used CXT-FIT from Parker and van Genuchten (1984b) to fit hydrodynamic dispersion coefficients to the simulated breakthrough curves. Not surprisingly, he found that the preferential flow model behaved very much like the convective-dispersive equation when saturated flow conditions and 100% mixing in all pore groups was adopted. He also found that even under unsaturated flow conditions, the dispersion coefficient calculated using the input parameters from the preferential flow model was highly correlated with the fitted coefficients from CXT-FIT. Figure 5 illustrates the relationship between the mixing coefficients expressed by equations. (28) and (29) and the fitted dispersion coefficients. Clearly, as expected, the mixing coefficients are inversely proportional to the hydrodynamic dispersion.

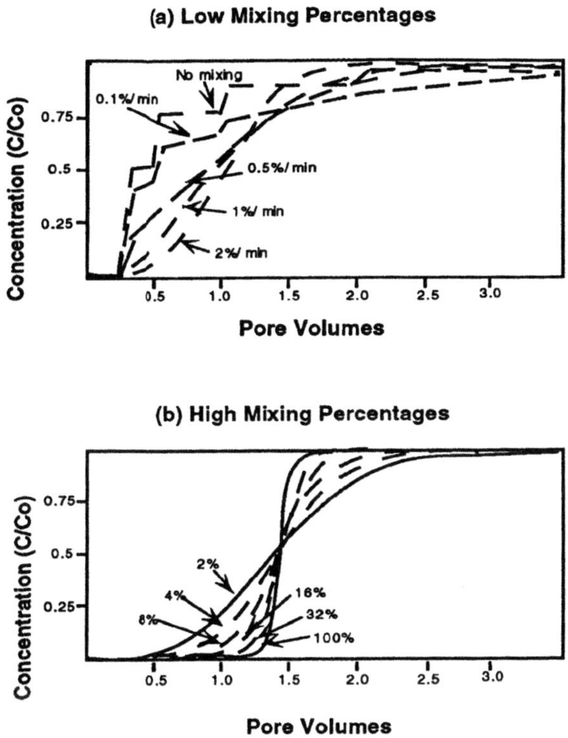

Figures 4a and 4b. Influence of different mixing percentages on the breakthrough curves for a simulated soil column.

6.0 Results And Discussion

The application of the preferential transport model is illustrated on a number of field and laboratory experiments. The first was a hillslope experiment conducted at the Cornell University's Turkey Hill Research Site near Ithaca, New York (Steenhuis and Muck, 1988). The plot is 23 m wide and 109 m long with an area of about 0.25 ha. The soil is a silt loam with a clay pan of compact till at 30 cm depth. The surface slope is a uniform 8%. A hay-grass mixture covered the plot. At the lower end of the plot both the volume and quality of the subsurface water and surface water were measured. A sprinkler irrigation system used water in a nearby creek to irrigate the plot. The outflow data for an experiment, carried out in December of 1978 in which nitrate-N and chloride-Cl was surface applied to saturated soil, is used here. The temperatures during this experimental period were just above freezing and consequently denitrification was extremely small. Evapotranspiration was negligible. The artificial precipitation averaged 0.6 cm/h and was maintained for over 80 hours. The amount of nitrate in the irrigated water was negligible but the concentration of chloride in the creek water ranged from 14 mg/l to 16 mg/l. Therefore, we took the initial concentration of nitrate in the soil to be zero and that for chloride at 15 mg/l. Before applying the chloride and nitrate, steady state drainage conditions were achieved by irrigating for a 24 hour period. The average measured saturated conductivity was 12 m/d and saturated moisture content was 52.5%. A field capacity of 46% was reported by Steenhuis and Muck (1980). The residual moisture content was 0.0%.

The mathematical formulation presented in this paper was developed for vertical percolation through the soil. However, Stagnitti, et al., (1991) demonstrated that with very little modification to the mathematical formulation presented here, the preferential flow model could also be applied to hillslope experiments. The major changes to the model include modifying the flux for downslope transport and incorporation of surface runoff. The reader is referred to that paper for further details. The applied solutes were only added initially to the macropore and surface runoff groups and then allowed to mix slowly into the micropores and immobile pore group as experimental measurements indicated that when the chemicals were surface applied to the saturated soil, the chemicals did not mix readily with the immobile water (Richard and Steenhuis, 1988).

For this experiment for equations (14) to (20), we took, f = 2 and N = 2 (two mobile pore groups and one immobile group). We added two additional pore groups, one for surface runoff and the other a macropore group. The moisture limiting values for the immobile pore group was 0 to 0.352 and for the mobile pore groups, they were 0.353 to 0.485 and 0.486 to 0.525. The limiting moisture contents for the macropore group was 0.525 to 0.675. If the local moisture content after accounting for precipitation exceeded 0.675 at any time, then the additional moisture was added to the surface runoff group. We also adopted an initial moisture profile which is consistent with steady-state drainage with an irrigation rate of 0.6 cm/h. Water and solutes in the surface runoff moved at a rate 32 times faster than the slowest moving group. Therefore, γ_p equalled 1, 2, 4 and 32, respectively. We chose our small time step in eq. (12) to be a quarter of an hour. This was similar to the time period in which measurements were taken. Therefore, $\Delta x_1 = 1.52$ m. How much of the solutes should be blended per time step? There is no *a priori* method of determining this. We decided to try four possibilities for α_p. The first was no mixing, ie., $\alpha_p = 0\%$ for all p. In subsequent trials we took 2%, 5% and 10% solutes blended per quarter hour per group. In each case, values for η_p were chosen to conserve mass (see Stagnitti, et al., (1991) for further details). The case for 2% gave the best fit and is presented in Figure 6.

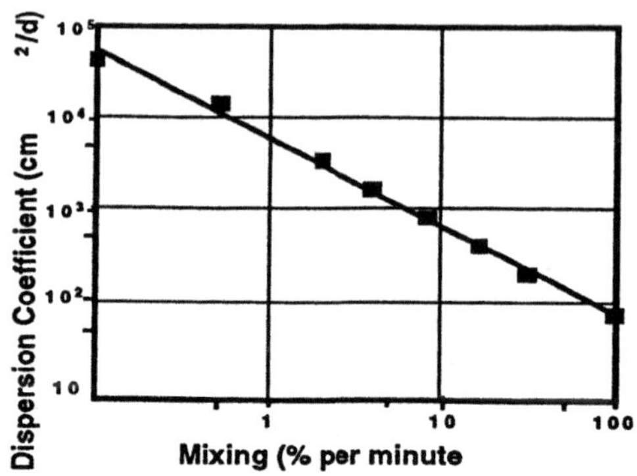

Figure 5. Relationship between the mixing coefficients for the preferential flow model and the hydrodynamic dispersion coefficients of CXT-FIT. (In each case the same value for α_p was adopted for each pore group and η_p was determined from conservation of mass within each cell per time step.

The predicted solute loss for both chloride and nitrate is in very good agreement with the observed loss. It is indeed remarkable that the one set of parameters fit both data sets extremely well. However, this experiment alone may not be very convincing given the number of degrees of freedom the model has for parameter estimation. The following two sets of laboratory experiments further illustrates the applicability of the model but in these cases the range of values that the parameters may adopt is much narrower.

Anderson and Bouma, (1977a,b) conducted a set of laboratory experiments in undisturbed soil columns to determine the dispersion coefficients for soils consisting of blocky and coarse prismatic structures. In these experiments, three similar columns were subject to four different flow treatments using chloride as a tracer. Each soil core was 10 cm in diameter and the length ranged from 43 to 52 cm. Three replications of each experiment were studied. The first two experiments examined saturated flow conditions. The columns were initially saturated with water and then a 300 ppm solution of potassium chloride was ponded on the top of each column at shallow depth (Experiment 1). After the effluent concentration reached the input concentration, the columns were then flushed with water and drained leaving the larger pores filled with air. Another continuous dose of 300 ppm of chloride concentration was applied by ponding (Experiment 2). This was then followed by another flushing and draining and then a pulse application of 1 cm per day irrigation by ponding (Experiment 3).

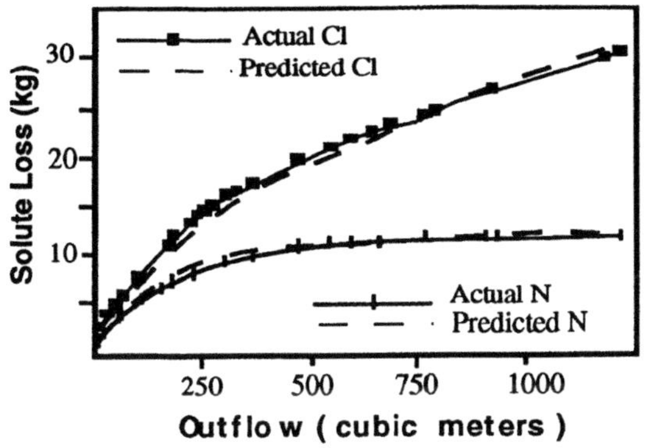

Figure 6. Predicted solute loss (broken lines) using the preferential solute transport model plotted with observed loss (solid lines) for application of chloride (squares) and nitrate (crosses).

The final experiment consisted of ponding water on a gypsum crust giving a steady, continuous infiltration of approximately 1 cm/d (Experiment 4). Steenhuis, et al., (1991) and Nijssen, et al., (1991) analysed these experiments by conducting two simulations. For the first simulation, N = 7 (ie. seven mobile pore groups) and f = 2. The velocity v_p (eq. 5) of the slowest moving group was 1.85 cm/d and the fastest pore-group was 118.6 cm/d. The simulation time step was 3 minutes and solutes were blended at the following rates α_p = 0, 0.1, 0.15, 0.5, 0.75, 1.5, 1.75, and 2.0 % per 3 mins. Although the mixing coefficients were large for the larger pore-groups, the overall mixing in the small pores is greater due to the longer retention time of solutes in those pores. In the second simulation, only the shape of the hydraulic conductivity function was modified by making the fastest moving pore-group transport water and solutes at a rate 256 times faster than the slowest moving pore-group rather than 64 times as in simulation 1 (ie. a factor of 4) and thus enhancing macroporous flow. All other parameters remained unaltered.

The predicted breakthrough curves for the two simulations plotted with the observed breakthrough curves are presented for each of the four experiments in figure 7. The range of values for the observed breakthrough curves provides an indication of the experimental variation between the columns. In all cases, except for Experiment 2, the predicted breakthrough curves for both simulations fell within the range of observed values. For Experiment 3, the predicted pulse application showed large variation within a day. These were not observed in the original data set because the concentrations were only measured daily whereas our model simulations were conducted on a 3-minute interval. If we averaged our predicted concentrations on a daily basis a very similar breakthrough curve would result. The second experiment clearly shows the significant effect of macroporous flow on the breakthrough curve. The almost instantaneous response in the effluent concentration was attributed by Anderson and Bouma to flow through the larger air-filled pores. Our second simulation which enhanced the effect of the macroporous flow more closely matched the observed values than the first simulation. It is very difficult to match the breakthrough curves for Experiments 1 and 2 at the same time and it is likely that different conductivity functions for drained columns and saturated columns should be applied due to swelling of the soil and consequent narrowing of the cracks and voids (Steenhuis, et al., 1991). These results are pleasing in that the model was able to simulate a wide variety of breakthrough curves without changing the physical properties for each experiment. In contrast, Anderson and Bouma tried to fit dispersion coefficients to the data sets using the convective-dispersion equation but found that the coefficients varied considerably and were highly dependent on the flow regime.

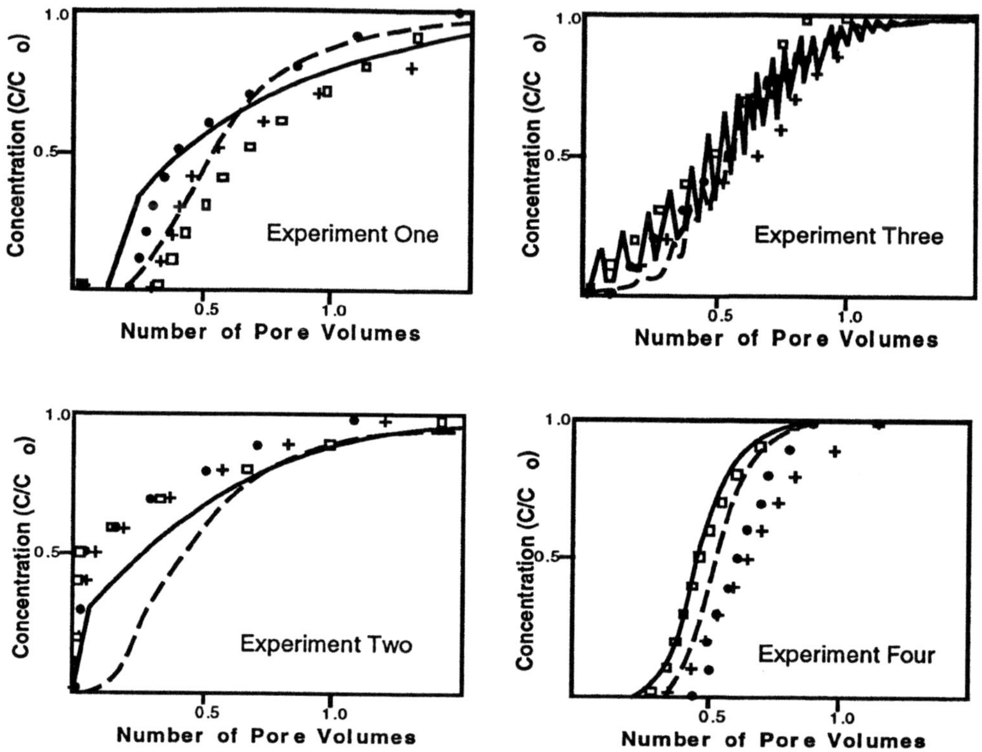

Figure 7. Predicted and observed breakthrough curves for the Anderson and Bouma, 1977a and 1977b data. The solid lines represent model predictions for the second simulation and the dashed lines represent model predictions for the first simulation. The symbols (squares, pluses and astricts) represent the observed data for three different replications (soil columns).

In more recent experiments conducted by Kluitenberg and Horton (1990), three undisturbed soil cores containing macropores were taken from the A horizon of a glacial-till comprising of a fine loam mix. Each column had a diameter of 18 cm and lengths of 33, 33 and 35 cm. Labelled A, B and C, the columns had saturated hydraulic conductivities ranging from 58, 28 to 12 cm/h respectively and an average saturated moisture content of 0.44. At the beginning of the experiment, all columns were saturated for a 12 h period with a 0.01 N solution of calcium sulphate solution. Then, a 0.05 N solution of calcium chloride solution was ponded with a constant head on each column (Experiment 1). After about two pore volumes of this solution had infiltrated, the columns were extensively leached with the 0.01 N solution of calcium sulphate solution and then allowed to drain for a further 12 hour period. Next, 200 ml of a 0.05 N solution of calcium chloride solution was applied to the surface as a pulse (Experiment 2). Once this solution had infiltrated, a 0.01 N solution of calcium sulphate solution was ponded on the surface and leached and drained once more. Then, an application of a 0.25 N solution of calcium chloride solution was dripped on the surface during a 6 to 8 minute interval during which no ponding occurred (Experiment 3). After 15 minutes, the columns were again ponded with the calcium sulphate solution.

For these experiments, four mobile pore-groups and an immobile pore group were selected and the velocities of solute and moisture flow for the mobile pore groups were 19.5, 39.1, 78.1 and 1250 cm/h respectively; the latter representing macroporous transport (see table 1). The "small" time step was 1/64 mins (see eq. (12)). Mixing was facilitated by blending approximately 0.8% of solutes in the immobile group and 0.004% of solutes in the mobile pore-groups in each time step (Nijssen, et al., 1991). Once again, this strategy results in more mixing in the slower moving groups than the faster ones due to the longer residence times in the soil. All other input parameters such as flow rates, duration of chloride pulse, initial moisture status etc. were determined from the actual data. Figures 8 to 10 illustrate the simulated and observed breakthrough curves for columns A, B, and C. Since Kluitenberg and Horton (1990) observed that air entrapment caused variation in the flow rate, the concentrations of the effluent are plotted against pore volume. Figures 8a, 9a and 10a show breakthrough curves for each column for the first experiment where the chloride was ponded at the surface of the soil for approximately two pore volumes. The model reasonably predicts the observed data except for the initial breakthrough. The most remarkable feature for the observed and predicted breakthrough curves is the difference between the pulse and drip applications (Experiments 2 and 3) which are shown in figures 8b, 9b and 10b. For the drip application, the column remained unsaturated, whilst for the pulse, the solution was applied in a single slug. The resulting breakthroughs are quite different as one might expect. For all columns, the pulse application resulted in a sharp, early peak whilst for the drip experiment, the breakthroughs were smaller and the peak arrived later. The model, without changing the soil and solute properties, was able to predict these differences quite well. Therefore, unlike the convective-dispersive equation, the model is sensitive to the rate of effluent application.

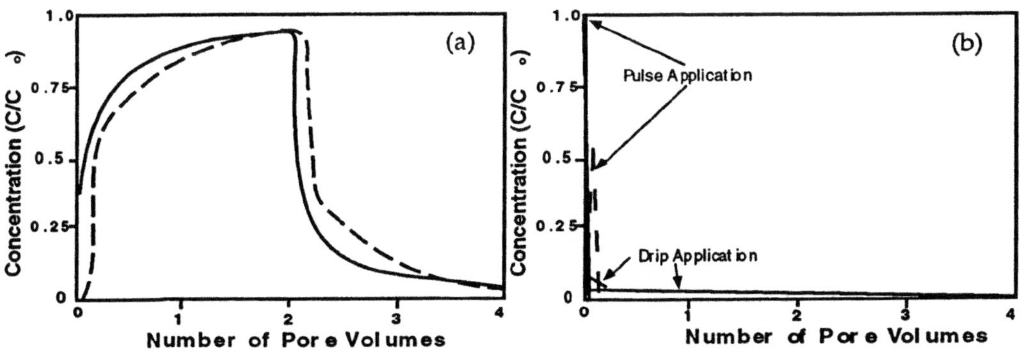

Figure 8. Predicted (dashed lines) and observed breakthrough (solid lines) curves for column A. (a) Ponded Application, and (b) Pulse and Drip Application. Data from Kluitenberg and Horton (1990).

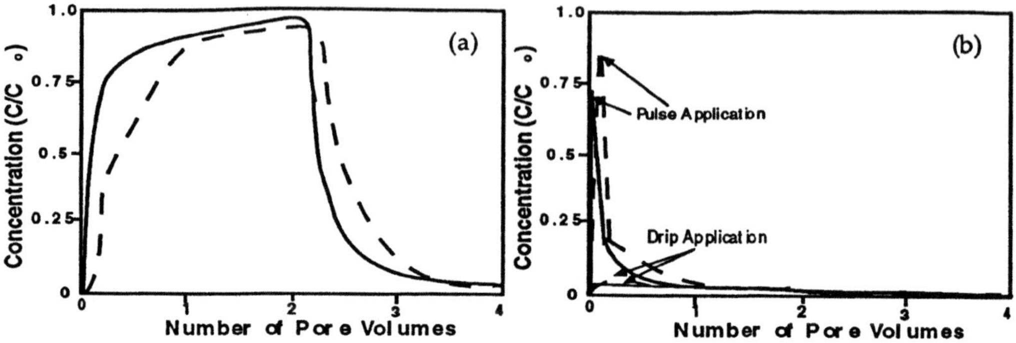

Figure 9. Predicted (dashed lines) and observed breakthrough (solid lines) curves for column B. (a) Ponded Application, and (b) Pulse and Drip Application. Data from Kluitenberg and Horton (1990).

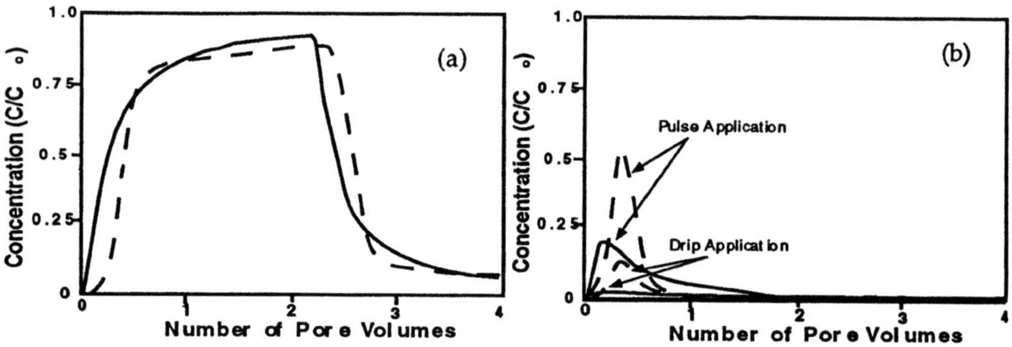

Figure 10. Predicted (dashed lines) and observed breakthrough (solid lines) curves for column C. (a) Ponded Application, and (b) Pulse and Drip Application. Data from Kluitenberg and Horton (1990).

7. Summary And Conclusions

Local scale heterogeneities in natural soils are a major concern in modelling solute movement to groundwater. The effect of preferential flow on groundwater quality is much the same whether it occurs in structured soils or homogeneous sandy soils. In fact, preferential flow seems to be most common in soils at both ends of the spectrum. In dense clay soils, cracks provide the only channels through which flow can occur, and in most coarse soils (when rainfall intensity is low), instabilities resulting in preferential flow are a near certainty. Preferential flow is especially important for chemicals that are toxic in the parts per billion or trillion range. For these chemicals, modelling contaminant transport with an averaged flux or averaged volume approach as advocated by the use of the convective-dispersion equation may result in gross inaccuracies.

We have developed a simple multi-domain, preferential flow model that numerically solves the simplified subsurface transport problem. The model which has assumptions based on Darcy's law, unit gradient theory and the conservation of mass, results in a one-dimensional partial differential equation which is solved by the method of characteristics. The model is based on the concept that the soil layer can be represented by a series of interconnecting pore groups in which mixing of moisture and solutes occur. These groupings can be simulated by taking various piece-wise linear approximations of the unsaturated hydraulic conductivity curve. The numerical scheme is greatly simplified by taking the pore-group velocities as a simple integer multiple of the slowest moving group. The exchange of water and solutes between the pore-groupings, in this paper was assumed to be a linear process but could easily be modified if more precise experimental data were available. The model was tested on a number of previously published and recently conducted experiments and the predicted outflows were in good qualitative and quantitative agreement with observed breakthroughs. In many cases, the preferential flow model offered more physically plausible explanations for the transport phenomenon than the convective-dispersion model. It can be used to predict a rich variety of flow patterns by changing initial and boundary conditions and is not restricted to assumptions of steady-state flow.

The presence of preferred paths in the soil presents not only significant difficulty for modelling but also for monitoring and measurement. Greater emphasis on experimental research directed to understanding the role of surface infiltration rates and chemical and physical processes in and between macropores is required to improve current models of vadose zone solute transport. Recent experimental findings show, unfortunately, that there are no simple means of estimating the size and the number of subsurface point measurements required to achieve representative samples. A better understanding of preferential flow will reveal factors that need to be considered in order to more accurately explore the likelihood that groundwater will be affected by agricultural chemicals, landfills, and industrial wastes. Ultimately, this should lead to sounder management practices that will diminish a serious threat to one of humanity's most precious resources.

8. Appendix: Derivation of Pore-Group Moisture and Conductivity

Select tangents to the hydraulic conductivity $k(\theta)$ such that

$$f = \frac{[\Delta k/\Delta\theta]_p}{[\Delta k/\Delta\theta]_{p-1}} \tag{A1}$$

where $[\Delta k/\Delta\theta]_p$ represents the gradient of the tangent line to $k(\theta)$ for the pth pore-group. Therefore, in general,

$$[\Delta k/\Delta\theta]_p = k_s\, f^{(p-N)} / [\, n\, (\theta_s - \theta_r)\,]; \qquad p = 1 \,..\, N \tag{A2}$$

and

$$[\Delta k/\Delta\theta]_0 = 0 \tag{A3}$$

The tangent point $(M_p^{(t)}, K_p^{(t)})$ for the pth pore-group (ie. the intersection point of the piece-wise linear approximation to the conductivity curve defined by eq. (16) is given by

$$M_p^{(t)} = \theta_r + (\theta_s - \theta_r)\, f^{(a+1)(N-p)} \tag{A4}$$

and

$$K_p^{(t)} = k_s f^{a(N-p)} \tag{A5}$$

Therefore, the linear equation with gradient $[\Delta k/\Delta\theta]_p$ that passes through $(M_p^{(t)}, K_p^{(t)})$ is given by

$$k(\theta_p) = \frac{k_s}{n}\left[\frac{(\theta_p-\theta_r)}{(\theta_s-\theta_r)} f^{(p-N)} + (n-1) f^{a(N-p)}\right]; \quad p = 1 .. N; \; M_{p-1} \le \theta_p \le M_p \tag{A6}$$

The intersection of the tangent line for the pth group with the (p+1) group leads to equations (14) and (15).

9. References

Anderson, J.L., and Bouma, J., Water movement through pedal soils: I. Saturated flow, *Soil Sci. Am. Proc.*, 37, 408 - 413, 1977a.

Anderson, J.L., and Bouma, J., Water movement through pedal soils: II. Unsaturated flow, *Soil Sci. Am. Proc.*, 37, 419 - 423, 1977b.

Barcelona, M.J. and Morrison, R.D., Sample collection, handling and storage: Water, soils and aquifer solids, in *Methods for Ground Water Quality Studies, Proc. Nat. Workshop*, Nelson, D.W. and Dowdy, R.H., Eds., Agricultural Research Division, University of Nebraska, Lincoln, Nebraska, 49 - 62, 1988.

Barry, D.A. and Sposito, G., Application of the convective-dispersion model to solute transport in finite soil columns. *Soil Sci. Soc. Am. J.*, 52(1): 3 - 9, 1988.

Barry, D.A., Modelling contaminant transport in subsurface: Theory and computer programs, in *Modelling Chemical Transport in Soils: Natural and Applied Contaminants*, ed. Ghadiri, H. and Rose, C. W., Lewis Pub.: 105 - 144, 1993a.

Barry, D.A., Parlange, J.-Y., Sander, G.C. and Sivapalan, M., A class of exact solutions for Richards' equation. *J. Hydrol.*, 142: 29 - 46, 1993b.

Charbeneau, R.J., Groundwater contaminant transport with adsorption and ion exchange chemistry: Method of characteristics for the case without dispersion, *Water Resources Res.*, 17, 705 - 713, 1981.

Beasly, R.S., Contribution of subsurface flow from the upper slopes of forested watersheds to channel flow, *Soil Sci. Soc. Am. J.*, 40, 955 - 957, 1976.

Biggar, J.W. and Nielsen, D.R., Spatial variability of the leaching characteristics of a field soil, *Water Resources Res.*, 12, 78 - 84, 1976.

Boll, J., Selker, J.S., Nijssen, B.M., Steenhuis, T.S., Van Winkle, J., and Jolles, E., Water quality sampling under preferential flow conditions, in *Lysimeters for Evapotranspiration and Environmental Measures*, Proc. American Society of Civil Engineers Int. Symp. Lysimetry, Allen, R.G., Howell, T.A., Pruitt, W.O., Walter, I.A., and Jensen, M.E., Eds., American Society of Civil Engineers, New York City, 1991, 290 - 298.

Bouma, J., Using morphometric expressions for macropores to improve soil physical analysis of field soil, *Geoderma*, 46, 3 - 11, 1990.

Everts, C.J. and Kanwar, R.S., Quantifying preferential flow to a tile line with tracers, *American Society of Agricultural Engineers*, Paper No. 88-2635, St. Joseph, MI, 15pp, 1988.

Freeze, R.A. and Cherry, J.A., *Groundwater*, Prentice Hall, 684pp, 1979.

Gerke, H.H. and van Genuchten, M.T., A dual-porosity model for simulating preferential movement of water and solutes in structured porous media, *Water Resources Res.*, 29, 305-319, 1993.

Gish, T.J., Helling, C.S., and Mojasevic, M., Preferential movement of atrazine and cyanazine under field conditions, *Trans, Am. Soc. Agr. Eng.*, 34(4), 1699 - 1705, 1991.

Glass, R.J., Steenhuis, T.S., and Parlange, J.-Y., Wetting front instability as a rapid and far-reaching hydrologic process in the vadose zone, *J. Contam. Hydrol.*, 3, 207 - 226, 1988.

Glass, R.J., Parlange, J.-Y., and Steenhuis, T.S., Wetting front instability, 1. Theoretical discussion and dimensional analysis, *Water Resources Res.*, 25, 1187 - 1194, 1989.

Glass, R.J., Steenhuis, T.S., and Parlange, J.-Y., Wetting front instability, 2. Experimental determination of relationships between system parameters and two-dimensional unstable flow field behaviour in initially dry porous media, *Water Resources Res.*, 25, 1195 - 1207, 1989.

Glass, R.J., Steenhuis, T.S., and Parlange, J.-Y., Mechanism for finger persistence in homogeneous, unsaturated, porous media: Theory and verification, *Soil Sci.*, 148, 60 - 70, 1989.

Glass, R.J., Parlange, J.-Y., and Steenhuis, T.S., Immiscible displacement in porous media: Stability analysis of three-dimensional, axisymmetric disturbances with application to gravity-driven wetting front instability, *Water Resources Res.*, 27, 1947 - 1956, 1991.

Hegert, G.W., Bouldin, D.R., Klausner, S.D., and Zwerman, P.J., Phosphorus concentration-water flow interaction in the effluent from manured land, *J. Env. Qual.*, 10, 338, 1981.

Hill, D.E. and Parlange, J.-Y., Wetting front instability in layered soils, *Soil Sci. Soc. Am. J.*, 36, 697 - 702, 1972.

Hillel, D. and Baker, R.S., A descriptive theory of fingering during infiltration into layered soils, *Soil Sci.*, 146, 51 - 56, 1988.

Hoa, N.T., A new method allowing the measurement of rapid variations on the water content in sandy porous media, *Water Resources Res.*, 17, 41 - 48, 1981.

Kluitenberg, G.J. and Horton, R., Effect of solute application method on preferential flow transport of solutes in soil, *Geoderma*, 46, 59 - 71, 1990.

Kung, K-J.S., Preferential flow in a sandy vadose soil: 1. Field observations, *Geoderma*, 46, 51 - 58, 1990a.

Kung, K-J.S., Preferential flow in a sandy vadose soil: 2. Mechanisms and implications, *Geoderma*, 46, 59 - 71, 1990b.

Lauren, J.G., Wagenet, R.J., Bouma, J., and Wîsten, J.H.M., Variability of saturated hydraulic conductivity in a Glossaquic Hapludalf with macropores, *Soil Sci.*, 145, 20 - 28, 1988.

National Research Council, *Groundwater Models: Scientific and Regulatory Applications,* National Academy Press, Washington D.C., 303pp, 1990.

Nijssen, B.M., Steenhuis, T.S., Kluitenberg, G.J., Stagnitti, F., and Parlange, J.-Y., Moving water and solutes through the soil: Testing of a preferential flow model, Paper presented at the National Symposium on Preferential Flow, Chicago, IL, December, 16-17, 223-232, 1991.

Parker, J.C. and van Genuchten, M. Th., Flux-averaged and volume-averaged concentration in continuum approaches to solute transport, *Water Resources Res.,* 20(7): 866 - 872, 1984a.

Parker, J.C. and van Genuchten, M. Th., Determining transport parameters from laboratory and field tracer experiments, Bulletin. 84-3, Virginia Agricultural Experimental Station, Blacksburg, Virginia, 1984b.

Parlange, J.-Y. and Hill, D.E., Theoretical analysis of wetting front instability in soils, Soil Sci., 122, 236 - 239, 1976.

Parlange, J.-Y., Steenhuis, T.S., Glass, R.J., Richard, T.L., Pickering, N.B., Waltman, W.J., Bailey, N.O., Andreini, M.S., and Throop, J.A., The flow of pesticides through preferential paths in soils, *New York's Food & Life Science Quarterly*, 18, 20 - 23, 1988.

Parlange, J.-Y., Steenhuis, T. S. and Stagnitti, F., Percolation, in *The Encyclopedia of Soil Science and Technology*, Finkl, C., Ed., Van Nostrand Reinhold, 1993.

Philip, J.R., Stability analysis of infiltration, *Soil Sci. Soc. Am. Proc.*, 39, 1042 - 1049, 1975.

Raats, P.A.C., Unstable wetting fronts in uniform and nonuniform soils, *Soil Sci. Soc. Am. Proc.*, 37, 681 - 685, 1973.

Richard, T.L. and Steenhuis, T.S., Tile drain sampling of preferential flow on a field scale, in Rapid and Far Reaching Hydrologic Processes in the Vadose Zone, Germann, P.F, Ed., *J. Contam. Hydrol.*, 3, 307 - 325, 1988.

Richards, L.A., Capillary conduction of liquids through porous mediums, *Physics*, 1: 318 - 333, 1931.

Selker, J.S., Leclerq, P., Parlange, J.-Y., and Steenhuis, T.S., Fingered flow in two dimensions, Part 1, Measurement of matric potential, *Water Resources Res.*, 1991a.

Selker, J.S., Parlange, J.-Y., and Steenhuis, T.S., Fingered flow in two dimensions, Part 2, Predicting finger moisture profile and measure of conductivity, *Water Resources Res.* , 1991b.

Selker, J.S., Steenhuis, T.S., and Parlange, J.-Y., Wetting front instability in homogeneous sand soils under continuous infiltration, *Soil Sci. Soc. Am. J.* , 1991c.

Shaffer, K.A., Fritton, D.D., and Baker, D.E., Drainage water sampling in a wet dual pore system, *J. Env. Qual.,* 8, 241 - 246, 1979.

Shuford, J.W., Fritton, D.D., and Baker, D.E., Nitrate nitrogen and chloride movement through undisturbed field soil, *J. Env. Qual.*, 6, 255 - 259, 1977.

Skopp, J., Gardner, W.R. and Tyler, E.J., Solute movement in structured soils: Two-region model with small interaction. *Soil Sci. Soc. Am. J.,* 45, 1147 - 1152, 1981.

Sposito, G. and Jury, W.A., The lifetime probability density function for solute movement in the subsurface zone, *J. Hydrol.*, 102: 503 - 518, 1988.

Stagnitti, F., Steenhuis, T.S., Parlange, J.-Y., Nijssen, B.M., and Parlange, M.B., Preferential solute and moisture transport in hillslopes, in *Challenges for Sustainable Development,* Proc. Int. Hydrology and Water Resources Symp., Institution of Engineers, Barton, Australia, 919 - 924, 1991.

Starr, J.L., Parlange, J.-Y., and Frink, C.R., Water and chloride movement through layered field soil, *Soil Sci. Soc. Am. J.*, 50, 1384 - 1390, 1986.

Steenhuis, T.S. and van der Molen, W.H., The Thornthwaite-Mather procedure as a simple engineering method to predict recharge, *J. Hydrol.*, 84, 221 - 229, 1986.

Steenhuis, T.S., Pacenka, S., and Porter, K.S., MOUSE: A management model for evaluating groundwater contamination from diffuse surface sources aided by computer graphics, *Appl. Agr. Res.*, 2, 277 - 289, 1987.

Steenhuis, T.S. and Muck, R.E., Preferred movement of non-adsorbed chemicals on wet, shallow, sloping soils. *J. Env. Quality,* 17, 376 - 384, 1988.

Steenhuis, T.S. and Parlange, J.-Y., Preferential flow in structured and sandy soils, *Engin. Quarterly,* Cornell University, 25(1): 7 -14, 1990a.

Steenhuis, T.S., Kung, K.J.S, and Cathles, L.M., Finding layers in the soil: Ground-penetrating radar as a tool in studies of groundwater contamination, *Engin. Quarterly,* Cornell University, 25(1): 15 -19, 1990b.

Steenhuis, T.S., Parlange, J.-Y. and Andreini, M.S., A numerical model for preferential solute movement in structured soils, *Geoderma,* 46, 193-208, 1990.

Steenhuis, T.S., Nijssen, B.M., Stagnitti, F., and Parlange, J.-Y., Preferential solute movement in structured soils: Theory and application, in *Challenges for Sustainable Development,* Proc. Int. Hydrology and Water Resources Symp., Institute of Engineers, Australia, 925 - 931, 1991a.

Steenhuis, T.S. and Parlange, J.-Y., Preferential flow in structured and sandy soils, National Symposium on Preferential Flow, Chicago, IL, December, 16-17, 12-21, 1991b.

van Dam, J.C., Hendrickx, J.M.H., van Ommen, H.C., Bannink, M.H., van Genuchten, M.Th., and Dekker, L.W., Water and solute movement in a coarse textured water repellent field soil, *J. Hydrol.*, 120, 359, 1990.

van der Molen, W.H., Desalinisation of saline soils as a column process, *Soil Sci.*, 81, 19 - 27, 1956.

van Es, H.M., Steenhuis, T.S., Geohring, L.D., Vermeulen, J., and Boll, J., Movement of surface-applied and soil-embodied chemicals to drainage lines in a well-structured soil, in *Proc. Nat. Symp. Preferential Flow,* American Society of Agricultural Engineers, St. Joseph, MI, 59, 1991.

Ward, R.C., On the response to precipitation of headwater streams in humid areas, J. Hydrol., 74, 171 - 189, 1984.

Wilson, L.G., *Monitoring in the vadose zone: A review of technical elements and methods,* Environmental Monitoring Systems Laboratory, Office of Research and Technology, U.S. Environmental Protection Agency, Las Vegas, NV, EPA-600/7-80-134, 168pp, 1980.

A PERSPECTIVE ON TIME-CONTINUOUS SPACE-DISCRETIZED SOLUTIONS OF CONVECTION-DISPERSION EQUATION IN VARIABLY SATURATED POROUS MEDIA

Sita Ram Singh[1] and S. K. Kamra[2]

ABSTRACT

Three time-continuous space discretized semi-analytic approaches to solve one and two-dimensional convection-dispersion equations have been critically examined. The methods are: (i) eigen function finite element method, (ii) Laplace transform finite element method, and (iii) Laplace transform finite analytic method. In the first approach the space discretization of convection-dispersion equation by Galerkin's finite element method yields a system of time dependent inhomogeneous ordinary differential equations that is solved by eigenvalue-eigenvector method of Euler. The second approach involves a simple application of Laplace transform to eliminate the temporal derivative before the application of Galerkin's procedure and subdivision of the domain into finite elements. The resulting transformed system of algebraic equations is solved for nodal concentration in Laplace p space, and thereafter the numerical Laplace inversion algorithm is employed to recover the nodal concentration in physical time domain. In the third approach, the temporal derivative is eliminated using Laplace transform, and the resulting steady state like convection-dispersion equation is solved analytically over small subdomains by assigning constant value to the parameters of the differential equation. The solution in the transformed domain is numerically inverted to get concentration as a function of time. Out of the three semi-analytic methods the Laplace transform finite analytic method is most accurate and has the potential of application to other areas as well.

INTRODUCTION

Analysis of the simultaneous transport of water and solutes in variably saturated natural geologic media is of paramount importance in understanding the processes of degradation and reclamation of land and water resources. Convection-dispersion equation describes adequately the transport of solutes in unsaturated and saturated soil-aquifer system. Solutions of this equation, subject to appropriate initial and boundary conditions, can predict solute concentration in space and time with reasonable accuracy. These solutions may be obtained using analytic, numeric, and semi-analytic techniques. Analytic solutions, generally developed for simplified and idealised situations, are exact, precise, and easy to compute.

---

1. Water Technology Centre For Eastern Region, Bhubaneswar, India, 751 016
2. Central Soil Salinity Research Institute, Karnal, India, 132001

V. P. Singh and B. Kumar (eds.), Water-Quality Hydrology, 153–173.

They are probably the most efficient alternative to get the first estimate especially when the data necessary for identification of the system is sparse and uncertain.

Many of the situations encountered in the analysis of water and solute transport in geologic formations, like soil and aquifers, involve irregular domain boundaries and heterogeneous material properties; consequently, the analytic solutions of the governing partial differential equations is intractable. In such situations one resorts to numerical techniques like finite difference and finite element methods that employ some form of descretization of the space and time variables of the partial differential equaions. The descretizations transform the partial differential equation into a set of approximate algebraic equations that are subsequently solved using the methods of linear algebra. Time derivative in conventional finite difference and finite element methods is discretized by finite difference technique and the solution is advanced through each time step to obtain hydraulic head and solute concentration fields for any desired time. It may lead to large computer storage and time requirements, and unbound truncation and round off errors, particularly for predictions at large time and distance from the source.

An alternative approach is to discretize the space domain only and leave the time variable continuous. An exact in-time analytic integration of the spatially discretized partial differential equation is performed that results in a semi-analytic solution. This analytical solution gives an explicit expression for hydraulic head or solute concentration at each nodal point as a function of time. The computer code can be implemented to calculate the concentration or head at any nodal point at a desired time. Hence long term predictions at a point can be obtained without marching through each time step, and without necessarily solving for all other points in the flow domain. This results in a lower computation time and fewer truncation errors, and provides convenience for data retrieval. Since the semi-analytic method is applied after the spatial discretization, it may not reduce the numerical errors associated with the space discretization.

In this paper we cast some light on the time-continuous space-discretized semi-analytic methods of solving convection-disperson equation to simulate solute transport in variably saturated natural geologic porous media without claiming completeness. To the authors' knowledge the following three type of semi-analytic methods have been developed.

(i) Eigenfunction finite element method.

(ii) Laplace transform finite element method.

(iii) Laplace transform finite analytic method.

We focus on the essence of these solution techniques and provide a clear perspective of their limitations and range of application.

GOVERNING EQUATIONS

The solute transport in a porous medium is the algebraic sum of (i) dispersive and diffusive flux in response to concentration gradients, and (ii) convective flux with the bulk water. Thus:

$$J_i = -\sum_{j=1}^{3} \theta\, D_{ij}\, \partial C/\partial X_j + q_i\, C \tag{1}$$

flux with respect to fixed axis	flux resulting from dispersion	flux resulting from bulk flow of water

where i, j = 1,2,3 stand for X, Y, Z, respectively; J_i is the mass of solute flux per unit area of the porous medium in direction i, $(M/L^2/T)$; D_{ij} is dispersion coefficient, (L^2/T); q_i is Darcy's specific discharge in direction i, (L/T); C is concentration of solute in the solution, (M/L^3); and θ is volumetric moisture content, (L^3/L^3), which gives the cross sectional area for dispersion in unit area of the porous medium. With the help of equation (1), a generalised form of the specific solute transport vector **J** is written as below.

$$J = J_x \mathbf{1x} + J_y \mathbf{1y} + J_z \mathbf{1z} \tag{2}$$

where **1x**, **1y**, and **1z** are, respectively, unit vectors in X, Y, and Z directions of a cartesian coordinate system; and Jx, Jy and Jz are components of the mass flux vector **J** $(ML^{-2}T^{-1})$ in X, Y, and Z directions, respectively.

The principle of conservation of mass in conjunction with the solute transport law yields the convection-dispersion equation:

$$\begin{aligned}\theta R_f \frac{\partial C}{\partial t} + C\frac{\partial \theta}{\partial t} = & \left[\frac{\partial}{\partial X}\left(\theta D_{xx}\frac{\partial C}{\partial X} + \theta D_{xy}\frac{\partial C}{\partial Y} + \theta D_{xz}\frac{\partial C}{\partial Z}\right)\right. \\ & + \frac{\partial}{\partial Y}\left(\theta D_{yx}\frac{\partial C}{\partial X} + \theta D_{yy}\frac{\partial C}{\partial Y} + \theta D_{yz}\frac{\partial C}{\partial Z}\right) \\ & \left. + \frac{\partial}{\partial Z}\left(\theta D_{zx}\frac{\partial C}{\partial X} + \theta D_{zy}\frac{\partial C}{\partial Y} + \theta D_{zz}\frac{\partial C}{\partial Z}\right)\right] \\ & - \left[\frac{\partial}{\partial X}(q_x C) + \frac{\partial}{\partial Y}(q_y C) + \frac{\partial}{\partial Z}(q_z C)\right] \\ & - S(X, Y, Z, t)\end{aligned} \tag{3}$$

where in addition to the terms defined earlier, C is the i-th specy solute concentration in liquid (ML^{-3}); S(X,Y,Z,t) is the rate of withdrawal (or production) of solute $(ML^{-3}T^{-1})$ per unit volume of the porous medium; and R_f is retardation factor given as below.

$$R_f = 1 + (\epsilon\, k_d\, N\, C^{N-1}) \,/\, \theta \tag{4}$$

where ϵ is the bulk density of the porous medium, (ML^{-3}); and k_d and N are constants of the Freundlich isotherm that gives the ith specy adsorption on solid. k_d is called the distribution coefficient of the solute. These constants depend upon several factors including (i) the solute specy, and (ii) the nature of the porous medium.

In steady unsaturated flows, $\partial\theta/\partial t=0$; consequently, the second term in the left hand side of (3) does not occur in such flows. When the porous medium is fully saturated, θ is replaced by porosity, n. In nondeformable saturated porous medium, $\partial n/\partial t=0$. Hence for transient as well as steady saturated flows equation (3) becomes:

$$
\begin{aligned}
nR_f\frac{\partial C}{\partial t} = {} & [\frac{\partial}{\partial X}(nD_{xx}\frac{\partial C}{\partial X} + nD_{xy}\frac{\partial C}{\partial Y} + nD_{xz}\frac{\partial C}{\partial Z}) \\
& + \frac{\partial}{\partial Y}(nD_{yx}\frac{\partial C}{\partial X} + nD_{yy}\frac{\partial C}{\partial Y} + nD_{yz}\frac{\partial C}{\partial Z}) \\
& + \frac{\partial}{\partial Z}(nD_{zx}\frac{\partial C}{\partial X} + nD_{zy}\frac{\partial C}{\partial Y} + nD_{zz}\frac{\partial C}{\partial Z})] \\
& - [\frac{\partial}{\partial X}(q_x C) + \frac{\partial}{\partial Y}(q_y C) + \frac{\partial}{\partial Z}(q_z C)] \\
& - S(X,Y,Z,t)
\end{aligned} \qquad (5)
$$

Equations (3) and (5) are the desired differential equations; the former for the partially saturated porous medium, and the latter for the saturated one. Reduction of these equations for two-dimensional solute transport in X-Y plane is straight forward. It is achieved by truncating all the Z components in equations (3) and (5). For example, equation (5) reduces to:

$$
\begin{aligned}
nR_f\frac{\partial C}{\partial t} = {} & [\frac{\partial}{\partial X}(nD_{xx}\frac{\partial C}{\partial X} + nD_{xy}\frac{\partial C}{\partial Y}) \\
& + \frac{\partial}{\partial Y}(nD_{yx}\frac{\partial C}{\partial X} + nD_{yy}\frac{\partial C}{\partial Y})] \\
& - [\frac{\partial}{\partial X}(q_x C) + \frac{\partial}{\partial Y}(q_y C)] \\
& - S(X,Y,t)
\end{aligned} \qquad (6)
$$

Similarly in absence of transverse dispersion, equation (3) in one-dimensional vertical unsaturated zone becomes:

$$
\theta R_f\frac{\partial C}{\partial t} = \frac{\partial}{\partial Z}(\theta D_{zz}\frac{\partial C}{\partial Z}) - \frac{\partial}{\partial Z}(q_z C) - S(Z,t) \qquad (7)
$$

Let D_L and D_T be longitudinal and transverse dispersion coefficients, respectively. They are related to longitudinal, α_L, and transverse, α_T, dispersivities as below (Sheidegger, 1961).

$$
D_L = \alpha_L\, q/n \qquad (8)
$$

and

$$
D_T = \alpha_T\, q/n \qquad (9)
$$

where q is the magnitude of the Darcy's specific discharge vector **q**, i.e.

$$
q = (q_x^2 + q_y^2 + q_z^2)^{1/2} \qquad (10)
$$

Dispersion coefficients, D_{ij}, for isotropic porous medium were given by Sheidegger (1961) as below.

$$D_{xx} = [D_L \, q_x^2 + D_T \, q_y^2 + D_T \, q_z^2] \; / \; q^2 \tag{11}$$

$$D_{yy} = [D_T \, q_x^2 + D_L \, q_y^2 + D_T \, q_z^2] \; / \; q^2 \tag{12}$$

$$D_{zz} = [D_T \, q_x^2 + D_T \, q_y^2 + D_L \, q_z^2] \; / \; q^2 \tag{13}$$

$$D_{xy} = [(D_L - D_T) \; q_x \, q_y] \; / \; q^2 \tag{14}$$

$$D_{xz} = [(D_L - D_T) \; q_x \, q_z] \; / \; q^2 \tag{15}$$

$$D_{yz} = [(D_L - D_T) \; q_y \, q_z] \; / \; q^2 \tag{16}$$

In isotropic porous medium the dispersion coefficient $D_{ij}=D_{ji}$; hence, D_{yx}, D_{zx} and D_{zy} are assigned the value of D_{xy}, D_{xz}, and D_{yz}, respectively.

The dispersion coefficients for solute transport in isotropic two-dimensional porous medium (X,Y plane) are easily obtained by substituting $q_z=0$ in equations (10) through (16) and truncating D_{zz}. Thus:

$$D_{xx} = [D_L \, q_x^2 + D_T \, q_y^2] \; / \; q^2 \tag{17}$$

$$D_{yy} = [D_T \, q_x^2 + D_L \, q_y^2] \; / \; q^2 \tag{18}$$

$$D_{xy} = [(D_L - D_T) \; q_x \, q_y] \; / \; q^2 \tag{19}$$

in which

$$q = (q_x^2 + q_y^2)^{1/2} \tag{20}$$

While analysing hydrodynamic dispersion in vertical Z direction, the transverse dispersion, D_T, is many times neglected. In such a case (13) becomes:

$$D_{zz} = D_L \tag{21}$$

Equations (11) to (16), (17) to (19), and (21) are employed to compute dispersion coefficients for the differential equations (3) or (5), (6) and (7), respectively.

An examination of equations (3) and (7) reveals that specific discharge q_i and volumetric moisture content θ are required before the transport equation could be solved in an unsaturated flow domain. Similarly the specific discharge q_i and porosity n are needed to solve equations (5) and (6) in saturated domain. It may be emphasized here that the magnitudes of dispersion coefficients, D_{ij}, and retardation factor, R_f depend upon q_i and θ in unsaturated flows, and q_i and n in saturated flows. Thus the transport equation is coupled to the flow equation. Hence, it is necessary to solve the flow equation which for a saturated-unsaturated three-dimensional medium is written as below (Narasimhan and Witherspoon, 1977).

$$m_c \frac{\partial h}{\partial t} = \frac{\partial}{\partial X}\left(K\frac{\partial h}{\partial X}\right) + \frac{\partial}{\partial Y}\left(K\frac{\partial h}{\partial Y}\right) + \frac{\partial}{\partial Z}\left(K\frac{\partial h}{\partial Z} + 1\right) + Q \qquad (22)$$

where m_c is the specific fluid mass capacity, (L^{-1}), and is a function of the specific storage coefficient and specific moisture capacity of the porous medium; h is the pressure head (L); K is the hydraulic conductivity (LT^{-1}); and Q is the fluid source or sink term.

Solution of the flow equation is required only once in case of steady state flow simulation ($\partial h/\partial t=0$), and repeatedly in the case of transient simulation. Whatever is the flow case, the specific discharge is obtained from Darcy's law as below.

$$q_i = -K \frac{\partial h}{\partial X_i}[h+Z] \qquad (23)$$

ESSENCE OF THE EIGENFUNCTION FINITE ELEMENT TECHNIQUE

The two steps involved in eigenfunction semi-analytic approach are: (i) descretization of the space domain using finite element or finite difference techniques, and (ii) time-continuous solution of the resulting system of time dependent ordinary differential equations employing eigenvalue eigenvector method of Euler. Guymon (1970) applied such a semi-analytic method to the one-dimensional convection-dispersion equation in saturated zone using variational finite element approach and a special transformation to transform the asymmetric coefficient matrix into a symmetric one to yield a real eigen system. The approach was extended to the two-dimensional solute transport equation without mixed partial derivatives (Guymon et al., 1970) and later on to a transport equation with mixed partial derivatives (Nalluswami et al., 1972). A continuous-in-time eigenvalue solution to the spatially discretized two-dimensional solute transport equation in steady ground water flow using complex matrix analysis has been developed by Hwang et al. (1984). A similar analysis but employing real arithmetic has been presented by Umari and Gorelick (1986). Singh et al. (1989) developed an one-dimensional semi-analytic solute transport model employing eigenvalue eigenvector method and real arithmetic for a saturated-unsaturated soil-aquifer profile. Kamra et al. (1991a,b) developed a semi-analytic two-dimensional solute transport model for a tile-drained soil-aquifer system employing real arithmetic, and applied it extensively in studying desalinization resulting from tube drainage. The outline of the method is discussed for one-dimensional convection-dispersion equation(7). The method is easily extended to higher dimensions.

AUXILIARY CONDITIONS

In order to solve equation (7) one needs to specify initial and boundary conditions. The flow domain extends from Z=0 at the land surface to Z=L at the basis layer. The water table is located at Z=DW. Application of the principle of mass conservation yields the following boundary condition at the land surface:

$$RC_e = -\theta D \frac{\partial C}{\partial Z} + qC \qquad (24)$$

in which R is the rate of infiltration (-R for rate of exfiltration) of water having concentration C_e . Sometimes, instead of assigning the general flux-type boundary condition given by (24), one assigns a Dirichlet-type boundary condition

$$C(0,t) = C_e(t) \qquad (25)$$

at the land surface. Here $C_e(t)$ is the concentration of solutes in the inflowing water.

Solute concentration, C_e, in water vapour is zero. Thus, when evaporation takes place from the land surface, boundary condition (24) becomes:

$$-\theta D \frac{\partial C}{\partial Z} + qC = 0 \qquad (26)$$

We substitute C and q in place of C_e and R respectively, in (24) to get the following boundary condition at the water table beneath the soil profile:

$$\frac{\partial C}{\partial Z} = 0 \qquad (27)$$

When there is evaporation from the water table we substitute the concentration of the ground water, C_g, in place of C_e in (24) to get the following boundary condition at the bottom of the soil profile:

$$-\theta D \frac{\partial C}{\partial Z} + qC = RC_g \qquad (28)$$

Some authors (Van Genuchten, 1987) use the concentration of ground water, say $C_g(t)$, as the boundary condition at the soil aquifer interface. Thus one gets:

$$C(Z,t)\big|_{Z=DW} = C_g(t) \qquad (29)$$

where DW is the depth of ground water below the land surface.

When the basis layer is impervious to flow, q and R are zero. Their substitution in (24) yields equation (27) as the boundary condition.

The concentration of solutes in the soil profile at t = 0, as obtained from sample measurements, is specified as the initial condition for the differential equation (7); i.e.

$$C(Z,0) = C_0(Z) \qquad (30)$$

Thus equations (7), (24), (25), (26), (27), (28) and (30) complete the formulation of the boundary-value problem of solute transport through a soil profile. Both the downward and upward fluxes with infiltrating and evaporating water has been considered. It may be remembered that during the upward flow of water, R is to be replaced by (-R) in all equations.

FINITE ELEMENT DISCRETIZATION OF SPACE DOMAIN

Consider the following linear operator for the one-dimensional solute transport in partially saturated porous media:

$$L(C) = \theta R_f \frac{\partial C}{\partial t} - \frac{\partial}{\partial Z} [\theta D \frac{\partial C}{\partial Z} - qC] \tag{31}$$

Thus (7) becomes:

$$L(C) + S(Z,t) = 0 \tag{32}$$

We approximate the solution C(Z,t) by C°(Z,t) given by

$$C^0(Z,t) = \sum_{j=1}^{NN} C_j(t)\ N_j(Z) \tag{33}$$

where $C_j(t)$ and $N_j(Z)$ are, respectively, unknown coefficient and one-dimensional basis function associated with node j, whose total number is NN. Here, out of several choices linear Lagrange polynomials are selected as basis functions. Over an element, (Z_i, Z_{i+1}), there are two nonzero linear basis functions, $N_i(Z)$ and $N_{i+1}(Z)$, where

$$N_i(Z) = (Z_{i+1} - Z) / (Z_{i+1} - Z_i) \qquad for \quad Z_i \le Z \le Z_{i+1} \tag{34a}$$

$$N_{i+1}(Z) = (Z - Z_i) / (Z_{i+1} - Z_i) \qquad for \quad Z_i \le Z \le Z_{i+1} \tag{34b}$$

As C° is an approximator to the solution, its substitution in (32) leaves a residual. In Galerkin's finite element method, we make the residual orthogonal to each basis function. It forces the residual to zero in an average sense. Thus

$$\langle\ [L(C^0) + S(Z,t)],\ N_i(Z)\ \rangle = 0 \tag{35}$$

for i = 1,2,3,..., NN.
Substituting (31) and (33) into (35) and integrating term by term, and dividing the region into finite elements, one gets a system of NN equations which could conveniently be written in matrix form as below.

$$[AM]\ \{dC/dt\} = [\ [DM] + [EM]\ ]\ \{C\} + \{F\} \tag{36}$$

where [AM], [DM] and [EM] are square matrices of order NN, and {dC/dt}, {C} and {F} are column vectors of size (NNx1). Typical elements of the matrices are given below.

$$AM_{ij} = \sum_{e=1}^{NE} \int_e \theta R_t N_i N_j dZ \tag{37}$$

$$DM_{ij} = - \sum_{e=1}^{NE} \int_e \theta D (dN_i/dZ)(dN_j/dZ)\, dZ \tag{38}$$

$$EM_{ij} = \sum_{e=1}^{NE} \int_e q (dN_i/dZ) N_j dZ \tag{39}$$

$$F_i = F_{si} - F_{wi} \tag{40}$$

$$F_{si} = - [(- \theta D \frac{\partial C}{\partial Z} + qC)\ N_i(Z)] \Big|_{Z=0}^{Z=L} \tag{41}$$

$$F_{wi} = \sum_{e=1}^{NE} \int_e S(Z,t) N_i(Z)\, dZ \tag{42}$$

where the subscript i corresponds to the ith row, and j corresponds to the jth column; e isa subdomain of the flow domain called an element; and NE is number of elements. Equation (41) gives the solute flux at the boundaries of the flow domain; Eq.(42) is due to salt removal by any means from the flow domain.

APPLICATION OF BOUNDARY CONDITIONS

Effect of all the boundary conditions is incorporated in Eq.(41). When the flux of water is downward, the boundary condition (24) is of the inflow type (concentration known) and the boundary condition (27) is of the outflow type (concentration unknown). Substituting (24) and (27) in (41), one gets:

$$F_{si} = RC_e\ N_i(0) - q(L)\ C(L,t)\ N_i(L) \tag{43}$$

for i=1,2,3,...,NN.
Since the basis function is nonzero only at the node with which it is attached and is zero at all other nodes; therefore, F_{si} becomes:

$$F_{s1} = RC_e \tag{44}$$

$$F_{s2} = F_{s3} = \ldots = F_{sNN-1} = 0$$

$$F_{sNN} = - q(L)\ C_{NN}(t) = [BM]\ \{C\} \tag{45}$$

where [BM] is a square matrix of order NN having all entries zero except the entry in the NNth row and NNth column, which is (-q(L)). Thus after incorporating the boundary conditions, one gets:

$$[AM]\ \{dC\ /\ dt\} = [\ [DM] + [EM] + [BM]\ ]\ \{C\} + \{F\} \tag{46}$$

in which AM_{ij}, DM_{ij}, and EM_{ij} are given by (37), (38) and (39), respectively; and $F_1 = F_{s1} = RC_e$ as given by (44), and $F_2=F_3= \ldots = F_{NN} = 0$ for S(Z,t)=0.

When the flow of water is upward, the boundary condition at the water table is of the inflow type and that of the land surface is of the outflow type. With the help of (26) and (28) equation (41) becomes:

$$F_{s1} = F_{s2} = \ldots = F_{sNN-1} = 0 \quad and \quad F_{sNN} = - RC_g \tag{47}$$

In this case the system of ordinary differential equations is described by (36) where F_{si} is defined by (47).

TIME-CONTINUOUS SOLUTION OF THE MATRIX DIFFERENTIAL EQUATION

The Galerkin space discretization of the one, two, and three-dimensional convection-dispersion equation yields the following system of ordinary differential equations (Singh, 1989):

$$[AM]\ \{dC\ /\ dt\} = [GM]\ \{C\} + \{F\} \tag{48}$$

with the initial condition

$$\{C(0)\} = \{C_0\} \tag{49}$$

where the square matrices [AM] and [GM] are of size (NNxNN) and the column vectors {dC/dt} , {C} , $\{C_0\}$ and {F} are of size (NNx1). The coefficient matrix [GM] contains the effects of dispersion, convection and outflow boundaries and is asymmetric. In one-dimensional simulation of solute transport it is given by [[DM]+[EM]] in (36) and [[DM]+[EM]+ [BM]] in (46). The column vector $\{C_0\}$ represents the initial solute concentration given by (30) at finite element nodes. Multiplication of equation (48) by $[AM]^{-1}$ gives:

$$\{dC\ /\ dt\} = [H]\ \{C\} + [AM]^{-1}\ \{F\} \tag{50}$$

in which

$$[H] = [AM]^{-1}\ [GM] \tag{51}$$

is an asymmetric matrix.

The solution of (50), the details of which are given elsewhere (Singh, 1989), is given below.

$$\{C\} = e^{[H]\,t}\,\{\{C_0\} + [GM]^{-1}\{F\}\} - \{[GM]^{-1}\{F\}\} \tag{52}$$

in which

$$e^{[H]\,t} = [Z]\,[QD]\,[Z]^{-1} \tag{53}$$

is matrix exponential; [Z] is a square matrix whose columns are the eigenvectors of the matrix [H]; and [QD] is a tridiagonal matrix whose elements are given below.

$$QD_{ii} = e^{\tau_i t} \tag{54a}$$

$$QD_{jj} = QD_{j+1,j+1} = e^{a_j t}\,Cos\ b_j t \tag{54b}$$

$$QD_{j,j+1} = e^{a_j t}\,Sin\ b_j t \tag{54c}$$

$$QD_{j+1,j} = -\,e^{a_j t}\,Sin\ b_j t \tag{54d}$$

All other elements of QD are zero (54e)

where τ_i is the i-th real eigenvalue, and a_j and b_j are real and imaginary parts of the j-th complex eigenvalue, τ_j, of the matrix [H]. In equation (54) i could assume any integer value from 1 to NN and j from 1 to (NN-1). Thus [QD] is of size (NNxNN).

The matrix exponential evaluated at any time, t, is a square matrix of real numbers. Its substitution in (52) results into a column vector which represents the solute concentration at time t at all the finite element nodes. It may be noted that the proposed algorithm uses real arithmetic following the evaluation of eigenvalues and eigenvectors and is much simpler to comprehend and programme on digital computers.

ESSENCE OF THE LAPLACE TRANSFORM FINITE ELEMENT TECHNIQUE

The Laplace Transform finite element method is implemented in four steps: (1) A Laplace transformation of the partial differential equation is performed to eliminate time, (2) the space variables in transformed partial differential equation are discretized using finite element (or finite difference) method, (3) the resulting system of simultaneous algebraic equations is solved that determines the transformed vector of the unknown concentrations in the Laplace space, and (4) a numerical inversion is performed on the transformed vector of unknown concentrations using an efficient Laplace inversion algorithm. Yu et al.(1987) applied the Laplace transform finite element scheme to the problem of one-dimensional nonreactive solute transport under uniform flow conditions and a time-invariant input concentration. Sudicky (1989) used Laplace transform together with the Galerkin finite element procedure for direct solution of advection-dispersion equation in two-dimensions with generalised boundary conditions and a spatially variable velocity field. His method is applied here on one-dimensional convection-dispersion equation for ease of understanding.

LAPLACE TRANSFORMATION

We define the Laplace transform, $\hat{C}(Z,p)$, of $C(Z,t)$ as

$$L[C(Z,t)] = \hat{C}(Z,p) = \int_0^\infty C(Z,t)\, e^{-pt} dt \tag{55}$$

where L[.] is the Laplace transform operator and p is the transform variable that, in general, is complex valued.

We have the choice of applying the Laplace transformation either before or after the spatial discretization. So far the authors have made the former choice(Sudicky, 1989; Moridis and Reddell, 1991) while using Laplace transform finite element or finite difference technique to solve ground water flow and mass transport problems. Applying Laplace transformation to (7) to eliminate time, one gets:

$$\theta\, R_f [p\hat{C} - C(Z,0)] = \frac{\partial}{\partial Z}(\theta D_{zz}\frac{\partial \hat{C}}{\partial Z}) - \frac{\partial}{\partial Z}(q_z \hat{C}) + \hat{S}(Z,p) \tag{56}$$

Equation(56) resembles the steady state form of the convection-dispersion equation. We also take Laplace transform of the boundary conditions (24), (25), (26), (27), (28) and (29) that are needed while developing Galerkin finite element solution of (56).

DISCRETIZATION OF SPACE VARIABLE

The Galerkin finite element technique, as described earlier, has been extensively used in mass transport analysis because of the ease with which the heterogeneous material properties and irregular domain boundaries can be accomodated. In this method we approximate the solution $\hat{C}(Z,p)$ by a trial solution $\hat{C}°(Z,p)$ as

$$\hat{C}^{o}(Z,p) = \sum_{j=1}^{NN} \hat{C}_j(p)\; N_j(Z) \tag{57}$$

where $N_j(Z)$ is basis function associated with node j, NN is the total number of nodes, and $\hat{C}_j(p)$ is unknown coefficient representing p space concentration at node j. Note that (57) is the Laplace transform of (33). Following the Galerkin finite element procedure we substitute (57) in (56), orthogonalize the residual to each basis function, divide the domain into finite elements, and incorporate the transformed boundary conditions to get a linear system of algebraic equations

$$[\,[A] + p[B]\,]\,\{\hat{C}\} = \{\hat{E}\} \tag{58}$$

in which [A] is an advective-dispersive transport matrix, [B] is the mass matrix, and $\{\hat{E}\}$ is the vector containing all the known quantities including the sink term and natural boundary conditions.

SOLUTION OF LINEAR SYSTEM OF ALGEBRAIC EQUATIONS

The system of linear algebraic equations(58) can easily be solved using any standard method of linear algebra provided we assign numerical values to the Laplace transform variable p. We prescribe several values to p, say p_k, k=1,2,...,K, and solve the system (58) for each p_k to get the transformed vector $\{\hat{C}(p_k)\}$ of unknown concentration. Corresponding to each p_k we numerically compute Laplace inverse of (57) to obtain

$$C^{o}(Z,t) = \sum_{j=1}^{NN} L^{-1}\, \hat{C}_j(p_k)\, N_j(Z) = \sum_{j=1}^{NN} C_j(t)\, N_j(Z) \qquad (59)$$

NUMERICAL INVERSION OF THE TRANSFORMED VECTOR

Numerous algorithms have been developed for the numerical inversion of Laplace transforms. Davies and Martin (1979) have presented a critical review of some of the methods with regards to their computational accuracy, efficiency, and ease of implementation. In the field of ground water, while Sudicky (1989) selected Crump (1976) algorithm, Moridis and Reddell (1991) employed Stehfest (1970) algorithm to numerically invert the Laplace transform vector for unknown concentration and pressure, respectively. Crump (1976) used Fourier series approximation of the inverse function and trapezoidal integration of the Laplace inversion integral to develop the following formula to invert the dependent variable, say $\hat{C}_j(p_k)$ from p space to the variable, say $C_j(t)$, in the physical time domain:

$$C_j(t) = \frac{1}{T} e^{(p_0 t)} \left[\frac{1}{2}\hat{C}_j(p_0) + \sum_{k=1}^{2N+1} \left(Re(\hat{C}_j(p_k))\cos\left(\frac{k\pi t}{T}\right) - Im(\hat{C}_j(p_k))\sin\left(\frac{k\pi t}{T}\right)\right)\right] + E \qquad (60)$$

where 2T is the period of the Fourier series approximating the inverse function on the interval [0, 2T]; Re and Im denote the real and imaginary parts, respectively, of the complex $\hat{C}_j(p_k)$ values; and E is an error term arising from the truncation of summation after 2N+1 terms. Crump (1976) performed extensive tests on a variety of functions and found that $T=0.8\ t_{max}$, where t_{max} is the maximum simulation time. This approximation yielded accurate results for broad range of times for which inversion were computed. He gave the following formula to compute k distinct values of the transform variable p_k:

$$p_k = p_0 + ik\pi / T \qquad (61)$$

where p_0 is the real-valued parameter and $i=\sqrt{(-1)}$. For real life transport problems, the parameter p_0 is approximated as below (Sudicky, 1989).

$$p_0 = -\ln(E) / (1.6\, t_{max}) \qquad (62)$$

where E is the absolute error tolerance, which in mass transport problems could be of the order of (10^{-6}). Sudicky (1989) used epsilon algorithm of Mac donald (1964) to accelerate the convergence of (60). A scruitiny of (60) reveals that the numerical inversion can be performed for any selected node j. Thus, one need not to solve (60) for all the nodes unless it is necessary.

ESSENCE OF THE LAPLACE TRANSFORM FINITE ANALYTIC METHOD

Three steps are involved in the space-time accurate Laplace transform finite analytic method. These are: (i) elimination of the temporal derivative by taking the Laplace transform of the original partial differential equation, (ii) solution of the transformed equation by the finite analytic method, and (iii) application of an efficient numerical Laplace inversion algorithm to recover the solution from Laplace transformed domain to the physical time domain.

Li et al. (1992) proposed the method and applied it in developing consistantly space-time accurate transient numerical solution of the one-dimensional advection-diffusion equation. The solution is free of spurious oscillations and numerical diffusion for all practical values of Peclet number and for a wide range of grid spacing. Their method is demonstrated by solving the following boundary value problem formulated by them:

$$\frac{\partial C}{\partial t} + q(z)\frac{\partial C}{\partial z} = \frac{\partial}{\partial z}\left(D(z)\frac{\partial C}{\partial z}\right) - K(z)\,C + S(z,t) \tag{63}$$

$$C(z,0) = f(z)\,; \quad C(0,t) = g_0(t)\,; \quad C(l,t) = g_l(t) \tag{64}$$

where in addition to all the terms defined earlier, K is a first order decay coefficient; f(z) is initial condition; $g_0(t)$ and $g_l(t)$ are the boundary conditions; and l is the length of the solution domain.

LAPLACE TRANSFORM

Taking the Laplace transform of (63) and (64) to eliminate time, we get

$$P\hat{C} + q(z)\frac{\partial \hat{C}}{\partial z} = \frac{\partial}{\partial z}\left(D(z)\frac{\partial \hat{C}}{\partial z}\right) - K(z)\,\hat{C} + \hat{S}(z,p) + f(z) \tag{65}$$

and

$$\hat{C}(0,p) = \bar{g}_0(p)\,; \qquad \hat{C}(l,p) = \bar{g}_l(p) \tag{66}$$

Equation (65) is like a "Steady State" advection-diffusion equation in which the initial condition is appearing with the sink term, making the equation nonhomogeneous. It will be solved by finite analytic method.

FINITE ANALYTIC SOLUTION OF THE TRANSFORMED EQUATION

In finite analytic method, the solution domain is discretized into small subdomains in which the parameters of differential equation could approximately be taken as constant. It greatly simplifies the differential equation that is easily solved on modified simple subdomains.

In order to illustrate the algorithm, we discretize the domain $0 \leq z \leq l$ into N subdomains: $0 = Z_0 < Z_1 < \ldots < Z_{i-1} < Z_i < Z_{i+1} < \ldots < Z_{N-1} < Z_N = l$. We define the interval $Z_{i-1} < Z_i < Z_{i+1}$ to be the local element associated with node Z_i. Thus each local element spans three consecutive computational nodes Z_{i-1}, Z_i, Z_{i+1}. Over this element all coefficients in (65) are approximated as locally constant. Thus over $Z_{i-1} < Z < Z_{i+1}$,

$$q(Z) \approx q_i; \ f(Z) \approx f_i; \ D(Z) \approx D_i; \ \hat{S}(Z,p) \approx \hat{S}_i; \ K(Z) \approx K_i \qquad (67)$$

where q_i, f_i etc. are evaluated at $Z=Z_i$. The analytic solution of the differential equation (65) with constant coefficient over the domain $Z_{i-1} < Z < Z_{i+1}$, becomes

$$\hat{C}(z,p) = A_1 \, e^{[r_1(z-z_i)/\Delta z_i]} + A_2 \, e^{[r_2(z-z_i)/\Delta z_i]} + \left(\frac{\hat{S}_i + f_i}{K_i + p}\right) \qquad (68)$$

where $\Delta Z_i = Z_i - Z_{i-1}$, and r_1 and r_2 are given by

$$r_1 = \frac{P_{ei}}{2}\left[1 + \sqrt{1 + \frac{4D_i}{q_i^2}(K_i + p)}\right] \qquad (69)$$

$$r_2 = \frac{P_{ei}}{2}\left[1 - \sqrt{1 + \frac{4D_i}{q_i^2}(K_i + p)}\right] \qquad (70)$$

A_1 and $A_2$are two arbitrary constants of integration; and $P_{ei}=(q_i \Delta Z_i / D_i)$, is the local Peclet number. We impose the boundary conditions

$$\hat{C}(Z_{i-1},p) = \hat{C}_{i-1} \qquad (71)$$

$$\hat{C}(Z_{i+1},p) = \hat{C}_{i+1} \qquad (72)$$

on (68) and get the following expressions for A_1 and A_2:

$$A_1 = \frac{(e^{r_2\rho_i}\,\hat{C}_{i-1} - e^{-r_2}\,\hat{C}_{i+1})}{(e^{r_2\rho_i - r_1} - e^{r_1\rho_i - r_2})} - \left[\frac{(e^{r_1\rho_i} - e^{-r_1})}{(e^{r_2\rho_i - r_1} - e^{r_1\rho_i - r_2})}\right]\left[\frac{(\hat{S}_i + f_i)}{(K_i + p)}\right] \qquad (73)$$

$$A_2 = \frac{(-e^{r_1\rho_i}\,\hat{C}_{i-1} + e^{-r_1}\,\hat{C}_{i+1})}{(e^{r_2\rho_i - r_1} - e^{r_1\rho_i - r_2})} + \left[\frac{(e^{r_1\rho_i} - e^{-r_1})}{(e^{r_2\rho_i - r_1} - e^{r_1\rho_i - r_2})}\right]\left[\frac{(\hat{S}_i + f_i)}{(K_i + p)}\right] \qquad (74)$$

where $\rho_i = \Delta Z_{i+1}/\Delta Z_i$

Equations (68) through (74) reveal the linear dependence of concentration $\hat{C}(Z,p)$ on the end node concentrations $\hat{C}(Z_{i-1},p)$ and $\hat{C}(Z_{i+1},p)$ of the local element $Z_{i-1}<Z<Z_{i+1}$. Thus we should aim at finding the nodal concentrations to compute the concentration at any interior point of the domain. Evaluating (68) at the interior node Z_i, we obtain

$$\hat{C}_i = \alpha_i \hat{C}_{i-1} + \beta_i \hat{C}_{i+1} + \gamma_i \tag{75}$$

in which

$$\alpha_i = \frac{(e^{r_2 p_i} - e^{r_1 p_i})}{(e^{r_2 p_i - r_1} - e^{r_1 p_i - r_2})} \tag{76}$$

$$\beta_i = \frac{(e^{-r_1} - e^{-r_2})}{(e^{r_2 p_i - r_1} - e^{r_1 p_i - r_2})} \tag{77}$$

$$\gamma_i = [1 + \frac{e^{r_1 p_i} - e^{r_2 p_i} - e^{-r_1} + e^{-r_2}}{e^{r_2 p_i - r_1} - e^{r_1 p_i - r_2}}][\frac{(\hat{S}_i + f_i)}{(K_i + p)}] \tag{78}$$

Equation (75) is the desired finite analytic algorithm for solving the transformed differential equation (65).

NUMERICAL INVERSION OF THE LAPLACE TRANSFORM

The Crump (1976) algorithm described earlier was employed by Li et al. (1992) to recover C(Z,t) from $\hat{C}$(Z,p). The finite analytic algorithm remains accurate even for non uniform spatial grids so long the coefficients of differential equation could be assigned constant values. Moreover, the solution is analytically differentiable to any order.

ACCURACY OF THE SEMI-ANALYTIC SOLUTIONS

Accuracy of the semi-analytic methods have been studied by comparing the semi-analytic solutions with analytic ones. Traditionally the semi-analytic and analytic solutions are compared for different values of interstitial velocity, dispersion coefficient, element length, and size of time step combined through dimensionless Peclet and Courant numbers. As time is continuous in all the aforementioned semi-analytic solutions, therefore, the Courant number criterion is unnecessary while studying the accuracy of the solutions.

Singh (1989) compared the time-continuous eigenfunction solution of one-dimensional convection-dispersion equation(7) with an analytic solution in a finite domain. The deviation of the finite element-eigenvalue solution from the analytic one is due to the space descretization of the differential equation (7). L_2 norm (Prenter, 1975; p 6-11) as defined below was employed to measure the goodness of the semi-analytic solution.

L2 NORM:

$$(\frac{1}{\sqrt{L}})||C_a - C||_2 = (\frac{1}{\sqrt{L}})[\int_0^L [C_a(Z,t) - C(Z,t)]^2 \, dz]^{1/2} \qquad (79)$$

It gives average distance of the numeric solution C(Z,t) from the analytic solution $C_a(Z,t)$ at time t. The two solutions were compared at different Peclet numbers of 0.50, 0.75, 1.0, 1.25, 1.5, 1.75, 2.0, 2.25, and 2.5 (Fig.1). It is evident from the figure that (i) at low Peclet numbers the deviation in semi-analytic solution from the analytic one is negligibly small, (ii) the error reduces with increase in time at all Peclet numbers, and (iii) at low Peclet numbers the error trend is uniformly decreasing but at high Peclet numbers it is decreasing with oscillation.

Kamra et al. (1991) studied the accuracy of the time-continuous eigen function solution of the two-dimensional convection-dispersion equation and concluded that (i) as the finite element mesh is refined, the difference between analytical and semi-analytical results decreases, (ii) the semi-analytic error decreases with increasing values of longitudinal and transverse dispersion coefficients, and decreasing interstitial velocity, and (iii) the error in the semi-analytic solution is relatively more sensitive to variations in seepage velocity than to variations in the longitudinal and tranverse dispersion coefficients. The sensitivity is attributed to the fact that when linear basis functions are used in the finite element approximations, the velocity field is constant inside each element but discontinuous accross the boundary from one element to another. Discontinuities in the velocity field often create problems for an accurate representation of the convection term in the solute transport equation.

Sudicky (1989) analysed the accuracy of the Laplace transform finite element solution of one and two-dimensional advection-dispersion equations. Fig.2 depicts a comparison of one-dimensional semi-analytic solution with the analytic one for a grid Peclet number of 33.3. Even at this large Peclet number, the semi-analytic solution is seen to produce results that are essentially identical to those of the analytical solution. In two-dimensions he compared the semi-analytic results with the analytic one at a grid Peclet number of 20 in the direction of flow, and found the two results to be identical. Only at very early time a slight amount of undershoot and overshoot appeared in the semi-analytic solution. He attributed it to the coarseness of the grid.

Li et al. (1992) compared the space-time accurate Laplace transform finite analytic solution of the one-dimensional advection-dispersion equation with an analytic solution at time t=150 days for four different Piclet numbers ranging from 5 to 500 (Fig. 3). As is seen in the figures, the finite analytic solution falls exactly on the analytic solution. It is due to the fact that the constant parameters in the differential equation alleviates the introduction of any approximation in the finite analytic method. Had the parameters been variable, the finite analytic method would have incorporated approximations, and the two solutions could have differred. Inspite of it, this method is more accurate than the other two time-continuous methods.

SUMMARY AND CONCLUSIONS

The purpose of this paper was to prepare a state of the art report on the time-continuous space discretized solutions of convection-dispersion equation. The time-continuous

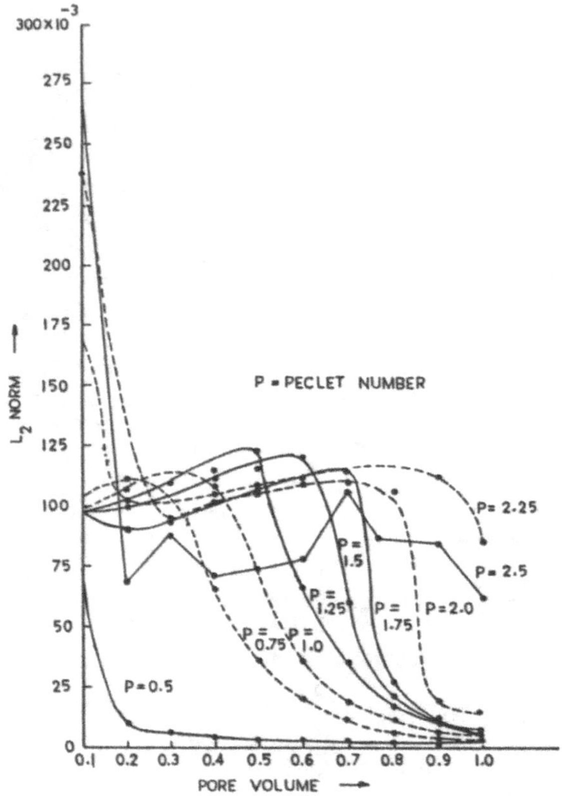

Fig.1. Average error in semi-analytic solution at different peclet numbers

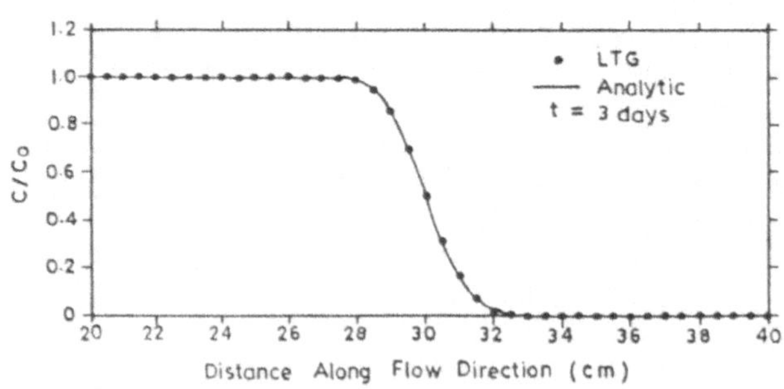

Fig.2. Comparison of laplace transform finite element solution with analytic solution at t = 3 days.
(Redrawn from Sudicky, 1989)

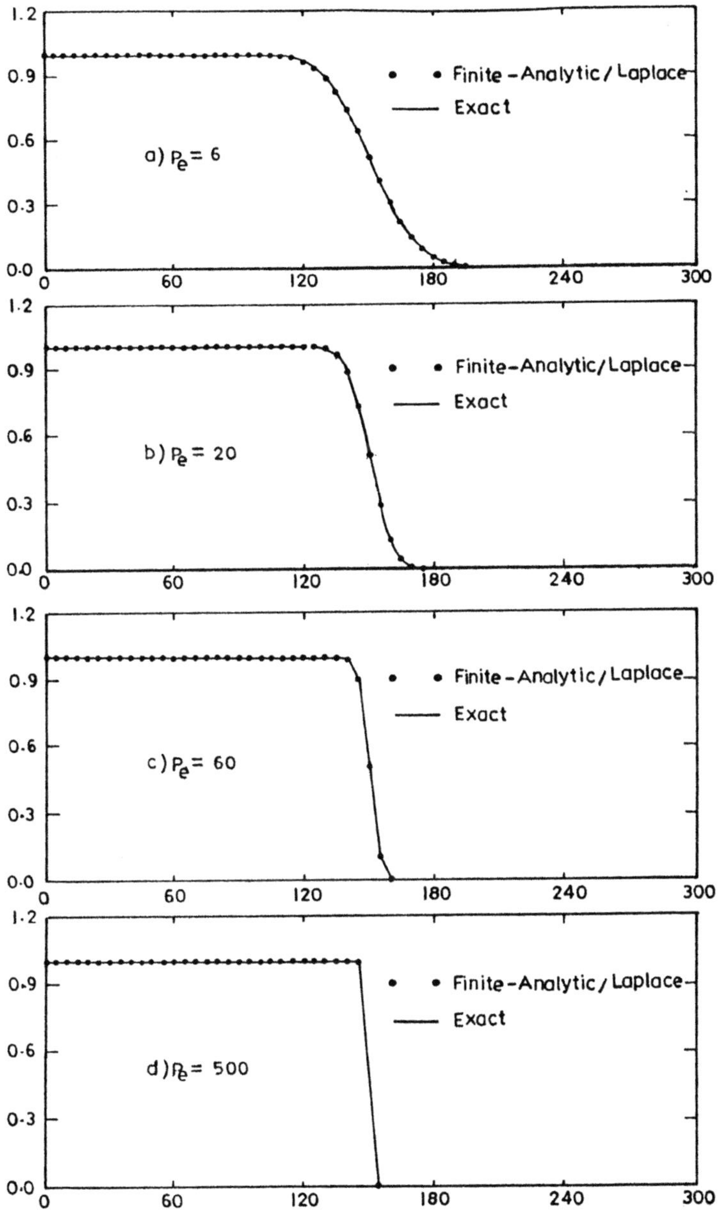

Fig.3. Spatial concentration distributions at t = 150 days for different Peclet numbers; vertical and horizontal axes are respectively normalized concentration and distance in m. (Redrawn from Li et al., 1992)

eigenfunction and Laplace tranform finite element methods have been used in developing semi-analytic solutions of one and two-dimensional convection-dispersion equations. To the authors knowledge the results from these two methods have not been compared so for. It may be noted that the eigenfunction method would require frequent evaluation of eigenvalues and eigenvectors if applied to a differential equation with time dependent coefficients. The Laplace transform method has advantage in solving such differential equations. Although the Laplace transform finite analytic method has so far been used to develop solution of only one-dimensional convection-dispersion equation, it can be extended to solve two- dimensional equation as well.

The following conclusions are drawn:

1. The time-continuous space discretized eigenfunction and Laplace transform solutions are accurate enough to simulate real life contaminant transport field problems.
2. The finite analytic solution matches with the analytic one for a wider range of Peclet numbers (upto 500), thereby ascertaining it to be the most accurate semi-analytic method. The technique has potential of solving various problems of surface and subsurface hydrology.

REFERENCES

Crump, K.S., 1976. Numerical inversion of Laplace transforms using a Fourier Series approximation, J. Assoc. Comput. Mach., Vol. 23 (1): pp. 89-96.

Davies, B., and Martin, B., 1979. Numerical inversion of the Laplace transform: A survey and comparison of methods, J. Comput. Phys., Vol. 33: pp. 1-32.

Guymon, G.L., 1970. A finite element solution of the one-dimensional diffusion-convection equation. Water Resources Research, Vol. 6 (1): pp. 204-210.

Guymon, G.L., Scott, V.H., and Hermann, L.,R., 1970. A general numerical solution of the two-dimensional diffusion-convection equation by the finite element method. Water Resources Research, Vol. 6 (6): pp.1611-1617.

Hwang, J.C., Cho, W.C., and Yeh, G.T., 1984. An eigenvalue solution continuous in time to the spatially discretized solute transport equation in steady groundwater flow. Water Resources Research, Vol. 20 (11): pp.1725-1732.

Kamra, S.K., Singh, S.R., Rao, K.V.G.K., and van Genuchten, M.Th., 1991a. A semidiscrete model for water and solute movement in tile-drained soils 1. Governing equations and solution. Water Resources Research, Vol. 27 (9): pp. 2439-2447.

Kamra, S.K., Singh, S.R., Rao, K.V.G.K., and van Genuchten, M.Th., 1991b. A semidiscrete model for water and solute movement in tile-drained soils 2. Field validation and applications. Water Resources Research, Vol. 27 (9): pp. 2449-2456.

Li, S.G., Ruan, F., and McLaughlin, D., 1992. A space-time accurate method for solving solute transport problems. Water Resources Research, Vol. 28 (9): pp. 2297-2306.

MacDonald, J.R., 1964. Accelerated convergence divergence , iteration, extrapolation, and curve fitting, J. Appl. Phys., Vol. (10): pp. 3034-3041.

Moridis, G.J., Reddell, D.L., 1991. The Laplace transform finite difference method for simulation of flow through porous media. Water Resources Research, Vol. 27 (8): pp. 1873-1884.

Nalluswami, M., Longenbaugh, R.A., and Sunanda, D.K., 1972. Finite element method for hydrodynamic dispersion equation with mixed partial derivatives. Water Resources Research, Vol. 8 (5): pp. 1247-1250.

Narasimhan, T.N., and Witherspoon, P.A., 1977. Numerical model for saturated-unsaturated flow in deformable porous media: I. Theory. Water Resources Research, Vol. 13(3): pp. 657-664.

Prenter, P.M., 1975. Splines and Variational Methods. John Wiley and Sons, New York. 323 pp.

Scheidegger, A.E., 1961. General theory of dispersion in porous media. J. Geophys. Res., 66(10): pp. 3273-3278.

Singh, S.R., 1989. Mathematical simulation of basinwide groundwater salinization. Terminal report of the PL 480 Scheme FG-IN-609, IN-ARS-200, Department of Soil and Water Engineering, P. A. U. Ludhiana, India.

Singh, S.R., Saini, A.K., and Kamra, S.K., 1989. Semi-analytic solution of one-dimensional dispersion-convection equation. Proc. International Workshop on Appropriate Methodologies for Development and Management of Groundwater Resources in Developing Countries, held at N.G.R.I., Hyderabad, India, Feb. 28 - March 4, 1989.

Stehfest, H., 1970. Numerical inversion of Laplace transforms-algorithm 368, Commun. Assoc. Comput. Mach., Vol. 13 (1): pp. 47-49.

Sudicky, E.A., 1989. The Laplace transform Galerkin technique: A time-continuous finite element theory and application to mass transport in groundwater. Water Resources Research, Vol. 25 (8): pp. 1833-1846.

Umari, A.M.J., and Gorelick, S.M., 1986. The problem of complex eigensystems in the semi-analytical solution for advancement of time in solute transport simulations: A new method using real arithmetic. Water Resources Research, Vol. 22 (7): pp. 1149-1154.

Van Genuchten, M.Th., 1987. A numerical model for water and solute movement in and below the root zone: Research report No. 121, United States Department of Agricultur, Agricultural Research Services, U.S. Salinity Laboratory River side, California.

Yu, Y.S., Heidari, M., and Wang, W.T., 1987. A semi-analytic method for solving unsteady flow equations, paper presented at Proceedings ASCE National Conference on Hydraulic Eng., Am. Soc. Civ. Eng., Williamsburg, Va., Aug.3-7,1987.

SECTION 5:

SOLUTE TRANSPORT IN GROUND WATER

USE OF TRACERS AND DYES TO ASSESS GROUND WATER CONTAMINATION POTENTIAL FOR GLACIAL TILL AQUIFERS

RAMESHWAR S. KANWAR
Professor of Agricultural & Biosystems Engineering
Iowa State University, Ames, IA 50011
U.S.A.

ABSTRACT. Studies were conducted to evaluate the effectiveness of chloride, bromide, lithium, and a fluorescent dye for tracing the movement of agricultural chemicals to ground water. In one study, bromide, nitrate, lithium, and rhodamine WT were applied with irrigation water to the field plot through a sprinkler system. In the second study, bromide, chloride, and rhodamine WT were mixed with two herbicides (cyanazine and alachlor) and were sprayed on the surface of the plot, which received rainfall with a rainfall simulator. In the third study, chloride as a tracer was used to trace the movement of NO_3-N in the vadose zone. During and after the experiments, groundwater samples were collected. These samples were analyzed for tracer and chemical concentrations. Based on adsorption coefficients in the literature and field experiments, rhodamine WT dye is roughly 20 to 40 times more strongly adsorbed than the commonly used pesticides. Experimental results also indicate that a combination of adsorbed tracers (fluorescent dyes) and non-adsorbed tracers (anions) have the potential for tracing the movement of water as well as adsorbed and non-adsorbed chemicals (nitrates and pesticides) to ground water. This study shows that macropores provide potential paths for rapid transport of chemicals to ground water and tracers offer a potential to investigate ground water contamination problems.

1. Introduction

The increased use of agricultural chemicals has contributed significantly to the agricultural productivity but has been the source of much controversy recently because of the perceived health risks posed by the presence of nitrate, pesticide, and other compounds in drinking water. This has resulted in several ground water quality legislation by several states in the U.S.A. High concentrations of nitrate-nitrogen (NO_3-N) in well water was first recognized as a health problem in 1945 when two cases of infant methemoglobinemia (bluebaby syndrome - a temporary blood disorder that reduces the ability of an infant's bloodstream to carry oxygen through the body) were reported in Iowa (Comly, 1945), and recently in South Dakota (Johnson et al., 1987). Some evidence exists that high NO_3-N ingestation is involved in the etiology of human cancer (Fraser et al., 1980; Foreman et al., 1985).

In recent years, the negative impacts of the use of pesticides to human health and the environment have been a source of concern. In addition to concern for the acute and chronic toxicity of pesticides, their potential as carcinogens and their presence in ground water sources have raised questions about their continued use in agriculture. Groundwater is the drinking water source for 50% of the U.S. population, whereas, 85% of the rural population relies on groundwater. Eighty-six percent of the total water resources of the U.S. are contained in groundwater aquifers (CAST, 1985).

Because of public concerns, water quality monitoring programs have focused on the presence of NO_3-N and pesticides in public water supplies. Recent studies in Iowa (Hallberg, 1986; Kross et al., 1993) and in other parts of the U.S. (Fairchild, 1987; Hallberg et al., 1989) have shown the evidence of the presence of NO_3-N and pesticides in ground water as a result of agricultural activities. Several studies have indicated the rapid transport of the surface applied chemicals to deeper soil depths and to shallow ground water systems (Czapar et al., 1993; Everts and Kanwar, 1993; Everts et al., 1989; Kanwar et al., 1985). This rapid movement of chemicals to shallow ground water is evidenced due to the preferential flow of solutes (Kanwar, 1991; Czapar et al., 1992; Kanwar at al., 1993; Olson and Kanwar, 1993; Everts

V. P. Singh and B. Kumar (eds.), Water-Quality Hydrology, 177–186.

and Kanwar, 1990). It is generally presumed that agricultural chemicals found in ground water are applied to the soil and move through the soil and into the aquifer by mass flow recharge of ground water.

A better understanding of ground water contamination mechanisms is needed to develop better management systems to protect groundwater aquifer systems. Research is needed to develop simple and rapid techniques to demonstrate how agricultural chemicals can move to aquifer systems. Several types of tracers and dyes have been used for tracing the movement of water and chemicals to groundwater (Everts and Kanwar, 1990, 1993; Aldous and Smart, 1988; Olson and Kanwar, 1993; Omoti and Wild, 1979; Smart and Laidlow, 1977; Trudgill et al., 1983; Wright et al., 1964), but relatively few studies show the significance of the use of tracers in the area of groundwater contamination.

An excellent review of the physical properties of fluorescent dyes has been given by Smart and Laidlow (1977). Fluorescent dyes can be detected in water using a filter fluorometer. An important consideration in selecting a dye for water and/or chemical tracing is the amount of dye adsorbed onto soil or aquifer materials (Smart and Laidlow, 1977). Dye adsorption can seriously restrict the use of a particular dye for some chemical tracing to ground water systems. Selection of a dye with adsorption characteristics similar to the organic chemical of interest could serve as a valuable tool (surrogate) for studying the movement of pesticides to groundwater. Fluorescent dyes offer the advantages of low detection levels and low analytical costs. If a particular dye could be used as a reliable tracer for pesticides, it would save significant costs in analytical analyses. It costs about $100 for one water sample for pesticide analysis, whereas the cost of analysis for dyes is less than one dollar. Omoti and Wild (1979) have used fluorescent dyes successfully for tracing paths of solute transport. Everts and Kanwar (1990, 1993) have used rhodamine WT, a florescent dye, for herbicide leaching to ground water. The sorption coefficients (K_{oc}) for rhodamine WT ranged from 1400 to 3700 compared to K_{oc} values of 148 for atrazine and 400 for alachlor in aquifer material.

Chloride and bromide have been used as conservative tracers for tracing NO_3-N to ground water. These tracers are useful in studying NO_3-N transport to groundwater because they are not adsorbed by the soil and do not undergo any biological transformations. Tracer studies are especially very important to study to the preferential movement of chemicals through soil (as a porous media) to shallow and deeper ground water systems as a function of the watershed characteristics and geologic conditions. Several studies have been conducted at Iowa State University in the last five years on the use of tracers and dyes to understand the leaching of agricultural chemicals to ground water. Therefore, the objective of this paper is to present summaries of some of the major tracer studies to determine the contamination potential of glacial till aquifers of Iowa due to agricultural activities. Seventy-five percent of the total land area of Iowa is currently being used for agricultural production.

2. Methods and Materials

2.1 RAINFALL SIMULATION STUDIES

2.1.1 Ames, Iowa site. A field tracer experiment was conducted at the Iowa State University Agricultural Engineering Research Center near Ames, Iowa (Everts and Kanwar, 1993; Everts et al., 1989). The soils at this site are artificially drained in the Clarion-Nicollet-Webster Association. These glacial till soils have a vertical conductivity of 0.014 to 0.40 cm/hr (Kanwar et al., 1985) and cover about 1.2 million hectares in Iowa and are under intensive agricultural use (receiving herbicides on 98% of corn and soybean acres in Iowa). The experimental site near Ames had nearly mature standing corn in a conventionally tilled field and was drained by a single 10.2 cm diameter clay drain tile line installed at a depth 1.1 m. This tile line was intercepted by a 1.2 m diameter steel culvert 1.8 m deep at the lower end of the field. A V-notch weir and stage recorder were installed to monitor tile flows and collect water samples for tracer analysis. Three tracers were used in this study. Bromide was used as a non-adsorbed tracer, and lithium and rhodamine WT were used as adsorbed tracers. In addition, nitrate was also added to the mix of tracers applied. All these chemicals were mixed with water in two 20-m^3 tanks and were applied to the field plot through a sprinkler system on the standing corn. Concentrations of tracers used in the applied irrigation water were chosen to be at least 1000 times larger than their detection limits. The concentrations of bromide, nitrate, lithium, and rhodamine WT in the irrigation water were 236, 234, 4.0 and 1.4 mg/l respectively, whereas the detection limits for these chemicals are 0.05, 0.01, 0.002, and 0.00001 mg/l, respectively. A total of 115.62 m^3 (5.6 cm) of irrigation water mixed with four tracers was applied over the plot area in two irrigations (2 days apart) at an average rate of 0.7 cm/hr. The number of sprinkler risers gave a uniform water distribution to the plot. Maximum irrigation rate was 30 mm/hr. Details about the sprinkler system used for irrigation are given by Everts et al. (1989). Water samples were collected frequently from tile outflows to calculate tracer concentrations in the subsurface drainage water.

2.1.2 Kanawha, Iowa site. The soils at this site are also derived from glacial till material and have properties similar to the Ames site. For the study conducted at the experimental site in Kanawha, Iowa, three tracers (Bromide, chloride, and rhodamine WT) were mixed with two herbicides (cyanazine and alachlor). These chemicals were applied to the soil surface at a rate of 2.8 kg/ha and 2.2 kg/ha to an area of 4.5 m^2 (1.5 m x 3 m) located directly over a drain tile line. Eight such drains were laid out in the form of wheel spokes draining into a 1.2 m dia steel culvert sump at the hub. These tile lines (3.4 m long) were installed in 1952 at an average depth of 1.2 m. During the experiment, galvanized metal borders (5 cm under the ground and 15 cm above the ground) were installed to border the 1.5 m x 3.0 m test plot areas after chemicals were sprayed on the surface. Test plots were irrigated 24 hours after chemical application with a rainfall simulator. The rainfall simulator consisted of three sprinkler heads, an electronic control device for controlling rainfall intensity, a water pump, a water tank and two portable electronic generators. For a proper rainfall pattern, the sprinkler nozzles were operated at a pressure of 40 kPa to give an application rate of 1 L/s. Rainfall of 5.2 cm and 7.8 cm was applied over the plot area in one hour and 1.5 hours, respectively, and the movement of tracers and herbicides to subsurface drain lines were monitored. The rainfall intensities used in the simulator correspond to recurrence interval in Iowa of 10 and 25 years, respectively. Tile outflows were monitored by collecting one-liter samples during the rainfall event and until the tile flow almost ceased. Tile effluent was analyzed for herbicides at the Iowa Hygienics Laboratory in Iowa City, Iowa and tracers were analyzed in the Chemistry Laboratory of the Department of Agricultural and Biosystems Engineering, Iowa State University, Ames, Iowa.

2.1.3 Large borehole cavity experiment. To better understand the mechanisms of chemical transport through the vadose zone and into groundwater, a large borehole cavity was installed at the Agricultural Engineering Research Center near Ames, Iowa. The borehole cavity consists of a 3.05 m diameter (having a hole augered to a depth of 3.3 m) galvanized steel culvert installed vertically to maintain the structure of the hole. In order to minimize the dewatering effect of the cavity, the perimeter of the culvert was backfilled with a bentonite and sand mixture. Also, a concrete floor and anchors were installed at the bottom of the hole to offset the buoyancy force of the water table and prevent water from entering the structure. Upon completion of the cavity structure, horizontal suction lysimeters were installed to sample the soil water. The soil water samplers used were 0.91 m by 0.06 m diameter porous stainless steel cylinders to minimize the adsorption of chemicals to the sampling material. Since many agricultural pesticides are likely to be adsorbed, it is imperative to use inert sampling materials to prevent bias in the water samples. Pore sizes in the samplers used were 0.2 micron in diameter with a bubbling pressure of 0.6 to 0.7 bars. A horizontal drilling machine was used to auger a horizontal hole to accommodate the soil water samplers. First, one 0.064 m hole was augered to a distance of 3.66 m and a schedule 40 PVC pipe was installed to maintain the hole. Then, a smaller (0.06 m) hole was drilled for an additional

1.18 m from the end of the PVC pipe where the sampler was to be located. The larger PVC pipe was installed so that the samplers could be removed and reinstalled if necessary. The sampler was fitted with an adaptor to connect to standard 0.051 m schedule 40 threaded well casing, which fit inside the larger PVC pipe and extended back into the borehole cavity. The well casing facilitated removal of the samplers and also provided space for the teflon sampling tube. The annulus between the PVC pipe and well casing was sealed with O-rings to prevent leakage into the borehole.

The suction lysimeters were installed at 0.6, 1.2, 1.8, 2.4, and 3.0 m depths at five locations around the perimeter of the cavity for a total of 25 lysimeters. The samplers were offset in a spiral fashion so that no sampler was located directly below a shallower sampler. This installation procedure provided a comprehensive sampling network around the borehole cavity. Additionally, fifteen electronic tensiometer (transiometers) were installed at the three shallower depths of 0.6, 1.2, and 1.8 m to continuously monitor the soil-water pressure potential using a datalogger. Operation of the lysimeters consists of applying a vacuum to the system and soil water is pulled into collection bottles. When the water samplers are beneath the water table, a sample is readily available. Water must be extracted in the case of dry soil, but must be collected under low tension to ensure that only macropore water flow is collected. The suction regulator ensured that a maximum of 200 cm of water suction was applied to the lysimeters. The area around the borehole cavity has been under no-till conditions since 1984. Additionally, the area has been managed using a controlled traffic, five row implement system. In order to observe the flow of water and chemicals, a rainfall simulation experiment was conducted on the borehole cavity during June, 1992 using two conservative tracers and two agricultural herbicides.

A rainfall simulation experiment was conducted by placing a rotating boom sprinkler apparatus over the borehole. The cover over the borehole was designed to support the weight of the machine. Two canvas tanks holding 5000 gallons of water each were erected on site as a source of water supply. A gasoline powered centrifugal pump was used to deliver water from the tanks to the simulator. A gate valve and flowmeter were used to control the flow rate and achieve a rainfall rate of 15.25 mm/hr. Since this was the first rainfall experiment over the borehole, the low rainfall rate was used to prevent surface runoff and thus simplify analysis of the data.

The two conservative tracers used in the experiment were bromide and chloride. Bromide was applied over a 19 m x 19 m area at a rate of 175 kg/ha the night before the simulation by dissolving potassium bromide in water and applying the mix with a tractor mounted sprayer. Chloride was applied by dissolving potassium chloride in the sprinkler applied water. The average concentration of chloride in the sprinkler applied water was 170 ppm with a total of 7.0 kg of chloride applied during the simulation. The two agricultural herbicides applied at planting time were Dual and Sencore. The herbicides were applied at labelled rates of 2.24 and 0.56 kg/ha, respectively. Ideally, the rainfall simulation would have been conducted shortly after application of the herbicides, but was delayed until three weeks after application. There was no significant precipitation in the three weeks after rainfall simulation.

A target rainfall depth of 127 mm was applied to the soil. A tipping bucket rain gage connected to the datalogger was placed in one corner of the plot to obtain a continuous record of rainfall intensity and depth. The rainfall period for the simulation was 450 minutes. In order to evaluate tracer movement through the profile, samples were taken every 15 minutes, or as often as possible for conservative tracer analysis. Additionally, pesticide samples were taken hourly throughout the experiment. Samples for conservative tracer analysis were collected in disposable 15 mL test tubes while pesticide samples were collected in 950 mL glass jars with teflon lined lids. Prior to initiation of the experiment, water samples were collected for pesticide analysis. However, no samples were attainable from the 60 and 120 cm samplers due to lack of precipitation. Samples were collected for a period of fourteen and one-half hours, which was seven hours after the rainfall was completed. Additionally, samples were collected the next morning to determine chemical movement overnight.

3. Results and Discussion

3.1 RAINFALL SIMULATION EXPERIMENT NEAR AMES, IOWA.

Figures 1 and 2 show the relative concentrations of rhodamine dye and two tracers in tile drainage water during and after two irrigations. Relative concentrations were calculated as tracer concentrations in tile drainage water as a percentage of the concentration of tracer in the applied irrigation water. Figures 1 and 2 show that tracer concentrations in tile drainage water peaked close to 65 minutes after the start of first irrigation and 75 minutes after the start of the second irrigation. Tracer concentrations then declined steadily, approaching a steady relative concentration even after the tracer irrigation continued. The decline in relative concentration before the irrigation ended could be explained by a dual porosity flow mechanism. This approach was used by Everts and Kanwar (1990) to separate the hydrograph into matrix and preferential flow. Preferential flow would contribute tracer concentration to the tile line with minimum opportunity for adsorption. Also, figures 1 and 2 show identical timings for all tracers and dye in reaching their respective peak concentrations. The initial peaks after about 65 minutes indicate that tracers reached the tile line without much interaction with the soil solution through preferential flow. The tracer concentrations in tile water after the peaks had leveled off are the result of tracer solutions passing uniformly through the soil matrix and then reaching the tile line. Matrix flow should take a lot longer to reach the ground water than the preferential flow, and tracer concentrations in the ground water should be dominated by the matrix flow. This phenomenon of dual flow mechanism (preferential and matrix) is quite visible in figures 1 and 2 and these figures show nearly identical breakthrough curves for nitrate and bromide in the tile water. Nitrate and bromide showed similar relative concentrations of 29% and 26% for the first and second irrigations, respectively. This indicates that bromide is a satisfactory tracer for monitoring the leaching of nitrates to groundwater. The relative peak concentrations for lithium and rhodamine WT were 16% and 6% for the first irrigation, and 15% and 6% for the second irrigation, respectively. The lower peak concentrations for rhodamine WT and lithium indicate that they are strongly adsorbed tracers and offer the potential for use in tracing the leaching characteristics of pesticides. The rapid appearance of adsorbed tracers in the ground water indicate that soil structure and the occurrence of preferential flow can be as important as adsorption coefficients in evaluating the leaching potential of pesticides to groundwater.

3.2 RAINFALL SIMULATION EXPERIMENT AT KANAWHA, IOWA.

Figure 3 gives the tile flow and bromide concentration hydrographs. Time zero is the start of the rainfall simulation. This figure shows that tile peak flows and greatest bromide concentration in tile water occurred about the same time, at about 100 minutes after the beginning of rainfall. Figure 4 shows the concentrations of cyanazine, alachlor, and rhodamine WT dye in tile water as a function of time. This figure also shows that peak concentrations of two herbicides and rhodamine WT occurred at the same time; about 100 minutes after the beginning of rain. Results of this experiment also suggest that rhodamine WT could possibly be used as a tracer for these herbicides. Dye concentrations peaks ranged from 1 to 38 ug/L, whereas significantly greater concentrations of alachlor, cyanazine, and rhodamine WT were observed. In all instances, herbicide and dye concentration peaked less than 150 minutes after the start of the rainfall and then decreased rapidly over the next couple of hours, indicating a dual flow mechanism. Similar results have been obtained by Singh and Kanwar (1991), and Everts and Kanwar (1990, 1993). Figures 3 and 4 also suggest that from a groundwater-monitoring point of view, water samples should be collected for herbicide analysis immediately after rainfall in the saturated and unsaturated zone above it. Under a worst-case scenario, heavy rains, soon after herbicide application, may allow macropore flow to carry surface applied chemicals rapidly through the soil profile and into the ground water. Understanding the importance of preferential chemical transport will help develop the best management practices to protect groundwater from contamination in the future.

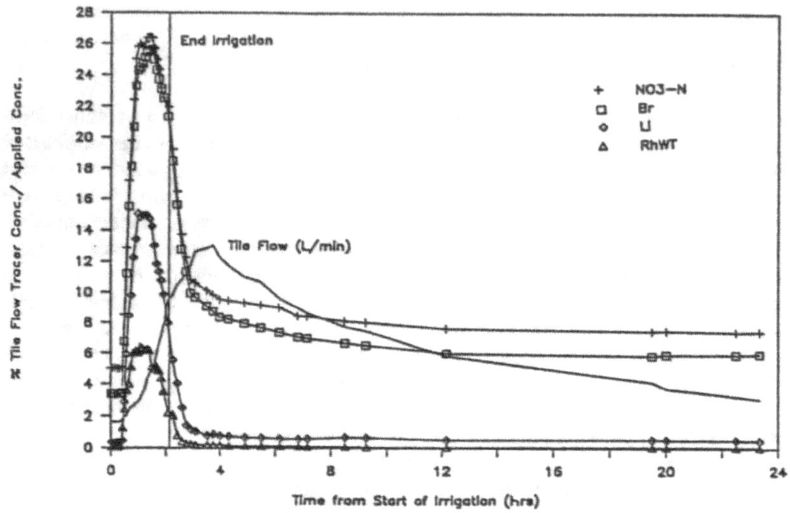

Figure 1. Relative concentration of tracers (Br, Li, NO_3-N) and dye (rh WT) in subsurface drainage water (shallow ground water) during first irrigation

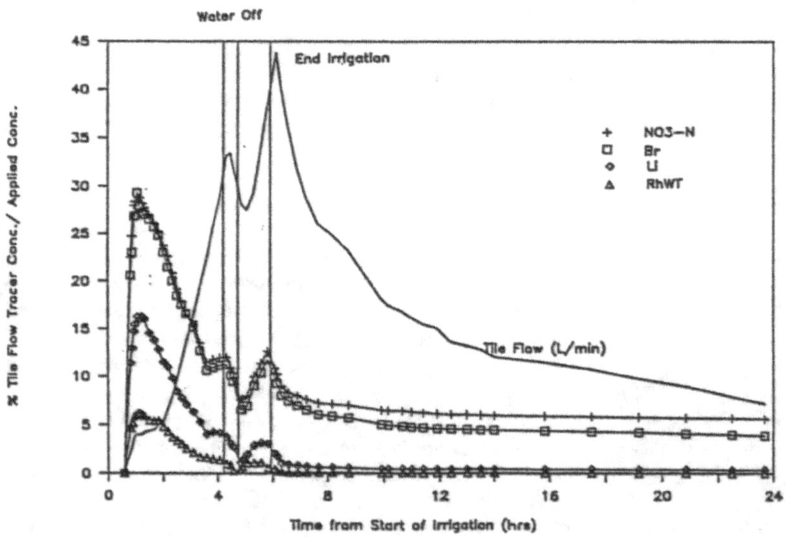

Figure 2. Relative concentrations of tracers and dye in subsurface drainage water during second irrigation

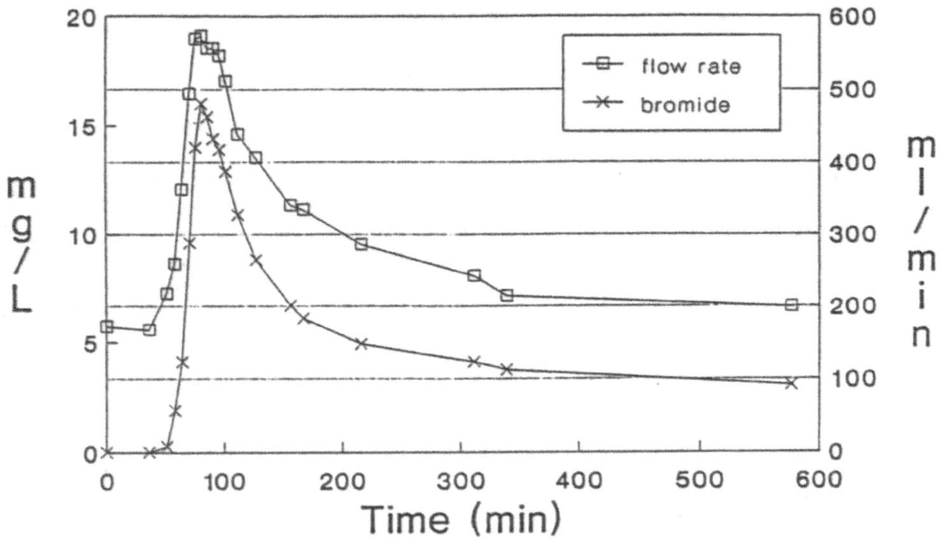

Figure 3. **Bromide concentration in subsurface drainage water and subsurface drainage flow rates**

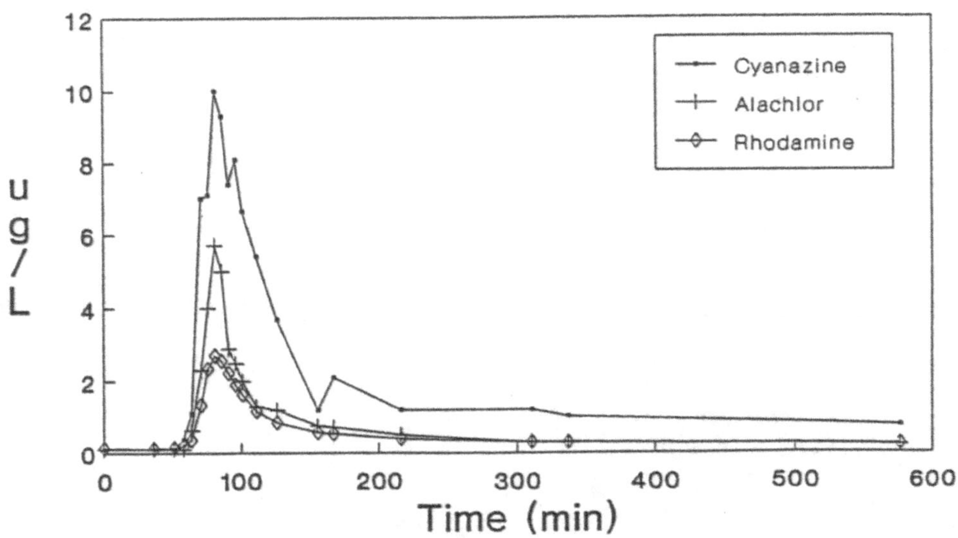

Figure 4. **Herbicide and rhodamine WT dye concentrations in the subsurface drainage water**

3.3 VADOSE ZONE MONITORING USING BOREHOLE CAVITY

Figures 5 and 6 give the chloride concentration hydrographs at 60 and 120 cm depths for areas A and B around the borehole cavity. The chloride concentration data for these two areas show an interesting tracer transport phenomena in the vadose zone. For area A, the concentration curve for the 60 cm depth resembles a classical breakthrough curve which can be described with the advective-dispersion equation. However, the chloride concentration data for the 120 cm depth of area A illustrate a two peak breakthrough curve with the first peak equal to the chloride concentration of the applied water. This suggest that preferential flow paths exist to the 120 cm depth. However, the concentration data in figure 6 suggest a strong evidence of preferential flow at 60 and 120 cm depths. Also, the chloride concentration data at the 120 cm depth indicates a two peak breakthrough, although the chloride concentrations peaks were lower than at the 60 cm depth. Again, figures 5 and 6 give the importance of using chloride as a tracer for tracing the movement of chemicals to groundwater.

4. Conclusion

A variety of tracers and dyes have been used to study the transport of NO_3-N and pesticides to shallow ground water systems. Chloride and bromide are favored because they are usually not adsorbed and react similarly in soils and aquifers to the nitrate anion without undergoing biological transformations. Results of this study show that chloride and bromide ions are good tracers for studying NO_3-N transport to ground water. Adsorbed tracers, such as fluorescent dyes used in field studies with pesticides, can reduce the number of samples required for pesticide analysis or eliminate the use of pesticides to study the transport of pesticides to the groundwater. Rhodamine WT dye used in this study is much more strongly adsorbed by soil than most commonly used pesticides and may not be the best tracer for all pesticides. More research is needed to identify different fluorescent dyes and other tracers (such as cations) with different adsorption properties that should resemble more closely the adsorption properties of agricultural chemicals. A comparison of alachlor, cyanazine, and rhodamine WT show that rhodamine dye is more strongly adsorbed than the other herbicides.

REFERENCES

Aldous, P.J., and P.L. Smart. 1988. Tracing ground-water movement in abandoned coal mined aquifers using fluorescent dyes. Ground Water 26(2):172-178.

Bowman, R.S., and R.C. Rice. 1985. Chemical tracers - their use in measuring deep percolation rates. In: Proceedings of the 3rd Percolation Symposium. Arizona Department of Water Resources, Phoenix, AZ, pp. 1-4.

Comly, H.H. 1945. Cyanosis in infants caused by nitrates in well water. J. Am. Med. Assoc. 129:112-117.

Council for Agricultural Science and Technology (CAST). 1985. Agriculture and groundwater quality. CAST Report no. 103. Ames, Iowa.

Czapar, G.F., R. Horton, and R.S. Fawcett. 1992. Herbicide and tracer movement in soil columns containing an artificial macropore. J. Environ. Qual. 21(1):110-115.

Czapar, G.F., R.S. Kanwar, and R.S. Fawcett. 1993. Herbicide and tracer movement to field drainage tiles under simulated rainfall conditions. Soil Tillage Res. (In press).

Everts, C.J. and R.S. Kanwar. 1993. Evaluation of rhodamine WT as an adsorbed tracer in an agricultural soil. J. Hydrol. (In press).

Everts, C.J., R.S. Kanwar, E.C. Alexander Jr., and S. Alexander. 1989. Comparison of tracer mobilities under laboratory and field conditions. Environ. Qual. 18:491-498.

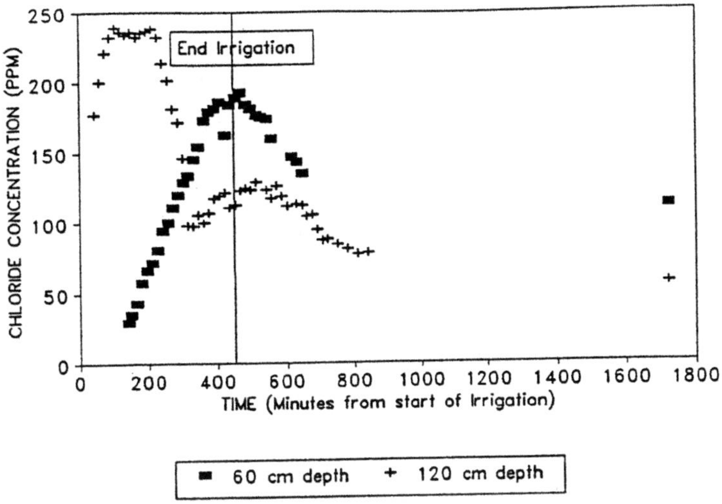

Figure 5. Chloride concentration in shallow ground water in area A

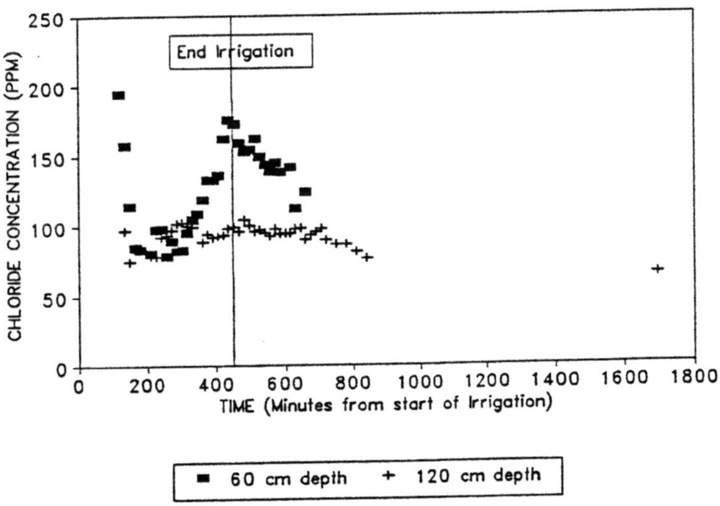

Figure 6. Chloride concentration in shallow ground water in area B

Everts, C.J., and R.S. Kanwar. 1990. Quantifying preferential flow to a tile line with tracers. Transactions of the ASAE 33(2):451-457.

Fairchild, D.M. 1987. A national assessment of groundwater contamination from pesticides and fertilizers. In groundwater quality and agricultural practices. Lewis Publishers, Inc., Chelsea, MI p. 273-294.

Forman, D., S. Al-Dabbagh, and R. Doll. 1985. Nitrates, nitrifies, and gastric cancer in Great Britain. Nature 313:620-625.

Fraser, P., C. Chilvers,V. Beral, and M.J. Hill. 1980. Nitrate and human cancer: a review of the evidence. Int. J. Epidemiol. 9:3-9.

Hallberg, G.R. 1986. Overview of agricultural chemicals in ground water. In: Agricultural Impacts on Ground Water, National Water Well Association, Wortherington, OH, pp. 1-67.

Hallberg, G.R. 1989. Pesticide pollution of groundwater in the humid united states. Agri. Ecosys. Environ. 26:299-367.

Isensee, A.R., C.S. Helling, T.J. Gish, P.C. Kearney, C.B. Coffman, and W. Zhuang. 1988. Groundwater residues of atrazine, alachlor, and cyanazine under no-tillage practices. Chemosphere 17:165-174.

Johnson, C.J., P.A. Bonrud, and T.L. Dosch. 1987. Fatal outcome of methemoglobinemia in an infant. J. Am. Med. Assoc. 257:2796-2797.

Johnson, C.J. and B.C. Kross. 1990. Continuing importance of nitrate contamination of groundwater and wells in rural Iowa. Am. J. Ind. Med. 18:449-456.

Jury, W,.A., D.D. Focht, and W.J. Farmer. 1987. Evaluation of pesticide groundwater pollution from standard indices of soil chemical adsorption and biodegradation. J. Environ. Qual. 16:422-428.

Kanwar, R.S., J.L. Baker, and J.M. Laflen. 1985. Nitrate movement through the soil profile in relation to tillage system and fertilizer application method. Transactions of the ASAE 28(6):1802-1807.

Kanwar, R.S. 1991. Preferential movement of nitrate and herbicides to shallow groundwater as affected by tillage and crop rotation. In: Proceedings of the National Symposium on Preferential Flow. American Society of Agricultural Engineers, pp. 328-337. Edited by T.J. Gish and A. Shirmohammadi.

Kanwar, R.S. and J.L. Baker. 1993. Tillage and chemical management effects on groundwater quality. In: Proceedings of National Conference on Agricultural Research to Protect Water Quality (In press).

Kross, B.C., G.R. Hallberg, D.R. Bruner, K. Cherryholmes, and J.K. Johnson. 1993. The nitrate contamination of private well water in Iowa. Am. J. Pub. Health 83(2):270-272.

Olson. D.I., and R.S. Kanwar. 1993. Tracer movement through a glacial till soil profile. In: Proceedings of National Conference on Agricultural Research to Protect Water Quality (In press).

Omoti, U. and A. Wild. 1979. Use of fluorescent dyes to mark the pathways of solute movement through soils under leaching conditions: 2. Field Experiments. Soil Sci. 128:98-104.

Singh, P. and R.S. Kanwar. 1991. Preferential solute transport through macropores in large undisturbed saturated soil columns. J. Environ. Qual. 20(1):295-300.

Smart, P.L. and I.M.S. Laidlow. 1977. An evaluation of some fluorescent dyes for water tracing. Water Resour. Res. 13:15-33.

Trudgill, S.T., A.M. Pickles, and K.R.J. Smettem. 1983. Soil-water residence time and solution uptake, 2. Dye tracing and preferential flow predictions, J. Hydrol. 62:279-285.

Wright, R.R. and M.R. Collings. 1964. Application of fluorescent tracing techniques to hydrological studies. J. Am. Water Works Assoc. 56(6):748-754.

CONTROL OF SEA WATER INTRUSION THROUGH BATTERY OF INJECTION WELLS

A. Mahesha and S.H. Nagaraja
Department of Civil Engineering
Indian Institute of Technology
Bombay-400 076.

Abstract. A vertically integrated quasi-three dimensional Galerkin finite element model is developed to study the transient effect of battery of injection wells on sea water intrusion into coastal aquifers. The model is derived based on sharp interface approach. Linear triangular elements are used to discretize the domain. The validity of the model is tested using existing theoretical solutions. Parametric studies are carried out using hypothetical data which cover wider practical range. A general relationship is established between the reduction of sea water intrusion and injection rate with time. A specific example is solved to illustrate the usefulness of the results.

1. Introduction

Artificial recharge of groundwater through injection wells is a widely used method worldwide to combat sea water intrusion. The method is suitable for confined or deep seated aquifers covered above and below by low permeable layers. Investigations in the past were carried out on theoretical (Hunt, 1985) and practical aspects of well recharge (Bruington and Seares, 1965; Bruington, 1969; Harpaz, 1971; Williams 1976; Schroeder et al, 1989). However, the transient effect of injection on sea water-fresh water interface behaviour is yet to be explored and present study considers this investigation through detailed parametric studies.

2. Materials and Methods

The definition sketch for a coastal confined aquifer is shown in Fig. 1. The initial seaward flow of fresh water QI establishes the equilibrium between sea water and fresh water with an initial intrusion length of L_o. The depth of aquifer below mean sea level is D and is covered above by an impermeable layer. The governing equations for fresh water and sea water flow can be expressed as (Bear, 1979):

V. P. Singh and B. Kumar (eds.), Water-Quality Hydrology, 187–196.

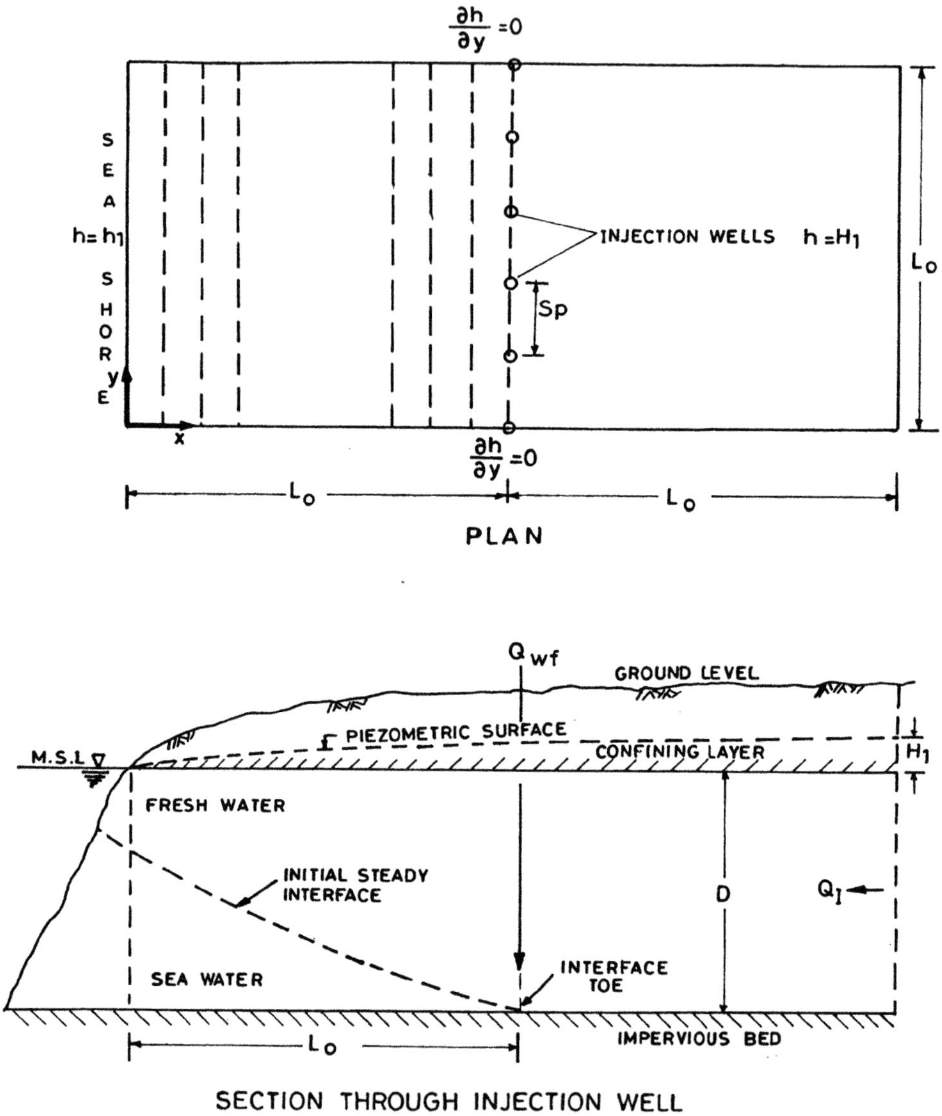

FIG. NO. 1: DEFINITION SKETCH FOR INJECTION WELL METHOD

$$\frac{\partial}{\partial x_i}\left(K_{fij}\, B_f\, \frac{\partial h_f}{\partial x_j}\right) + \frac{n}{\alpha}(1+\alpha)\frac{\partial h_s}{\partial t} - \left[n\left(\theta + \frac{1}{\alpha}\right) + S_f\right] \times \frac{\partial h_f}{\partial t} + Q_{wf} = 0 \quad (1)$$

$$\frac{\partial}{\partial x_i}\left(K_{sij}\, B_s\, \frac{\partial h_s}{\partial x_j}\right) - \left[\frac{n}{\alpha}(1+\alpha) + S_s\right]\frac{\partial h_s}{\partial t} + \frac{n}{\alpha}\frac{\partial h_f}{\partial t} = 0 \quad (2)$$

where K is the hydraulic conductivity; h is the vertically averaged piezometric head; B is the saturated thickness; n is the porosity; α is the excess density ratio; S is the storage coefficient; t is time; θ is a coefficient (= 0 for confined and = 1 for unconfined aquifers); Q_{wf} is the fresh water injection rate; x_i, x_j are lateral axes (= x, y); subscripts f amd s refer to fresh water and sea water respectively; $K_{ij} = 0$ $(i \neq j)$.

The continuity of pressure across the interface yields the following relationship between interface elevation and fresh water, sea water piezometric heads

$$z = -\frac{1}{\alpha}\left[h_s(1+\alpha) - h_f\right] \quad (3)$$

Using Galerkin's approach for the finite element formulation,

$$\int W_i\, R_e\, dA = 0 \quad (4)$$

where W_i = 1,2,3 are linear interpolation functions; R_e is the residual of Eq. (1) and (2). Linear triangular elements are used to discretize the domain. Finer subdivisions are adopted around the wells and towards sea coast to cope with the non-linearities in piezometric surface and interface elevations. The system of equations can be expressed in the matrix form as

$$[S]\left\{\frac{dH}{dt}\right\} + [C]\{H\} + \{F\} = 0 \quad (5)$$

where [S] is the storage matrix; [C] is the conductivity matrix; {F} is the flux vector and {H} is the state vector.

The matrices are symmetric and banded in nature. The time derivative is represented by finite difference scheme. A backward time stepping scheme is used for the above system of equations to minimise the interface oscillations. The element equations are assembled into global matrix and are stored in compact form. After applying suitable boundary conditions, the system of equations is solved iteratively by Newton -Raphson method.

The initial condition is assumed to be steady and is computed using steady state analytical solution (Vappicha and Nagaraja, 1975). The following boundary conditions are applied for fresh water and sea water

$$h_f(0,y,t) = h_f(0,y,0);\; h_f(L_a,y,t) = h_f\,(L_a,y,0) \qquad (6)$$

$$Q_f(x,0,t) = Q_f(x,y_{max},t) = Q_s(x,0,t) = Q_s(x,y_{max},t) = 0 \qquad (7)$$

$$h_s(0,y,t) = h_s(L_a,y,t) = 0 \qquad (8)$$

where L_a is the length of the aquifer perpendicular to the coast ($=2L_o$); y_{max} is the maximum dimension along y-axis.

The performance of the numerical model is tested with the available steady (Strack, 1976) and transient solutions (Sa da Costa and Wilson, 1979) for a single pumping well in a coastal phreatic aquifer. The position of interface toe in plan is compared in Fig.2. The results are in good agreement except minor derivation near the well due to the linear elements in the present model.

For the parametric studies, hypothetical cases are considered over wide practical range. Major parameters in the analysis are location of well battery, spacing of wells, rate of fresh water injection, motion of the interface and time. These parameters are expressed in non-dimensional form as:

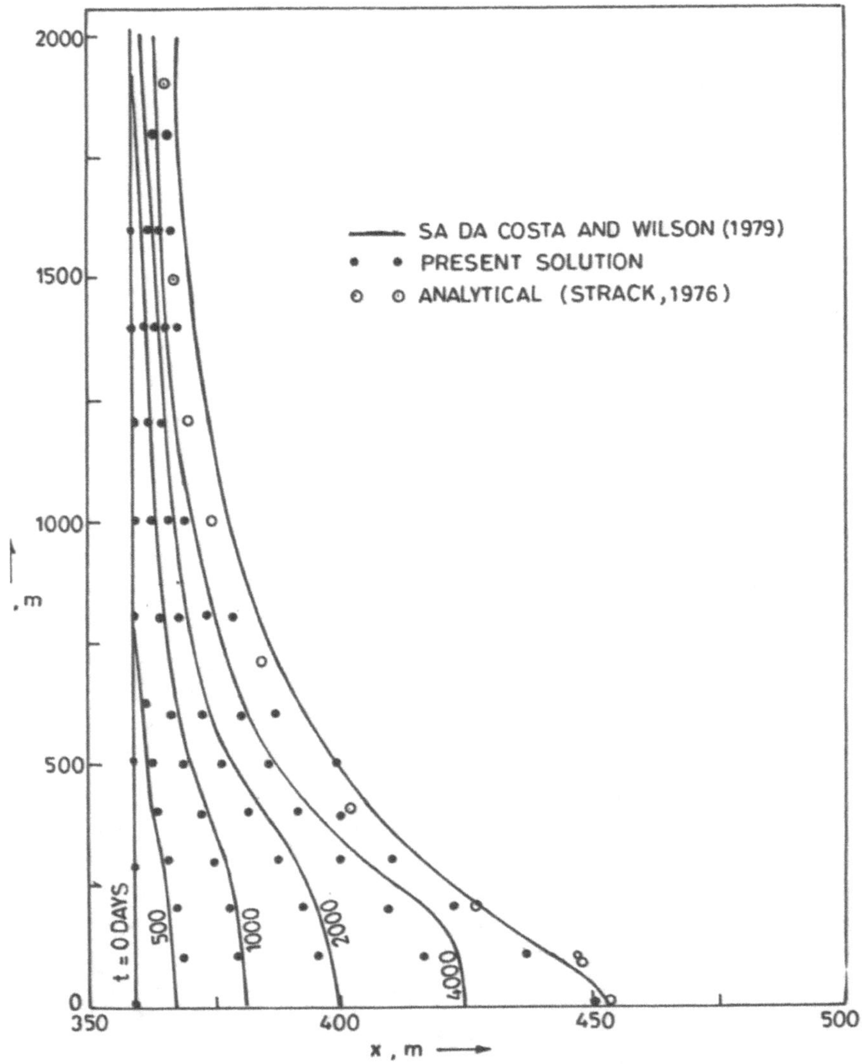

FIG. NO. 2 : TRANSIENT POSITION OF INTERFACE TOE DUE TO PUMPING WELL

$$W'_L = W_L | L_0 \tag{9}$$

$$S'_p = \frac{S_p}{L_0} \times 100 \tag{10}$$

$$Q'_f = \frac{Q_{wf}}{\alpha K D^2} \tag{11}$$

$$P = \frac{L_0 - L_t}{L_0} \times 100 \tag{12}$$

$$\delta_t = \frac{Tt}{nL_0^2} \tag{13}$$

where W_L is the distance to the location of injection well battery from the coast; L_o is the initial length of intrusion, S_p is the spacing of wells in the battery; Q_{wf} is the fresh water injection rate; L_t is the length of intrusion at time t; T is the transmissivity. The well battery is located at a distance of L_o from the coast. The injection rate has a range of Q'_f = 0.05 - 0.8 for the wells spaced at S_p' = 7.5 the battery.

3. Results and Discussion

The motion of the interface due to fresh water injection is tracked and the magnitude is expressed as percentage of initial length of intrusion. The overall reduction of intrusion is the average of reduction of intrusion at different sections running perpendicular to the coast. A relationship is established between the reduction of intrusion and rate and period of fresh water injection and is shown in Fig.3. The above relationship is useful in estimating the reduction of intrusion at any time t due to injection well battery located at a distance L_o from the coast with wells spaced at S_p' = 7.5. A numerical example is solved to illustrate the usefulness of the results. Fig.4 shows the transient motion of the interface due to injection wells spaced at 600m apart recharging at the rate of 1280m^3 /d. The reduction of intrusion is rigorous at initial stages which slows down with time as steady state is approached. The reduction of intrusion achieved after ten years of operation according to the above figure is around 14%. Better results can be achieved either by increasing the injection rate or reducing the spacing of wells. Further studies are on to analyse the effect of the location of the well battery and spacing of wells on the interface motion. As a field example, the injection test barrier created by the injection of treated sewage water in Magothy aquifer, Long Island, U.S.A. at the

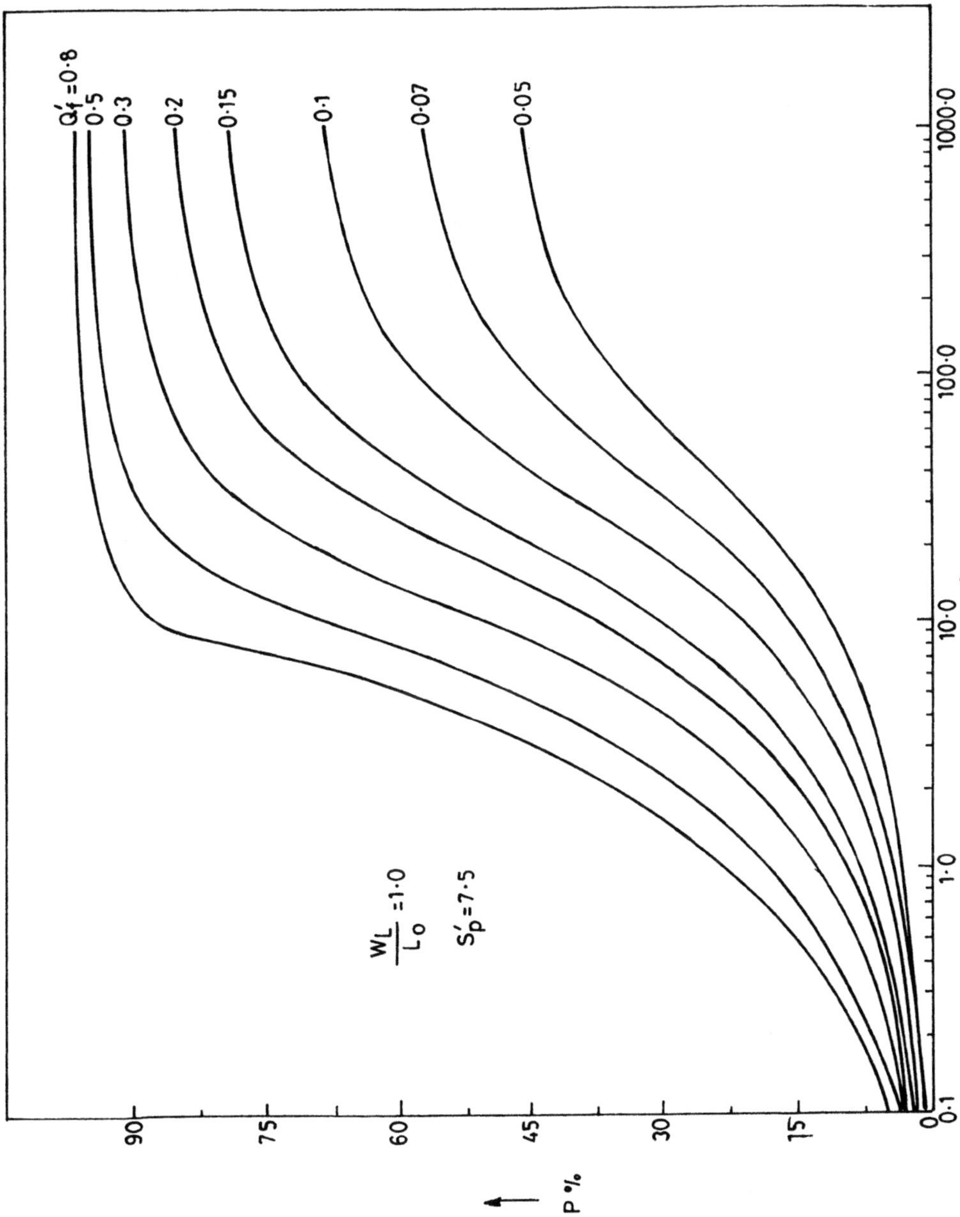

FIG. NO. 3: REDUCTION OF INTRUSION DUE TO FRESH WATER INJECTION

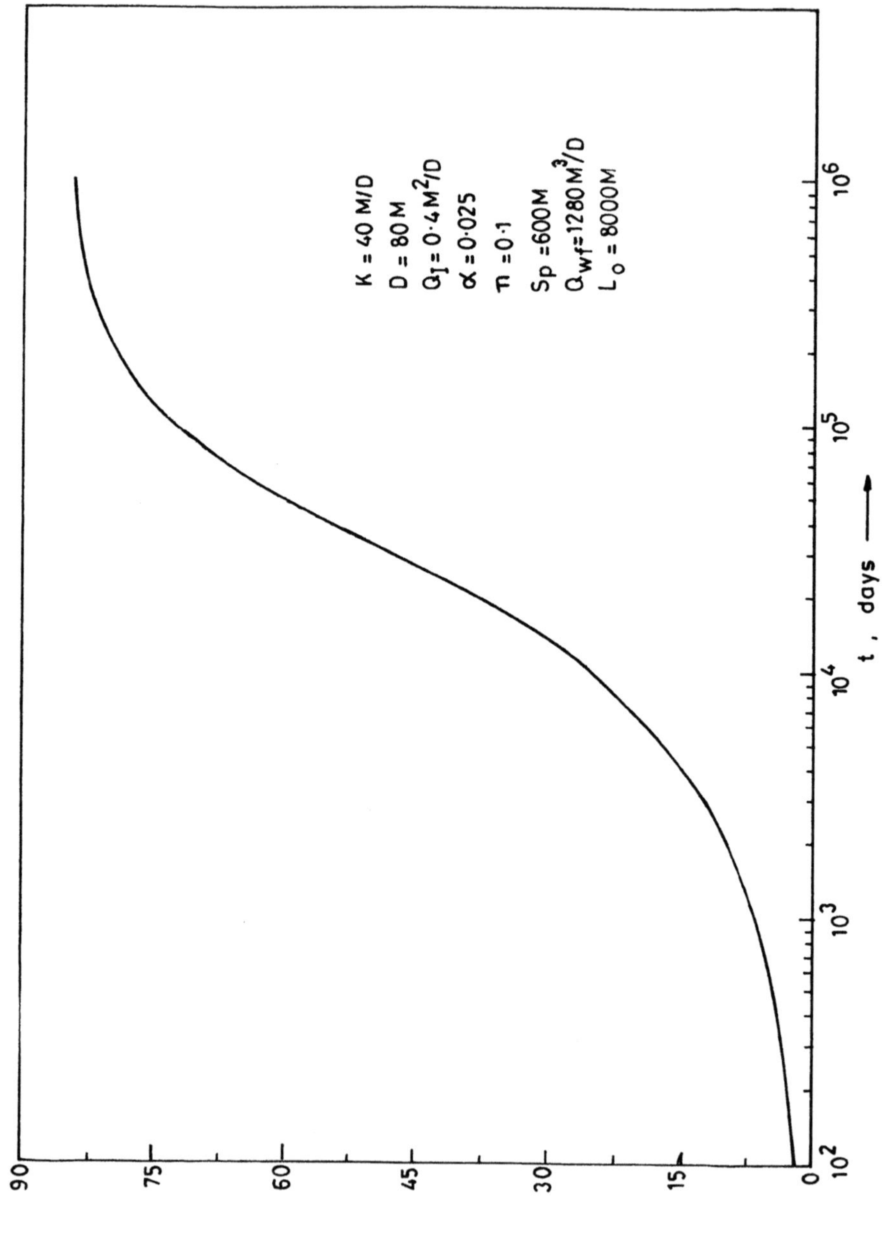

FIG. NO. 4 : REDUCTION OF INTRUSION FOR THE SOLVED EXAMPLE

rate of 1900 m^3/d for a period of 33 days showed positive results (Das, 1993). However, the costs prohibited injection into deeper parts of the aquifer. Another example is reported from Los Angeles County, U.S.A. (Das, 1993) where piezometric head maintained 1 to 3m above mean sea level with the injection of fresh water at the rate of 1500m^3 /d proved successful in preventing intrusion. Injection well experiments in intrusion affected coastal districts of Gujarat improved the situation significantly (Mistry et al, 1989, Raju et al, 1993). The detailed assessment of the situation are awaited.

4. Conclusion

Artificial recharge of coastal aquifers through battery of injection wells is an effective method to control sea water intrusion. Transient motion of the interface due to multiple injection wells is of utmost importance in coastal management schemes and needs detailed investigations. The numerical solution represented in the form of non-dimensional curves is useful in predicting transient motion of sea water wedge for wide ranged practical problems. Significant reduction of intrusion can be achieved through battery of injection wells over a reasonable period of time.

5. References

1. Bear, J., 1979. Hydraulics of Groundwater, McGraw-Hill, New York, N.Y., 567 pp.
2. Bruington, A.E. and Seares, F.D., 1965. 'Operating a Sea Water Barrier Project', J. Irrig. Drain Div., Proc. Am. Soc. Civ. Eng., 91 (IR1), 117-140.
3. Bruington, A.E., 1968. 'The Amelioration or Prevention of Salt Water Intrusion in Aquifers - Experience in Los Angeles County, California', Louisiana Water Resources Res. Inst. Bull. 3, Louisiana State Univ., Baton Rouge, Louisiana, U.S.A., 153-168.
4. Das, S., 1993. 'Hydrogeological Considerations for Recharging Coastal Aquifers', In Proc. Workshop on Artificial Recharge of Ground Water in Coastal Aquifers held at Bhubaneswar, Orissa, India during March 27-28, 1993, 1-34.
5. Harpaz, Y., 1971 'Artificial Groundwater Recharge by Means of Wells in Isreal', J. Hydraul Div., Proc. Am. Soc. Civ. Eng., 97 (HY12), 1947-1964.
6. Hunt, B., 1985. Some Analytical solution for Sea Water Intrusion Control with Recharge Wells', J. Hydrol, 80, 9-18.
7. Mistry, J.F., 1989. 'Measures to Overcome the Salinity Problems of Coastal Areas in Gujarat', Eds. C.P. Gupta, S. Ahmed, V.V.S.G. Rao and M.T. Rajan, Vol. II, Oxford and IBH, New Delhi, 829-840.
8. Raju,T.S. Agashe, Romani, S. and Rao, S.V.R., 1993. Manual on Artificial Recharge of Groundwater, Central Ground Water Board, New Delhi, India.

9. Sa da Costa A.A.G. and Wilson, J.L. 1979. A Numerical Model of Sea Water Intrusion in Aquifers, Rep. No. 247, Ralph M. Parsons Lab for Water Resour. Hydrodyn., Dept. Civ. Eng., M.I.T. Cambridge, Massachusetts, U.S.A. 245 pp.
10. Schroeder, D.J. Harley, B.M. and Mejia, P.E.C., 1989. 'Sea Water Intrusion Barrier Evaluated with 3-D Ground water Models', Water Eng. Mgmt, 136 (2), 26-29.
11. Strack, O.D.L., 1976. 'A Single Potential Solution for Regional Interface Problems in coastal Aquifers', Water Resour Res., 12 (1) 1165 - 1174.
12. Vappicha, V.N. and Nagaraja, S.H., 1975. 'Steady State Interface in an Unconfined Aquifer with a Vertical Outflow Face', - In Proc. Natl. Symp. on Hydrology held at Univ. of Roorkee, Roorkee, India during Nov. 13-15, 1975, H7-H11.
13. Williams D., 1976. 'The Dashte-Naz Groundwater Barrier and Recharge Project', Ground Water, 15 (1), 23-31.

Application of Monte Carlo Analysis to Saturated Subsurface Contamination Modelling

A . Ghosh Bobba[1] and Vijay P. Singh[2]
[1]National Water Research Institute
Burlington, Ontario, Canada.

[2]Department of Civil Engineering
Louisiana State University
Baton Rouge, Louisiana, U.S.A. 70803-6405

Abstract

The prediction and evaluation of contamination in saturated subsurface environment by using contaminant transport models is necessary to plan remedial actions at existing sites and may also be useful in siting and designing new land-based waste treatment or disposal facilities. Most models used to make such predictions assume that the system behaves deterministically. A variety of factors, however, introduce uncertainty into the model predictions. The factors include model, parameters, and geometric uncertainty. The Monte Carlo technique is used to evaluate the uncertainty, as illustrated by applying three analytical saturated subsurface contaminant transport models. The uncertainty analysis provides estimates of statistical reliability in model outputs of contaminant concentration and arrival time. Examples are provided which demonstrate that (a) confidence limits around predicted values of concentration and arrival time can be obtained, (b) the selection of probability distributions for input parameters affects the output variables, and (c) the probability distribution of the output variables can be different from that of the input variables, even when all input parameters have the same probability distribution.

Introduction

Various saturated subsurface transport models have been developed (Bobba and Singh, 1994) which can describe the fate and transport of contaminants in the saturated subsurface system. A generic modelling procedure for most deterministic models involves: (1) production of a simplified conceptual model of the system of interest with assumptions of geometry and transport mechanisms; (2) development of sets of Partial Differential Equations (PDE) based on mass balances or momentum or energy balances for a more detailed description of the saturated subsurface flow field with initial and boundary conditions; (3) development of analytical and/ or numerical solutions; (4) sensitivity analysis to determine the relative importance of parameters; (5) estimation of parameters from literature or from experiment; (6) comparison of computed results with observed data; and (7) application of mathematical models for prediction and parameter estimation.

Saturated subsurface contaminant models are mathematical constructs that integrate a number of complex phenomena such as water transport, reaction kinetics, and external loadings. There are two basic reasons for constructing mathematical representations of natural aquatic ecosystems. First, there is a need to increase the current level of understanding regarding the cause-effect relationships operative in all aquatic environments. Second, models provide a synthesis of

V. P. Singh and B. Kumar (eds.), Water-Quality Hydrology, 197–213.

understanding that is increasingly important in the policy arena. Various types of errors can be introduced at each modelling step. The errors are due to the simplification of model geometry as well as initial and boundary conditions, the incomplete description of transport mechanisms, and variability associated with parameters requiring measurements.

Burges and Lettenmaier (1975) identified two components of uncertainty or error associated with mathematical modelling. Type A error results from the selection of an incorrect model with correct (deterministic) parameters. Type A uncertainty can be subdivided into inappropriate model selection and inherent modelling error due to process aggregation. Type B error assumes the use of a perfect model with parameters which are characterized by a degree of uncertainty.

Type B error has been investigated using both first-order uncertainty analysis and Monte Carlo simulation. Jaffe and Parker (1984) conducted a first-order uncertainty analysis of a simple first order decay model. First-order uncertainty analysis provides estimates of the mean and standard deviation for the dependent variable as functions of the mean and standard deviation associated with each random independent variable. The description of the output distribution will only be accurate for the unlikely case that the distribution is normal. In addition, the accuracy of the first-order approximation deteriorates as the coefficient of variation (standard deviation divided by mean) for the random variables increases in magnitude. Burges and Lettenmaier (1975) provide an excellent review and subsequent application of first-order uncertainty analysis to a simplified Streeter-Phelps equation. Currently, uncertainty analyses of complex systems are being conducted using Monte Carlo simulation techniques. This type of techniques requires knowledge of the statistical distribution of each independent variable together with its mean and variance. The Monte Carlo technique involves repeated simulation of a single model, each run being parameterized by unbiased selection of values of the independent variables from their respective statistical distributions. The process ends when sufficient output has been obtained to yield a clear statistical description of dependent variables. Details regarding the use of Monte Carlo analysis with various statistical distributions are given by Sobal (1974).

Monte Carlo analysis provides a different view of the model uncertainty from that derived by first-order analysis. Although both methods yield similar estimates of average values (Burges and Lettenmaier, 1975), variance estimates can be quite different. Scavia et.al (1981) point out that first-order analysis estimates variability about typical components of the modeled population whereas Monte Carlo methods gives variance estimates of the population mean. That is, Monte Carlo results describe expected variability in the system. Additionally, analysis of Monte Carlo results allows determination of the probability density function associated with an output variable. This in turn provides significant insight into total system behaviour.

This paper presents results derived from Monte Carlo simulations undertaken to quantify the effect of input data uncertainty (Type B error) on the estimation of saturated subsurface contamination. The objective of the Monte Carlo simulation is to account for the uncertainties which are inherently present in deterministic models. Since the purpose of this paper is to present modelling methodology and to demonstrate generic pattern results, a hypothetical subsurface system was used to illustrate the magnitude of expected model output uncertainty. Specific results include, the probability distribution of model output and a comparison of uncertainty levels associated with highly reactive and relatively inert contaminants. In addition, the impact of improper contaminant loading characterization (Type A error) on model output uncertainty is investigated.

One Dimensional Contaminant Transport Model in Vertical Direction

A contaminant or leachate may enter subsurface directly from a landfill site, or a contaminated pond from an industrial site. The one dimensional model equation used here incorporated advection, dispersion, adsorption, and first order decay (radioactive decay,hydrolysis and biodegradation) is as follows:

$$R_d \frac{\partial C}{\partial t} = D_z \frac{\partial^2 C}{\partial z^2} - V_z \frac{\partial C}{\partial z} - R_d k_e C \qquad (1)$$

with $C(z,t=0) = 0$
$C(z=0,t) = C_0 \exp(-k_d t)$
$\partial C(z=\infty,t)/\partial z = 0$

where, C is contaminant concentration (mg/l), C_0 is concentration strength at the source site (mg/l), V_z is the seepage velocity (m/day), D_z is dispersion coefficient (m^2/day), R_d is retardation coefficient (dimensionless), k_e is effective first order decay constant in subsurface system (1/day), k_d is decay rate constant at the surface, z=0, (1/day), Z is the distance in vertical direction (m), and t is the time (day). The solution of Equation (1) is given by van Genuchten and Alyes (1982) with initial and boundary conditions. The resulting solution is given as:

$$\frac{C(z,t)}{C_0} = \frac{1}{2}\exp\left(\frac{vz}{2D_z} - k_d t - 2ab\right) erfc\left(-at^{\frac{1}{2}} + \frac{b}{t^{\frac{1}{2}}}\right) +$$

$$\frac{1}{2}\exp\left(\frac{vz}{2D_z} - k_{dt} + 2ab\right) erfc\left(+at^{\frac{1}{2}} + \frac{b}{t^{\frac{1}{2}}}\right) \qquad (2)$$

where $a = \{k_e - k_d + v^2/(4D_z R_d)\}^{1/2}$

$b = 0.5z(R_d/D_z)^{1/2}$

erfc = complementary error function

Two Dimensional Horizontal Flow with a Slug Contaminant Source Model

The distribution of the contaminant concentration down gradient from a point discharge is considered in this model. The situation, which is conceptualized, is based on the assumption that the instantaneous discharge is uniformly distributed over the vertical dimension of the saturated subsurface system. In addition, concentration gradients at large distances ($x=\infty$, $y=\infty$) are zero. The contaminants, in the form of a vertical line source, are discharged into a saturated zone of depth B. Further, it is assumed that the saturated zone is infinitely extended with a uniform thickness.

The two dimensional partial differential equation governing the mechanisms of advection, dispersion, adsorption, effective first order decay, and instantaneous solute loading is given as:

$$R_d \frac{\partial C}{\partial t} = D_x \frac{\partial^2 C}{\partial x^2} + D_y \frac{\partial^2 C}{\partial y^2} - V_z \frac{\partial C}{\partial z} - R_d k_e C \tag{3}$$

with

$$\begin{aligned} C(x,y,t=0) &= 0 \\ C(x=0,y=0,t) &= M/n \text{ if } t=0 \\ &= 0 \text{ if } t>0 \\ \partial C(x=\infty,y,t)/\partial x &= 0 \\ \partial C(x,y=\infty,t)/\partial y &= 0 \end{aligned} \tag{4}$$

where M is the strength of discharge $C_0 Q'/B$, Q' is volume of contaminant discharge (m^3), B is the thickness of saturated zone (m), n is the porosity (dimensionless), D_x is longitudinal dispersion coefficient (m^2/day), D_z is transverse dispersion coefficient (m^2 /day), x is the distance in the direction of flow (m), and y is the distance in the direction normal to flow.

The solution to Equation (3) is presented as (Wilson and Miller, 1978)

$$\frac{C(x,y,t)}{C_0} = \frac{Q'}{4\pi n B t (D_x D_y)^{\frac{1}{2}}} \exp\left[-k_e t - \frac{(xR_d - vt)^2}{4R_d D_x t} - \frac{(yR_d)^2}{4R_d D_y t}\right] \tag{5}$$

As a special case, the time when the maximum concentration occurs is calculated by:

Two Dimensional Horizontal Flow with Continuous Pollutant Source Model

$$t_{max} = -\frac{R_d}{2\,(k_e R_d + \frac{v^2}{4D_x})}\,[1-\{1+(k_e R_d + \frac{v^2}{4D_x})\,(X^2 D_x + y^2 D_y)\}^{\frac{1}{2}}] \tag{6}$$

With the governing Equation (3), the two dimensional model for continuous discharge of a contaminant is derived by replacing Equation (4) with:

$$C(x=0,\ y=0,t) = C_0 Q/Bn, \qquad t \geq 0 \tag{7}$$

where Q is volumetric flow rate input to the saturated zone (m^3/day). The volume flow rate is estimated directly or computed from the product of area (m^2) of source and water recharge rate (m/day). If the first order decay of contaminant at the point (x=0,y=0) is considered, Equation (7) may be rewritten as:

$$C(x=0,y=0,t) = C_0\ Q\ \exp(-k_d t)/Bn, \qquad t \geq 0 \tag{8}$$

The solution of Equation (3) with the conditions expressed by Equations (7) and (8) is given as (Wilson and Miller, 1978):

$$\frac{C(x,y,t)}{C_0} = \frac{Q'}{4\pi nB\,(D_x D_y)^{\frac{1}{2}}}\ \exp\left(\frac{vx}{2D_x}\right)\, W\left(u, \frac{r}{B}\right) \tag{9}$$

where W(u,r/B) is the leaky well function of Hantush (1964), B is $2D_x/v$ (m), r is $\{\theta(x^2+y^2 D_x/D_y)\}^{1/2}$ (m), u is $r^2R_d/(4\theta D_x t)$ (dimensionless); θ is $1+2BR_d\ k_e/v$ (dimensionless).

The leaky well function, W(u,a) is of an infinite integral form and must be integrated, which is defined as:

$$W(u,a) = \int_u^{\infty} \frac{1}{w} \exp\left[-w - \frac{a^{\frac{1}{2}}}{4w}\right] dw \tag{10}$$

where a is r/B (dimensionless) and w is a dummy integration variable.

The well function can be integrated numerically by using the trapezoidal method, or the quadrature method. Tabulated values are reported by Bear and Verruijt (1987). For limited cases such as r/B larger than 1, approximate solutions are possible. However, this paper used the Gauss quadrature to more generally calculate the well function, since the quadrature method is easier and more convenient to use once it is computed. The infinite integral of the well function is replaced by Gauss quadrature formula of a finite form (Carnahan et. al, 1969).

$$W(u,a) = \frac{\underline{u}-u}{2}\sum_{i=1}^{NX}\left[W_i\left\{f\left(\frac{\underline{u}-u}{2}X_i + \frac{u+\underline{u}}{2}\right) + f\left(\frac{u-\underline{u}}{2}X_i + \frac{u+\underline{u}}{2}\right)\right\}\right] \quad (11)$$

where u is lower integration limit, $\underline{u}$ is upper integration limit, NX is number of quadrature points, x_i are locations of collocation points which are the roots to a family of orthogonal polynomials, and W_i are weighting factors. More information on the generation of X_i and W_i is available from Finalyson (1980). This paper utilized the quadrature constants generated by using the Jacobian polynomial. The quadrature method provided accurate results even with a collocation points used in the simulation. As a special solution at very large times, Equation (9) is simplified as:

$$\frac{C(x,y,t=large)}{C_0} = \frac{Q}{2nB(\pi v D_y r)^{\frac{1}{2}}}\exp\left[\frac{vx}{2D_x} - \frac{r}{B}\right] \quad (12)$$

Equation (12) may be used to calculate a Dilution Attenuation Factor (DAF). For a location of well, the value of C/C_0 or DAF is always constant deterministically. An allowable concentration of a contaminant at a discharge point is back calculated by dividing a Maximum Contamination Level (MCL) by the DAF.

Application of Monte Carlo Simulations

Uncertainty in the prediction of contaminant arrival time and concentration is examined by applying the Monte carlo method to three analytical models. The example of contamination of concern in this analysis is chloroform and its contamination scenario for the individual models. First, random input parameters used in the simulation are introduced. Here, two schemes of assigning probability distributions to the random input parameters are considered: a) normal distribution for all the random parameters, and b) a combination of different probability distributions. With the two schemes, each analytical model simulates the contaminant arrival time of a specified contaminant concentration and the concentration breakthrough curve at a location of interest. In each simulation run, 95% confidence limits for the response variables are calculated. Finally, the types of probability distributions for the response variables are identified by a Goodness-of Fit Test (GFT).

Input Data for Monte Carlo Simulation Runs

The following simulations use the parameters listed in Tables 1 and 2. Table 1 presents the non-random parameters used in the simulation. Table 2 provides the assumed values of parameters and the types of probability distributions selected for this simulation.

The contaminant-specific parameters are estimated based on the literature sources as far as possible. The parameter values of hydrolysis for chloroform are obtained from the USEPA document 'Screening hydrolytic of OSW chemicals. The partition coefficient is estimated by using Karickhoff et.al's (1979) equation with the information on the octanol/water partition coefficient from the USEPA (1982). The biodegradation decay rate constant is approximated from Table 3 since no adequate value is currently available. These reaction-related parameters are of course site specific, i.e., spatially-varying. However, it is assumed that they are non-random except for the biodegradation rate constant.

The subsurface system parameters must be estimated from field or laboratory data for the true prediction value of a model. However, the collection and the analysis of data are expensive and time consuming. This lack of real data leads to the difficulty in estimating parameters and identifying the probability distribution of individual input parameters. Against this difficulty, this paper assigns these parameters and their probability distributions. Different choices for these could have been made. As shown in Table 2, Scheme I assigns the normal distribution to all the random parameters. Scheme II arbitrarily mixes different probability distributions for the random parameters.

The extent to which a parameter is uncertain is expressed in terms of the Coefficient of Variation (CV) defined as the ratio of standard deviation to mean. Although the CV varies from parameter to parameter, this paper assumes that each CV is 10%. However, as noted in Table 2, the CV of the soil diameter is 100% for scheme II. The reason is that the mean and the standard deviation of the exponential distribution should be equal to each other. For certain soil physical parameters such as soil porosity, temperature, and pH, the assumption of 10% may not be unrealistic. Jury (1986) reports about 10% of CV for soil porosity and Ph.

Means and standard deviations resulting from the Monte Carlo simulation are listed in Table 4. In addition to the parameters listed in Table 2, means and standard deviations for the derived parameters are also shown in Table 4. The estimated means and standard deviations can be compared to the statistics of random parameters for each probability distribution in Table 2. Note that different realizations of probability distributions are possible which also lead to different results of the Monte Carlo simulations.

In the next section, uncertainty analysis using the input values described above are demonstrated. Each model computes the arrival time of chloroform and the time variation of concentration at a location of concern. Each simulation considers two schemes of probability realization. If possible, the goodness-of-fit test is applied to examine the type of probability density function for the arrival time and concentration at a specified time and location.

Results

Concentration Breakthrough Curves

One Dimensional Model:

Figure 1 shows the breakthrough curves with 95% confidence bands simulated by the one-dimensional vertical solute transport model. Chloroform concentrations at the water table, z=15m, are plotted versus time. The curves in Figure 1(a) were obtained using scheme I, while the curves in Figure (1b) used scheme II. At the time of 43 days, the confidence interval for 6 mg/l is approximately 3 to 9 mg/l as shown in Figure 1(a). In contrast, if scheme II is chosen, as in Figure 1(b) contamination arrives within 37 days. The reason is that the estimated velocity for scheme II is larger than the one for scheme I.

Two Dimensional Model with a Slug Source:

Figure 2 shows the transient transport of a contaminant slug at X=100, Y=0. For scheme I, maximum concentration occurs at about 50 days as shown in Figure 2(a), while for scheme II, arrival is at about 85 days as shown in Figure 2(b). The maximum concentrations for both cases never exceed 6 mg/l. The reason is that the contaminant plume is spread due to dispersion and adsorption. A portion of chloroform is also lost by the effective decay action, hydrolysis and biodegradation. Since the velocity and the dispersion coefficient for scheme II are higher than those for scheme I, the mass centre arrives earlier and the curve is spread more. Further, the curve in Figure 2a appears to be more bell shaped than the curve in Figure 2(b). It is possible that the section of non-normal probability distributions for some parameters may cause more skewed curves.

Two Dimensional Model with Continuous Contaminant Source:

Figure 3 shows the contaminant concentration profiles simulated by the two-dimensional model with a continuous source. As shown in Figure 3(a), the 6 mg/l level of chloroform arrives about 110 days after the spill started, and at this time the 95% confidence interval for the mean of 6 mg/l falls between 5.4 and 6.8 mg/l. However, the contaminant level of 6 mg/l does not appear in the scheme II simulation as shown in Figure 3(b). In both cases, a maximum concentration decreases slowly. This gradual decline is caused by the effective decay effect.

It should be noted that the estimated contamination arrival times in this paper may be unrealistically short for some sites. The hydraulic gradient of 10% used in the simulations may be to steep, and also the soil particle size of 0.1mm, classified as fine sand, produces relatively large values of hydraulic conductivity.

The goodness-of-fit test was used to identify a probability density function that appropriately describes the distribution of concentration and arrival time. The test can conduct null-hypothesis tests for ten probability density functions using the statistics of Chi-square and Kolmogorov-Smirnov. The significance level of testing the goodness-of-fit is internally controlled at maximum 25%. The following distributions may be used: (a) Normal distribution, (b) Gamma

Table 1: Non Random Basic Parameters

Parameter	Value	Parameter	Value
Ratio (D_x/D_y)	5.00	Soil particle density (g/cm^3)	2.65
Decay rate at surface (1/day)	0.00	Reference temp. for hydrolysis (^0C)	25.00
Acid catalyzed hydrolysis (l/Mole-day)	0.00	Reference temp. for bio-decay (^0C)	20.00
Base catalyzed hydrolysis (l/Mole-day)	5.52	Distance in x-direction (m)	15.00 100.00
Neutral hydrolysis (1/day)	7.20×10^{-8}	Distance in y-direction (m)	0.00
Partition coefficient (cm^3/g)	58.80	Maximum contaminant level (mg/l)	6.00 0.60

Table 3: Biodegradation rates under aerobic and anaerobic conditions

Organism concentration		Degradation (1/day)		Degradation (1/day)	
mg/l	No/l	aerobic	anaerobic	aerobic	anaerobic
10.000	10^7	50.000	5.0000	0.014	0.140
1.000	10^6	5.000	0.5000	0.140	1.400
0.100	10^5	0.500	0.0500	1.400	14.000
0.010	10^4	0.050	0.0050	14.000	140.000
0.001	10^3	0.005	0.0005	140.000	1400.000

Table 2: Probability density functions for random parameters and their assumed statistics

	Type of PDF		Statistics		
Parameter	Scheme I	Scheme II	Mean	Standard Deviation	Remark
Longitudinal dispersivity (m)	-	-	-	-	(1)
Concentration strength (mg/l)	Normal	normal	1000.00	100.000	
Soil particle diameter (cm)	Normal	Exponential	0.01	0.0010	(2)
Hydraulic gradient (dim.)	Normal	Normal	0.10	0.0100	
Fraction of organic content	Normal	Lognormal	0.005	0.0005	
pH	Normal	Triangular	6.200	0.6200	(3)
Porosity (dim.)	Normal	Triangular	0.400	0.0400	(4)
Aquifer temperature (°C)	normal	Triangular	15.000	1.5000	(5)
Aquifer thickness (m)	Normal	Uniform	15.000	1.5000	(6)
Discharge volume (m^3)	Normal	Normal	100.000	10.000	(7)
Biodegradation decay (1/day)	Normal	Normal	0.050	0.0050	

Remarks:
(1) For dispersion coefficient, the scale dependent method is used. (D_x =0.1x)
(2) Scheme II, diameter follows 2-parameter exponential (mean=standard deviation = 0.01)
(3) Scheme II, Ph follows (min=4, max=8, mode=6.2)
(4) Scheme II, porosity follows (min=0.30, max=0.50, mode=0.35)
(5) Scheme II, temperature ranges (min=10.00, max=15.00, mode=12.00)
(6) Scheme II, thickness follows uniform (min=10.00, max=20.00, mode=15.00) and no need for 1-D (7) No need for 1-D model.

Table 4: Probability density functions for random parameters and their assumed statistics

Parameter	SCHEME(I) Mean	SCHEME(I) Standard deviation	SCHEME(II) Mean	SCHEME(II) Standard deviation
A: Basic Random Parameters				
* Concentration strength (mg/l)	993.80	103.379	993.800	103.379
* Soil particle diameter (cm)	0.010	0.00105	0.00979	0.01119
* Hydraulic gradient (dim.)	0.09827	0.00978	0.09827	0.00978
* Fraction of organic content	0.00503	0.00048	0.00498	0.00005
* Ph	6.250	0.60829	6.13730	0.83170
* Porosity (dim.)	0.39917	0.04149	0.38074	0.03822
* Aquifer temperature (°C)	14.989	1.56670	12.31400	1.08690
* Aquifer thickness (m)	14.9800	1.56670	15.05200	2.88630
* Discharge volume (m^3)	100.4500	9.8440	100.4500	9.84440
* Biodegradation decay (1/day)	0.04963	0.00517	0.04963	0.00517
B: Derived Random Parameters				
* Pore water Velocity(m/day)	1.9054	0.77370	3.8293	10.5210
* Long. Dispersion Coefficient (m^2/d)	19.0540	7.73720	38.2930	105.2100
* Effective decay decay coeff.(1/day)	0.0352	0.00515	0.0913	0.0034
* Retardation Coefficient	2.2200	0.24489	2.1737	0.0232
* Hydraulic conductivity (m/day)	8.0028	3.94330	9.7880	19.9410

Table 5: Results of Goodness of Fit Test for the outputs of contamination arrival time and concentration

Scheme	Model	Response variable	Probability distribution
		Arrival Time	Beta-Kappa, Beta-P, Lognormal
		Concentration (t=2.5 day)	Weibull
	1-D	Concentration (t=5 day)	Weibull
		Concentration (t=7.5 day)	Beta-Kappa,Weibull
I: All Normal		Concentration (t=30day)	Not applicable
	2-DS	Concentration (t=60 day)	Not applicable
		Concentration (t=90 day)	Weibull, Lognormal
		Concentration (t=100 day)	Gamma, Beta-type,Lognormal
	2-DC	Concentration (t=200day)	Beta-Kappa,Lognormal
		Concentration (t=300 day)	Beta-Kappa, Lognormal
II:Mixed	1-D	Arrival Time	Beta-Kappa, Beta-P, Lognormal
	2-DS		Not applicable
	2-DC		Not applicable

Remarks:

1-D one dimensional model
2-DS Two dimensional model with slug source
2-DC Two dimensional model with continuous source

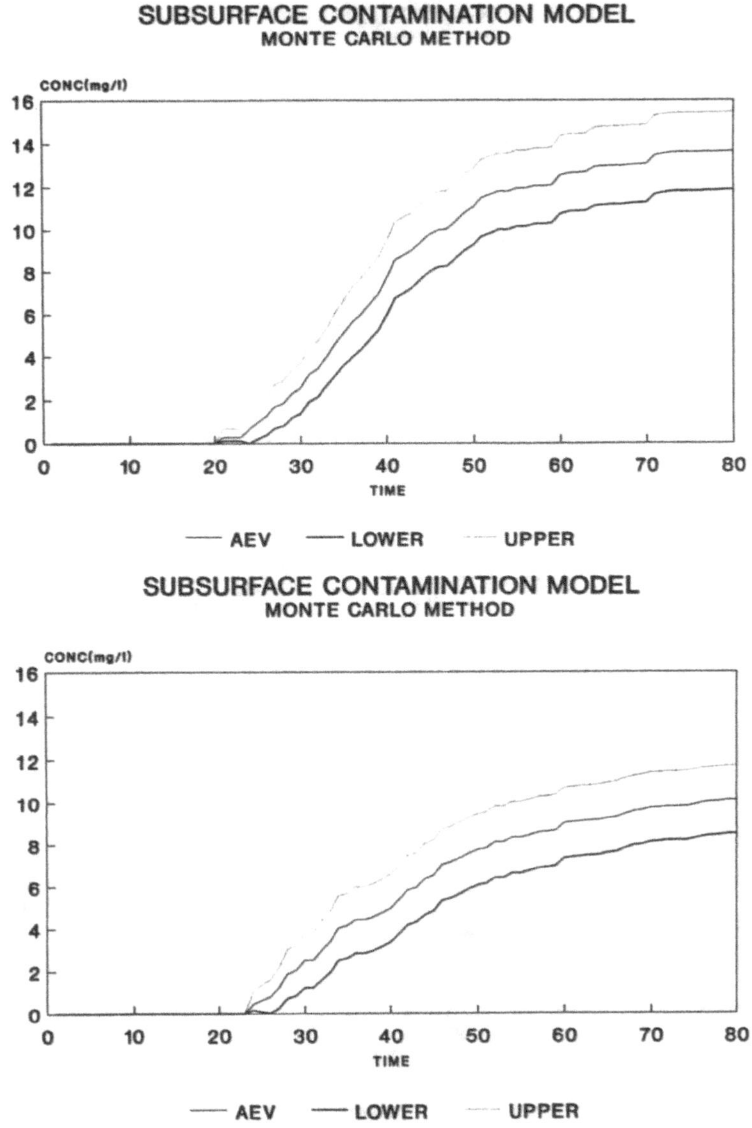

Figure 1: 95% confidence interval of contaminant breakthrough curves computed by one dimensional vertical transport model. (a) with scheme I of probability assignment; (b) with scheme II of probability assignment.

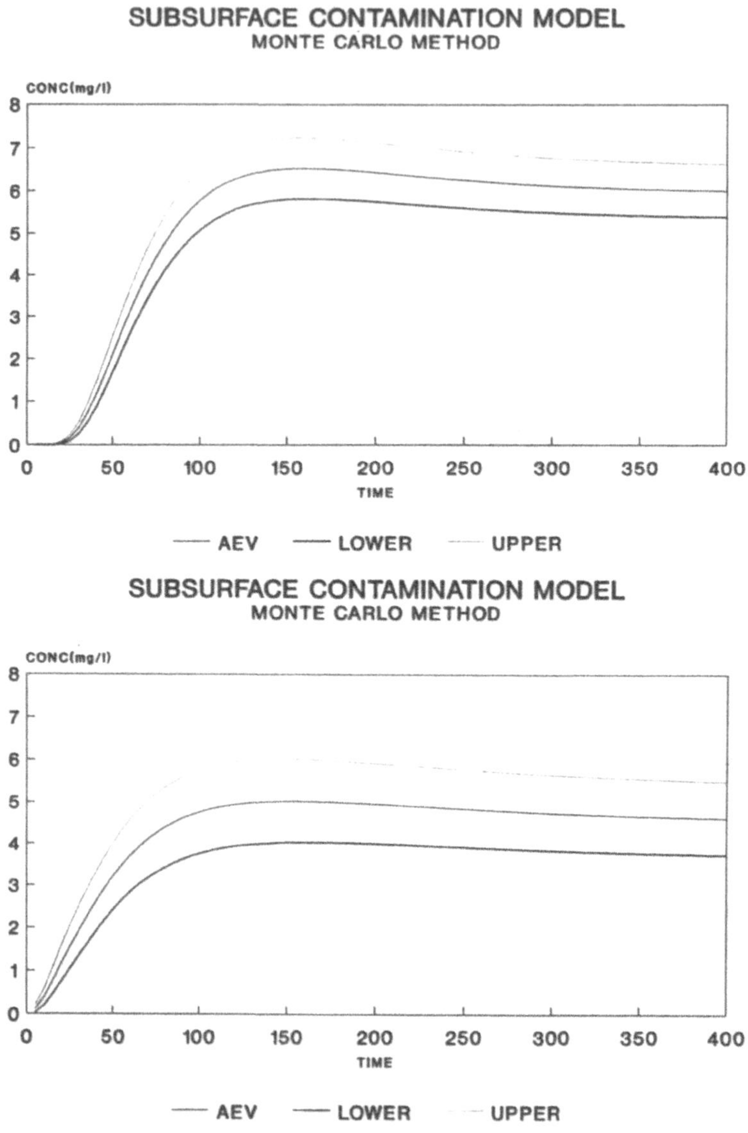

Figure 2: 95% confidence interval of contamination breakthrough curves computed by two dimensional horizontal model with slug source. (a) with scheme I of probability assignment, (b) with scheme II of probability assignment.

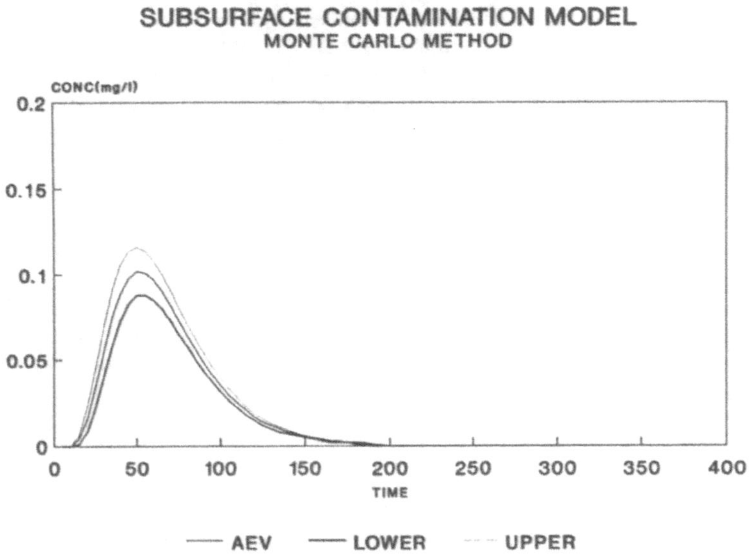

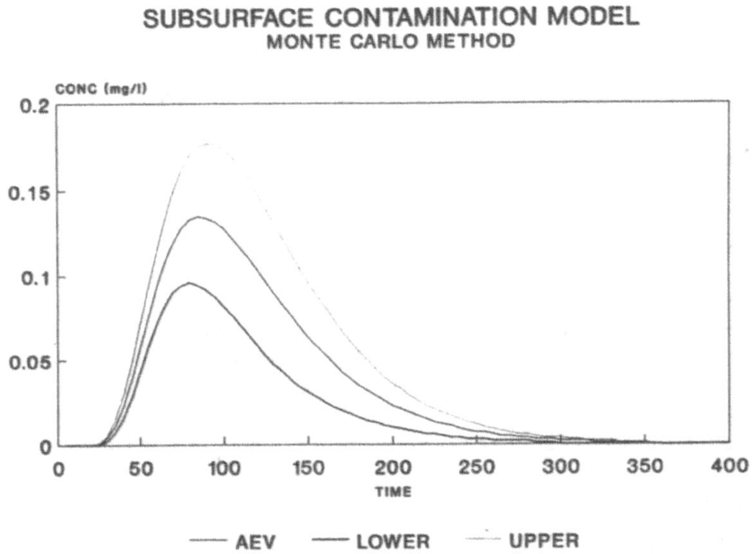

Figure 3: 95% confidence interval of contamination breakthrough curve simulated by two dimensional horizontal model with continuous source. (a) with scheme I of probability assignment, (b) With scheme II of probability assignment.

distribution,(c) Extreme value type-I distribution: The maximum values, (d) Beta distribution, (e) Beta-Kapa distribution, (f) Beta-P distribution, (g) Pareto distribution, (h) Lognormal distribution, (i) Log-Pearson Type-III distribution, (j) Log-Extreme value distribution: Log-Gumbel and Weibull. The type of probability density function of the response variable depends on the mathematical structure of the model and the probability functions assigned to input parameters. Although a normal distribution is assigned to all the random input parameters, an output variable may not follow the normal distribution.

Conclusions

This paper presents the application of Monte Carlo analysis with three existing analytical contaminant transport models. The models consider the transport mechanisms of advection, dispersion, adsorption, radioactive decay, hydrolysis, and biodegradation. The Monte Carlo technique can be used to examine the uncertainty in output variables, either contamination arrival time or concentration. In the example, two schemes of assigning probability distributions are used: normal for all input parameters and mixed distributions. The two schemes show different results which suggests that the probability function for each input parameter must be carefully selected. It is demonstrated that the type of probability distribution for arrival time or concentration is dependent on the mathematical structure of the model and the selected probability functions.

Based on the results of the Monte Carlo simulation demonstrated in this paper, the following are recommended for further study are, (a) the simulation used in this paper has not been validated against any independently measured values of concentrations and such validation step is essential, (b) for the Monte Carlo simulation, it is most important to identify an adequate probability density function for each input parameter. Sometimes, the scarcity of information requires assumptions of probability distributions for individual random parameters. The belief that hydraulic conductivity may follow lognormal distribution is useful. According to the Kozeny-Carman equation, hydraulic conductivity is a function of porosity and particle diameter. From this, at least the distributions of porosity and particle diameter may be identified. (c) Information on reaction rate constants, especially, biodegradation decay constants for organic chemical, is needed. (d) A input file should be prepared to store the rate data so that the required data can be accessed easily for use in a subsurface flow and contaminant transport model.(e) This model can analyze the uncertainty in contaminant arrival time. However, it has limits when a break through curve is not an ever increasing function. A better root finding scheme other than the bisection method should be selected or developed for inclusion into the model.

Acknowledgments

We thank Dean Jeffries and G. Padmanabhan for helpful comments on early drafts of this paper. The encouragement and support from John Carey is also appreciated.

References

Bear, J. and Verruijt, A. 1987. Modelling Groundwater Flow and Pollution, D. Reidel Pub., Norwell, Ma. U.S.A.

Bobba, A.G and Singh, V.P, 1994. Groundwater contamination: Modelling. Chapter, Environmental Hydrology. (In preparation)

Burges, S.B and Lettenmaier, D.P.,1975. Probabilistic methods in stream water quality management. Water Resources Bulletin, vol. 11, pp. 115-130.

Carnaham, B., Luther, H.A., and Wilkes,J.O., 1969. Applied Numerical Methods. John Wiley and Sons, New York.

Finlayson, B.A. 1980. Nonlinear Analysis in Chemical Engineering, McGraw-Hill, New York.

Freeze, R.A. and Cherry, J.A. 1979. Groundwater, Prentice-Hall, Inc., Inglewood Cliffs, Nj.

van Genuchten, M.Th. and Alves, W.J., 1982. Analytical solutions of the one-dimensional convective dispersive solute transport equation. U.S. Department of Agricultural, Technical Bulletin No. 1661.

Hantush, M.S.,1964. Hydraulics of wells. Advances in Hydrosciences. vol.1,pp.282-432.

Jaffe, P.R.,and Parker, F.L., 1984. Uncertainty analysis of first order decay models. Journal of Environmental Engineering Division, ASCE, vol.110, pp.131-140.

Jury, W.A. 1986. Spatial variability of soil properties. Vadose zone modeling of organic pollutants. edited by S.C. Hern and S.M. Melancon, Lewis Publishers, Inc., Chelsea, Michigan.

Karcickhoff, S.W., Brown, D.S., and Scott, T.A. 1979. Sorption of hydrophobic pollutants on natural sediments. Water Research, vol. 13, 241-248.

Scavia, D.,1981. Comparison of First Order Error Analysis and Monte Carlo simulation in time dependent Lake Eutrophication Models. Water Resources Research. vol. 17, pp.1051-1059.

Sobal, I.M., 1974. The Monte Carlo Method. The University of Chicago Press. Chicago. p.66.

Wilson, J.L., and Miller, P.J., 1978. Two dimensional plume in uniform groundwater flow. American Society of Civil Engineers, Journal of Hydraulics Division. vol. 104, pp. 503-514.

SECTION 6:

WATER QUALITY MODELING

WATER QUALITY MODELS FOR WATERSHED MANAGEMENT

J. R. Williams and J. G. Arnold
USDA-Agricultural Research Service
808 East Blackland Road
Temple, Texas 76502
USA

ABSTRACT. Several water quality models are available for use in assessing the effects of agricultural management on the environment. Since these models are being used to solve a variety of water quality problems, they vary considerably in structure and complexity. For example, simple screening models may be adequate and appropriate for identifying potential pollutant sources. However, more comprehensive models are needed in comparing agricultural management effects on chemical transport by runoff and sediment. Model requirements may also vary depending upon temporal and spatial scales, cost, and risk associated with proposed projects.

Some of the most widely used water quality models, particularly those most useful in agricultural management are described briefly. Two of the models, the Erosion-Productivity Impact Calculator (EPIC) and the Simulator for Water Resources in Rural Basins (SWRRB), are presented in more detail to serve as examples of field and watershed scale water quality models. These models were selected because they feature convenient and comprehensive agricultural and soil management components. For example, EPIC is useful in solving management problems involving crop varieties and rotations, tillage, furrow diking, irrigation, drainage, fertilization, pest control, weather variation, atmospheric CO_2 concentration, erosion (wind and water), water quality (nutrients and pesticides), manure handling, crop residue management, liming, and grazing. The model operates on a daily time step and is capable of simulating hundreds of years if necessary. It is also useful in solving short term (within growing season) management problems operating in a real time mode. The SWRRB model was designed for solving watershed scale problems like water supply and quality (nutrients and pesticides), pond and reservoir design, groundwater flow contributions, irrigation water transfer, and stream channel routing of sediment and agrichemicals. SWRRB also operates on a daily time step and allows watershed subdivision. Subdivisions are made to account for spatial variability of soils, land use, weather, and topography. This gives SWRRB the capability to estimate off-site impacts including channel and reservoir deposition and total water supplies. Example applications of EPIC and SWRRB to water quality problems are also described.

1. Simulation Models

Mathematical models are among the best tools available for analyzing water quality issues. They can project the consequences of alternative management, planning, or policy-level activities and substantially reduce the cost of managing water resources (Office of Technology Assessment,

V. P. Singh and B. Kumar (eds.), Water-Quality Hydrology, 217–241.

1982). Models are also useful for absolute predictions, relative response, developing designs, organizing concepts, improving the understanding of the system, and dealing with parameter uncertainties.

There is also an increasing concern over the off-site impacts of nonpoint source pollution including lake water quality and instream nutrient and toxic concentrations. To estimate off-site loadings, the ability to simulate large basins with heterogeneous soils, land use, and topography is required. Several basin scale water quality models have recently been developed that simulate spatial variability within a watershed. However, these models have been limited by several factors including: 1) computer speed; 2) computer memory; and 3) availability of inputs. These limitations have produced models falling into one of the following three categories:

- Continuous time models with natural subwatershed boundaries that require considerable lumping of subwatershed inputs;
- Single event models subdivided into grid cells that allow more spatial detail;
- Continuous time, spatial models that are so complex that obtaining required inputs inhibits their general use.

Several models were chosen as examples to illustrate the variety of model configurations. The models are placed into three categories for discussion--field scale, watershed scale event, and watershed scale continuous.

1.1 FIELD SCALE

1.1.1 *Creams*. The Chemicals, Runoff, and Erosion from Agricultural Management Systems (CREAMS) model (Knisel, 1980) is a physically based, daily simulation model that estimates runoff, erosion/sediment transport, plant nutrient, and pesticide yield from field-sized areas. The hydrologic component consists of two options. When only daily rainfall data are available to the user, the SCS curve number model is used to estimate surface runoff. If hourly or breakpoint rainfall data are available, an infiltration-based model is used to simulate runoff. Both methods estimate percolation through the root zone of the soil.

The erosion component maintains elements of the USLE, but includes sediment transport capacity for overland flow. A channel erosion/deposition feature of the model permits consideration of concentrated flow within a field. Impoundments are treated in the erosion component also. The plant nutrient submodel of CREAMS has a nitrogen component that considers mineralization, nitrification, and denitrification processes. Plant uptake is estimated, and nitrate leached by percolation out of the root zone is calculated. Both the nitrogen and phosphorus parts of the nutrient component use enrichment ratios to estimate that portion of the two nutrients transported with sediment. The pesticide component considers foliar interception, degradation, and washoff, as well as adsorption, desorption, and degradation in the soil. This method, like the nutrient model, uses enrichment ratios and partitioning coefficients to calculate the separate sediment and water phases of pesticide loss. The CREAMS model is applicable for a field having (1) a single land use; (2) relatively homogeneous soils; (3) spatially uniform rainfall, and (4) a single management system, such as terraces. Normally, a field is less than 100 ha.

CREAMS can estimate the impact management systems, such as planting dates, cropping systems, irrigation scheduling, and tillage operations, have on sediment and nutrient movement. The model is also useful in long-term simulations for pesticide screening of management systems.

1.1.2 *Gleams*. The Groundwater Loading Effects of Agricultural Management Systems (GLEAMS) model (Leonard et al., 1987) is a continuous simulation, field scale model which was developed as an extension of the CREAMS model. GLEAMS assumes that a field has homogeneous land use, soils, and precipitation. It consists of three major components: hydrology, erosion/sediment yield, and pesticide transport. GLEAMS was developed to evaluate the impact of management practices on potential pesticide leaching within, through, and below the root zone. It also estimates surface runoff and sediment losses from the field. GLEAMS can be used to assess the effect of farm level management decisions on water quality.

GLEAMS considers soil properties and weather characteristics in estimating management impacts, such as planting dates, cropping systems, irrigation scheduling, and tillage operations, have on pesticide movement. The model is also useful in long-term simulations for pesticide screening of soil/management systems.

The model tracks movement of pesticides with percolated water, runoff, and sediment. Upward movement of pesticides and plant uptake are simulated with evaporation and transpiration. Degradation into metabolites is also simulated for compounds that have potentially toxic products. Erosion in overland flow areas is estimated using a modified Universal Soil Loss Equation. Erosion in channels and deposition in temporary impoundments such as tile outlet terraces are considered in estimating sediment yield at the edge of the field.

1.1.3. *EPIC*. The EPIC model (Williams et al., 1984) was developed in the early 1980's to assess the effect of erosion on productivity. Since the 1985 RCA application, the model has been expanded and refined to allow simulation of many processes important in agricultural management (Sharpley and Williams, 1990).

EPIC is a continuous simulation model that can be used to determine the effect of management strategies on water quality. The drainage area considered by EPIC is generally a field-sized area, up to 100 ha, where weather, soils, and management systems are assumed to be homogeneous. The major components in EPIC are weather simulation, hydrology, erosion-sedimentation, nutrient cycling, pesticide fate, plant growth, soil temperature, tillage, economics, and plant environment control.

EPIC can be used to compare management systems and their effects on nitrogen, phosphorus, pesticides and sediment. The management components that can be changed are crop rotations, tillage operations, irrigation scheduling, drainage, furrow diking, liming, grazing, manure handling, and nutrient and pesticide application rates and timing.

1.1.4 *NLEAP*. Nitrate Leaching and Economic Analysis Package (NLEAP) (Shaffer, 1991) is a field scale computer model that was developed to provide a rapid and efficient method of determining potential nitrate-N leaching associated with agricultural practices. Basic information concerning farm management practices, soils, and climate are translated into N budgets and nitrate-N leaching indices. The model also estimates potential nitrate-N leaching below the root zone and to groundwater supplies, the potential off-site effects of leaching, and the economic impacts of leaching.

NLEAP uses a three-phase approach to determine leaching potential: an annual screening analysis and the more detailed, monthly and event-by-event analyses. The processes modeled include movement of water and nitrate-N, crop uptake, denitrification, ammonia volatilization, mineralization of soil organic matter, nitrification, and mineralization-immobilization associated with crop residue, manure, and other organic wastes.

The screening procedure uses a simplified annual water and nitrogen budget and is designed to give only a general estimate of potential leaching of nitrate-N. The monthly budget analysis calculates leaching with consideration for the seasonal and monthly effects of precipitation, temperature, evapotranspiration, and farm management. The event-by-event analysis provides the best estimate of nitrate-N leaching. Its water and nitrogen budgets track the impacts of each precipitation, irrigation, fertilization, and tillage event on potential nitrate-N leaching.

1.1.5. *PRZM*. Pesticide Root Zone Model (PRZM) (Carsel et al., 1984) is a one-dimensional, dynamic, compartmental model that can be used to simulate chemical movement in unsaturated zone within and immediately below the plant root zone. Hydrology and chemical transport are the major components. The hydrology component calculates runoff and erosion based upon the Soil Conservation Service curve number procedure and the Universal Soil Loss Equation, respectively. Evapotranspiration is estimated directly from pan evaporation or by an empirical formula if pan evaporation data is not available.

Pesticide application on soil or on the plant foliage are considered in the chemical transport simulation. Dissolved, adsorbed, and vapor-phase concentrations in the soil are estimated by simultaneously considering the processes of pesticide uptake by plants, surface runoff, erosion, decay, volatilization, foliar washoff, advection, dispersion, and retardation. PRZM considers pulse loads, predicts peak events, and estimates time-varying mass emission of concentration profiles, thus overcoming limitations of the more commonly used steady-state models.

1.1.6. *RUSTIC*. Risk of Unsaturated/Saturated Transport and Transformation of Chemical Concentrations (RUSTIC) (Dean et al., 1989) links three subordinate models in order to predict pesticide fate and transport through the crop root zone, and saturated zone to drinking water wells through PRZM, VADOFT, SAFTMOD.

VADOFT is a finite-element model for simulating moisture movement and solute transport in the vadose zone. The model simulates one-dimensional, single-phase moisture and solute transport in unconfined, variably saturated porous media. Transport processes include hydrodynamic dispersion, advection, linear equilibrium sorption, and first-order decay. VADOFT predicts infiltration or recharge rate and solute mass flux entering the saturated zone. Parent/daughter chemical relationships may be simulated.

SAFTMOD performs two-dimensional simulations in an areal plane or a vertical cross section. In addition, the model can also perform axisymmetric simulations. Both single (unconfined and confined) and leaky two-aquifer systems can be handled. Transport of dissolved contaminants may also be simulated within the same domain. Transport processes accounted for include hydrodynamic dispersion, advection, linear equilibrium sorption, and first-order decay. Parent/daughter chemical relationships may be simulated.

1.2 WATERSHED SCALE EVENT MODELS

1.2.1 *AGNPS*. Agricultural Nonpoint Source Pollution Model (AGNPS) was developed (Young et al., 1987) to analyze nonpoint source pollution in agricultural watersheds. AGNPS uses a distributed parameter approach by dividing a watershed into square grid areas called cells. Cell sizes are selected by the user (normally 0.4 to 16 ha) and the model can be applied on watersheds up to 20,000 ha. Runoff is calculated for each cell using the SCS curve number method (USDA Soil Conservation Service, 1972) and sediment yield for each cell is estimated with the Universal Soil Loss Equation (Wischmeier and Smith, 1978) adjusted for slope shape.

AGNPS can compare the effects of implementing various conservation alternatives within the watershed. Cropping systems, fertilizer application rates and timing, point source loads, contributions from feedlots, and the effect of terraced fields can be modeled.

The model partitions soluble nitrogen and phosphorus between surface runoff and infiltration. Chemical oxygen demand and soluble nutrient contributions from feedlots are transported with runoff. When the soluble pollutants reach concentrated flow, they are conservative and accumulate in the flow. Sediment transported nitrogen and phosphorus are also determined. A modified Universal Soil Loss Equation, adjusting for slope shape, predicts local sediment yield within the originating cell. An estimate of gully erosion occurring in a cell can be added to the total amount of sediment yield in the cell. Sediment and runoff routing through impoundment terrace systems are also simulated.

1.2.2 *ANSWERS*. Area Nonpoint Source Watershed Environment Response Simulation (ANSWERS) (Beasley and Huggins, 1982) uses a distributed parameter approach and is limited to a single storm event. ANSWERS consists of a hydrology model, a sediment transport model, and several routing components. The conceptual basis of the model was developed by Huggins and Monke (1966). Hydrology model components include rainfall interception, infiltration, surface detention, and surface retention. A sediment continuity equation is employed which describes the process of soil detachment, transport, and deposition (Foster and Meyer, 1972). A watershed being modeled is divided into a series of small independent elements. The size of the elements normally ranges from 1 to 4 hectares and watershed size is limited to approximately 10,000 ha because of computational and input preparation time. The use of small elements allows considerable spatial detail in representing topography, soils, and land use. However, the building of input files and interpreting output requires significant time and considerable knowledge of the model and its operation (Engel and Arnold, 1991).

1.3 WATERSHED SCALE, CONTINUOUS TIME MODELS

1.3.1 *HSPF*. The Hydrological Simulation Program - FORTRAN (HSPF) (Johansen et al., 1984) simulates watershed hydrology and sediment yield. Data requirements for HSPF are extensive (Donigian and Huber, 1990). HSPF is a continuous simulation model and requires continuous data (generally hourly rainfall is required) to drive the simulations. The watershed is divided into land segments and stream channel segments. HSPF has been applied to the 68,000 square mile watershed draining into Chesapeake Bay. Although the watershed can be subdivided, the land segments are normally large enough to require considerable lumping of inputs.

HSPF uses such information as the time history of rainfall, temperature, solar intensity, and parameters related to land use patterns, soil characteristics, and agricultural practices to simulate the processes that occur in a watershed. The initial result of an HSPF simulation is a time history of the quantity and quality of water transported over the land surface and through various soil zones down to the groundwater aquifers. Runoff flow rate, sediment loads, nutrients, pesticides, toxic chemicals and other quality constituent concentrations can be predicted. The model then takes these results and information about the receiving water channels in the watershed and simulates the processes that occur in these channels. This part of the simulation produces a time history of water quantity and quality at any point in the watershed.

1.3.2 *SWRRB*. Simulator for Water Resources in Rural Basins (SWRRB) was developed to predict the effect of alternative management decisions on water and sediment yields with reasonable accuracy for ungaged, rural basins (Arnold et al., 1990; Williams et al., 1985). The model was developed by modifying the CREAMS daily rainfall model (Knisel, 1980) for application to large, complex, rural basins. The major changes involved were (a) the model was expanded to allow simultaneous computations on several subwatersheds, and (b) components were added to simulate weather, return flow, pond and reservoir storage, crop growth, transmission losses, and sediment movement through ponds, reservoirs, streams, and valleys. SWRRB operates on a daily time step and is efficient enough to run for many years (100 or more). Since the model is continuous time, it can determine the impacts of management such as crop rotations, planting and harvest dates, and chemical application dates and amounts. Basins can be subdivided into subareas based on differences in land use, soils, topograhpy, and climate. Soil and associated chemical are then routed to the basin outlet. SWRRB has been validated on basins up to 500 square kilometers (Arnold and Williams, 1987). Since SWRRB allows a limited number of subareas, some lumping of inputs is required.

1.3.3 *ROTO*. A model called ROTO (Arnold, 1990) was developed to estimate water and sediment yields on large basins (several thousand Km^2). ROTO is a continuous time model operl'ating on a daily time step that accepts inputs from continuous time soil water balance models including SWRRB, EPIC, and GLEAMS. ROTO will also accept point sources and withdrawals, measured data, and output from another ROTO run. ROTO uses a command structure to route and add flows down the watershed through channels and reservoirs and operates on continuous (daily time step) allowing management decisions to be evaluated. ROTO also has several advantages over SWRRB including: 1) greater spatial detail; 2) improved watershed routing structure; and 3) greater flexibility, allowing input from several models, point sources, and measured data.

2. Example Models

One of the major strengths in the modern comprehensive water quality model is the ability to simulate processes and interactions that effect water quality. For example, runoff has a strong influence on water quality, but runoff is affected by many variables including soil and climatic conditions, plant growth, evapotranspiration, plant residue decay, tillage, etc. Continuous simulation models offer many advantages including water quality probability distributions, seasonal variations, and little dependence on initial condition estimates. However, it is important that the major processes and interactions are simulated realistically. The EPIC and SWRRB models were chosen to illustrate the complex process interactions contained in fairly comprehensive continuous simulation water quality models.

2.1 EPIC

The components of the field scale EPIC model can be placed into 10 major divisions for discussion--hydrology, weather, erosion, nutrients, pesticide fate, soil temperature, plant growth, tillage, plant environment control, and economics. A detailed description of the EPIC components was given by Williams et al. (1990). A brief description of each of the 10 components is presented here.

2.1.1. *Hydrology*

2.1.1.1. *Surface Runoff*. Surface runoff from daily rainfall is predicted using a procedure similar to the CREAMS runoff model, option one (Knisel 1980; Williams and Nicks, 1982). Like the CREAMS model, runoff volume is estimated with a modification of the SCS curve number method (USDA Soil Conservation Service, 1972). The curve number varies non-linearly from the 1 (dry) condition at wilting point to the 3 (wet) condition at field capacity and approaches 100 at saturation. The EPIC model also includes a provision for estimating runoff from frozen soil.

Peak runoff rate predictions are based on a modification of the Rational Formula. The runoff coefficient is calculated as the ratio of runoff volume to rainfall. The rainfall intensity during the watershed time of concentration is estimated for each storm as a function of total rainfall using a stochastic technique. The watershed time of concentration is estimated using Manning's Formula considering both overland and channel flow.

2.1.1.2. *Percolation*. The percolation component of EPIC uses a storage routing technique to predict flow through each soil layer in the root zone. Downward flow occurs when field capacity of a soil layer is exceeded if the layer below is not saturated. The downward flow rate is governed by the saturated conductivity of the soil layer. Upward flow may occur when a lower layer exceeds field capacity. Movement from a lower layer to an adjoining upper layer is regulated by the soil water to field capacity ratios of the two layers.

Percolation is also affected by soil temperature. If the temperature in a particular layer is 0°C or below, no percolation is allowed from that layer.

2.1.1.3. *Lateral Subsurface Flow*. Lateral subsurface flow is calculated simultaneously with percolation. A nonlinear function of lateral flow travel time is used to simulate the horizontal component of subsurface flow. The magnitudes of the vertical and horizontal flow components are determined by a simultaneous solution of the two governing equations.

2.1.1.4. *Evapotranspiration*. The model offers four options for estimating potential evaporation--Hargreaves and Samani (1985), Penman (1948), Priestley-Taylor (1972), and Penman-Monteith (Monteith, 1965). The Penman and Penman-Monteith methods require solar radiation, air temperature, wind speed, and relative humidity as input. If wind speed, relative humidity, and solar radiation data are not available, the Hargreaves or Priestley- Taylor methods provide options that give realistic results in most cases. The model computes soil and plant evaporation separately as described by Ritchie (1972).

2.1.1.5. *Snow Melt*. If snow is present, it is melted on days when the second soil layer temperature exceeds 0°C, using a linear function of air and snow pack temperature. Snow pack temperature is estimated using daily air temperature and soil surface temperature. Melted snow is treated the same as rainfall for estimating runoff and percolation, but rainfall energy is set to 0.0 and peak runoff rate is estimated assuming uniformly distributed rainfall for a 24 h duration.

2.1.2. *Weather*. The weather variables necessary for driving the EPIC model are precipitation and air temperature. If the Penman methods are used to estimate potential evaporation, solar radiation, wind speed and relative humidity are also required. Of course, wind speed and direction are also needed when wind erosion is simulated. If daily precipitation, air temperature, and solar radiation data are available, they can be input directly to EPIC. Otherwise, EPIC provides options for simulating various combinations of the five weather variables.

2.1.2.1 *Precipitation*. The EPIC precipitation model developed by Nicks

(1974) is a first-order Markov chain model. Thus, input to the model must include monthly probabilities of receiving precipitation if the previous day was dry and if the previous day was wet. Given the wet-dry state, the model determines stochastically if precipitation occurs or not. When a precipitation event occurs, the amount is determined by generating from a skewed normal daily precipitation distribution. The amount of daily precipitation is partitioned between rainfall and snowfall using average daily air temperature.

2.1.2.2. *Air Temperature and Solar Radiation.* The temperature-radiation model developed by Richardson (1981) was selected for use in EPIC because it simulates temperature and radiation that exhibit proper correlation between one another and rainfall. The residuals of daily maximum and minimum temperature and solar radiation are generated from a multivariate normal distribution. Details of the multivariate generation model were described by Richardson (1981). The dependence structure of daily maximum temperature, minimum temperature, and solar radiation was described by Richardson (1982a).

2.1.2.3. *Wind.* The wind simulation model was developed by Richardson (1982b) for use in simulating wind erosion with EPIC. The two wind variables considered are average daily wind velocity and direction. Average daily wind velocity is generated from a modified exponential distribution. Wind direction expressed as radians from north in a clockwise direction is generated from an empirical distribution specific for each location.

2.1.2.4. *Relative Humidity.* The relative humidity model simulates daily average relative humidity from the monthly average using a triangular distribution. Triangular coordinates are set to produce higher relative humidities on rainy days, lower values on dry days, and to preserve the long-term monthly average.

2.1.3. *Erosion*

2.1.3.1 *Water.* The EPIC water erosion model simulates erosion caused by rainfall and runoff and by irrigation (sprinkler and furrow). To simulate rainfall/runoff erosion, EPIC contains six equations--the USLE (Wischmeier and Smith, 1978), the Onstad-Foster modification of the USLE (Onstad and Foster, 1975), and three variations of MUSLE (Williams, 1975a). There is also a provision for inputting four coefficients for use in a MUSLE-like equation. Only one of the equations (user specified) interacts with other EPIC components.

The hydrology model supplies estimates of runoff volume and peak runoff rate. To estimate the daily rainfall energy in the absence of time-distributed rainfall, it is assumed that the rainfall rate is exponentially distributed. This allows simple substitution of rainfall rates into the USLE equation for estimating rainfall energy. The fraction of rainfall that occurs during 0.5 h is simulated stochastically. The soil erodibility factor is estimated as a function of soil texture and organic content. The crop management factor is evaluated with a function of above-ground biomass, crop residue on the surface, and the minimum C factor for the crop. Other factors of the erosion equation are evaluated as described by Wischmeier and Smith (1978). A nonlinear function of topsoil coarse fragment content is used to adjust the erosion estimates.

2.1.3.2 *Wind.* The Manhattan, Kansas wind erosion equation (Woodruff and Siddoway, 1965), was modified for use in the EPIC model. The original equation computes average annual wind erosion as a function of soil erodibility, a climatic factor, soil ridge roughness, field length along the prevailing wind direction, and vegetative cover. The new, more physically based model estimates daily potential wind erosion for a smooth, bare soil by integrating an erosion equation as a function of wind speed. Mean erodible particle diameter and surface water content are used to determine daily threshold wind speed. Potential erosion is adjusted using four factors based on soil properties, surface roughness, cover, and distance across the field in the wind direction.

2.1.4. *Nutrients*

2.1.4.1 *Nitrogen.* The amount of NO_3-N in runoff is estimated by considering the top soil layer only. The decrease in NO_3-N concentration caused by water flowing through a soil layer can be simulated satisfactorily using an exponential function. The average concentration for a day can be obtained by integrating the exponential function to give NO_3-N yield and dividing by volume of water leaving the layer (runoff, lateral flow, and percolation). Amounts of NO_3-N contained in runoff, lateral flow, and percolation are estimated as the products of the volume of water and the average concentration.

Leaching and lateral subsurface flow in lower layers are treated with the same approach used in the upper layer, except that surface runoff is not considered. When water is evaporated from the soil, NO_3-N is moved upward into the top soil layer by mass flow.

A loading function developed by McElroy et al. (1976) and modified by Williams and Hann (1978) for application to individual runoff events is used to estimate organic N loss. The loading function estimates the daily organic N runoff loss based on the concentration of organic N in the top soil layer, the sediment yield, and the enrichment ratio.

Nitrification, the conversion of ammonium N to nitrate N is estimated using a combination of the methods of Reddy et al. (1979a) and Godwin et al. (1984). The approach is based on the first-order kinetic rate equation of Reddy et al. (1979a). The rate constant is adjusted to account for variation in soil temperature, water content, and pH. The equation initiates nitrification when soil temperature exceeds 5°C and the rate constant increases linearly with temperature. Nitrification does not occur until soil water content exceeds wilting points (WP). The nitrification rate constant is a linear function of soil water between WP and 25% of the difference between field capacity (FC) and WP. Soil water contents between WP + 0.25 (FC-WP) and FC do not affect the rate constant. The nitrification rate constant declines linearly above FC and approaches zero at saturation. The nitrification rate constant is not affected for pH values between 7.0 and 7.4. The rate constant decreases linearly below pH 7.0 and approaches zero at pH 4.1. Similarly, the rate constant decreases linearly above pH 7.4 and approaches zero at pH 9.0.

Volatilization, the loss of ammonia N to the atmosphere is estimated simultaneously with nitrification using the method of Reddy et al. (1979b). Volatilization of surface applied ammonia is estimated as a function of temperature and wind speed. Depth of ammonia within the soil, cation exchange capacity (CEC) of the soil, and soil temperature are used in estimating below surface volatilization. The volatilization temperature function is the same as used in estimating nitrification. The surface volatilization rate constant increases nonlinearly with windspeed. The subsurface rate constant decreases nonlinearly with soil depth and linearly with CEC.

Denitrification, one of the microbial processes, is a function of temperature and water content. Denitrification is only allowed to occur when the soil water content is 5% of field capacity. The denitrification rate is estimated using an exponential function involving temperature, organic carbon, and NO_3-N.

The N mineralization model is a modification of the PAPRAN mineralization model (Seligman and van Keulen, 1981). The model considers two sources of mineralization: fresh organic N associated with crop residue and microbial biomass and the stable organic N associated with the soil humus pool. The mineralization rate for fresh organic N is governed by C:N and C:P ratios, soil water, temperature, and the stage of residue decomposition. The N associated with the soil humus pool is divided into two pools (active and stable). Mineralization occurs only in the active pool, but N is allowed to flow very slowly from the stable to the active pool. Mineralization is estimated as a function of organic N weight, soil water, and temperature.

Like mineralization, immobilization is simulated with a modification of the PAPRAN model. Immobilization is a very important process in EPIC because it determines the residue decomposition rate, and residue decomposition has an important effect on erosion. The daily amount of immobilization is computed by subtracting the amount of N contained in the crop residue from the amount assimilated by the microorganisms.

Crop use of N is estimated using a supply and demand approach. The daily crop N demand is estimated as the product of biomass growth and optimal N concentration in the plant. Optimal crop N concentration is a function of growth stage of the crop. Soil supply of N is limited by mass flow of NO_3-N to the roots. Actual N uptake is the minimum of supply and demand.

Fixation of N is an important process for legumes. Daily N fixation is estimated as a fraction of daily plant N uptake. The fraction is a function of soil NO_3 and water content and plant growth stage. No fixation occurs if the root zone NO_3 content is greater than 300 $kg \cdot ha^{-1} m^{-1}$. The fraction is allowed to increase to 1.0 as the root zone NO_3 content is lowered to 100 $kg \cdot ha^{-1} m^{-1}$. The fraction decreases linearly from 1.0 to 0 as soil water increases from 85% of field capacity to saturation. Below 85% of field capacity, the fraction reduces linearly to zero at wilting point. Also, fixation only occurs during the period between 15 and 75% of crop maturity.

To estimate the N contribution from rainfall, EPIC uses an average rainfall N concentration at a location for all storms. The amount of N in rainfall is estimated as the product of rainfall amount and concentration.

2.1.4.2. *Phosphorus*. The EPIC approach to estimating soluble P loss in surface runoff is based on the concept of partitioning pesticides into the solution and sediment phases as described by Leonard and Wauchope (Knisel, 1980). Because P is mostly associated with the sediment phase, the soluble P runoff is predicted using labile P concentration in the top soil layer, runoff volume, and a partitioning factor. Sediment transport of P is simulated with a loading function as described in organic N transport.

The P mineralization model developed by Jones et al. (1984) is similar in structure to the N mineralization model. Mineralization from the fresh organic P pool is governed by C:N and C:P ratios, soil water, temperature, and the stage of residue decomposition. Mineralization from the stable organic P pool associated with humus is estimated as a function of organic P weight, labile P concentration, soil water, and temperature. The P immobilization model also developed by Jones et al. (1984) is similar in structure to the N immobilization model.

The mineral P model was developed by Jones et al. (1984). Mineral P is transferred among three pools: labile, active mineral, and stable mineral. When P fertilizer is applied, it is labile (available for plant use). However, it may be quickly transferred to the active mineral pool.

Simultaneously, P flows from the active mineral pool back to the labile pool (usually at a much slower rate). Flow between the labile and active mineral pools is governed by temperature, soil water, a P sorption coefficient, and the amount of material in each pool. Flow between the active and stable mineral P pools is governed by the concentration of P in each pool and the P sorption coefficient.

Crop use of P is estimated with the supply and demand approach described in the N model. However, the P supply is predicted using an equation based on plant demand, labile P concentration, and root weight.

2.1.5. *Pesticide Fate.* GLEAMS technology for simulating pesticide transport by runoff, percolate, soil evaporation, and sediment was added to EPIC. Pesticides may be applied at any time and rate to plant foliage or below the soil surface at any depth. The EPIC simulated plant leaf-area-index determines what fraction of foliar-applied pesticide reaches the soil surface. Also, a fraction of the application rate (called application efficiency) is lost to the atmosphere.

Each pesticide has a unique set of parameters including solubility, half life in soil and on foliage, wash-off fraction, organic carbon adsorption coefficient, and cost. Pesticides on plant foliage and in the soil degrade exponentially according to the appropriate half lives. When rainfall greater than 2.5 mm occurs, pesticide is washed from the foliage to the soil surface according to the wash-off fraction. If the field capacity of the top soil layer (10 mm thick) is exceeded, percolation occurs. Pesticide losses are estimated as a function of water lost/storage volume and the adsorption isotherm (Kd) of the chemical. Leaching, runoff, and lateral flow loss amounts are determined by loss volumes and pesticide concentration. Pesticide leaching is estimated for each soil layer when percolation occurs. Soil layers with low storage volumes have high leaching potentials not only because percolation is greater, but also because storage volume displacement is greater (higher concentration). Pesticides with low Kd values and high solubility are transported rapidly with water. Conversely, high Kd value pesticides are adsorbed to soil particles and travel largely with the sediment. Pesticide concentration adsorbed to sediment is estimated as a function of pesticide and soil weight, Kd, and water storage capacity of the soil layer. Pesticide transport by sediment is calculated as a function of pesticide concentration, sediment yield, and an enrichment ratio.

2.1.6. *Soil Temperature.* Daily average soil temperature is simulated at the center of each soil layer for use in nutrient cycling and hydrology. The temperature of the soil surface is estimated using daily maximum and minimum air temperature, solar radiation, and snow, plant, and residue cover. Soil temperature is simulated for each layer using a function of damping depth, surface temperature, and mean annual air temperature. Damping depth is dependent upon bulk density and soil water.

2.1.7. *Crop Growth Model.* A single model is used in EPIC for simulating all the crops considered (corn, grain sorghum, wheat, barley, oats, sunflower, soybean, alfalfa, cotton, peanuts, potatoes, durham wheat, winter peas, faba beans, rapeseed, sugarcane, sorghum hay, range grass, rice, casava, lentils, and pine trees). Of course, each crop has unique values for the model parameters. Energy interception is estimated as a function of solar radiation and the crop's leaf area index. The potential increase in biomass for a day is estimated as the product of intercepted energy and a crop parameter for converting energy to biomass. The leaf area index is simulated with equations dependent upon heat units, the maximum leaf area index for the crop, a crop parameter that initiates leaf area index decline, and five stress factors.

Crop yield is estimated using the harvest index concept. Harvest index increases as a non-linear function of heat units from zero at planting to the optimal value at maturity. The harvest index may be reduced by water stress during critical crop stages (usually between 30 and 90% of maturity).

The fraction of daily biomass growth partitioned to roots is estimated to range linearly between two fractions specified for each crop at emergence and at maturity. Root weight in a soil layer is simulated as a function of plant water use within that layer. Root depth increases as a linear function of heat units and potential root zone depth.

The potential biomass is adjusted daily if one of the plant stress factors is less than 1.0 using the product of the minimum stress factor and the potential biomass. The water-stress factor is the ratio of actual to potential plant evaporation. The temperature stress factor is computed with a function dependent upon the daily average temperature, the optimal temperature, and the base temperature for the crop. The N and P stress factors are based on the ratio of accumulated plant N and P to the optimal values. The aeration stress factor is estimated as a function of soil water relative to porosity in the root zone.

Roots are allowed to compensate for water deficits in certain layers by using more water in layers with adequate supplies. Compensation is governed by the minimum root growth stress factor (soil texture and bulk density, temperature, and aluminum toxicity). The soil texture-bulk density relationship was developed by Jones (1983).

2.1.8. *Tillage*. The EPIC tillage component was designed to mix nutrients and crop residue within the plow depth, simulate the change in bulk density, and convert standing residue to flat residue. Other functions of the tillage component include simulating ridge height and surface roughness.

Tillage operations convert standing residue to flat residue using an exponential function of tillage depth and mixing efficiency. When a tillage operation is performed, a fraction of the material (equal the mixing efficiency) is mixed uniformly within the plow depth. Also, bulk density is reduced as a function of mixing efficiency, bulk density before tillage, and undisturbed bulk density. After tillage, bulk density returns to the undisturbed value at a rate dependent upon infiltration, tillage depth, and soil texture.

2.1.9. *Plant Environment Control*

2.1.9.1. *Drainage*. Underground drainage systems are treated as a modification to the natural lateral subsurface flow of the area. Simulation of a drainage system is accomplished by reducing the travel time in a specified soil layer.

2.1.9.2. *Irrigation*. The EPIC user has the option to simulate dryland or irrigated agricultural areas. Sprinkler or furrow irrigation may be simulated and the applications may be user specified or automatic. With the automatic option, the model decides when and how much water to apply. The user must input a plant water stress level or a soil water tension value to trigger automatic irrigation, the maximum volume applied per growing season, and the minimum time interval between applications.

2.1.9.3. *Fertilization*. The EPIC model provides two options for applying fertilizer. With the first option, the user specifies dates, rates, and depths of application of N and P. The second option is more automated--the model decides when and how much fertilizer to apply. The four required inputs are: (1) a plant stress level to trigger nitrogen fertilizer application, (2) the maximum N application per growing season, (3) the maximum N per application as a fraction of the maximum per growing season, and (4) the minimum number of days between applications. At planting time, the model takes a soil sample and applies enough N and P to bring the concentrations in the root zone up to the concentration level at which the soil contains the maximum per application amount. Additional N may be applied during the growing season.

2.1.9.4 *Lime*. The EPIC model simulates the use of lime to neutralize toxic levels of aluminum in the plow layer. Two sources, KCl-extractable aluminum in the plow layer and acidity caused by ammonia-based fertilizers, are considered. When the sum of acidity due to extractable aluminum and fertilizer N exceeds 4 $t \cdot ha^{-1}$, the required amount of lime is added and incorporated into the plow layer.

2.1.9.5. *Pests*. The effects of insects, weeds, and diseases are expressed in the EPIC pest factor. The pest factor is simulated as a function of minimum daily temperature, 30-day moving average rainfall, above ground growing biomass, and crop residue. At the end of the growing season, the pest factor is normalized to fall in the range between the minimum value for the particular crop and 1.0. Crop yields are estimated as the product of the simulated yield and the simulated pest factor.

2.1.9.6. *Grazing*. Livestock grazing is simulated as a daily harvest operation. Users specify daily grazing rate minimum grazing height, harvest efficiency, and date grazing begins and ends. Any number of grazing periods may occur during a year and the grazing schedule may vary from year-to-year within a rotation. Grazing ceases when forage height is reduced to the user specified cutoff value and resumes automatically when new growth exceeds the cutoff height if the grazing period has not expired.

2.1.10. *Economics*. The crop budgets are calculated using components from the Enterprise Budget Generator (Kletke, 1979). Inputs are divided into two categories: fixed and variable. Fixed inputs include depreciation, interest or return on investment, insurance, and taxes on equipment, land, and capital improvements (terraces, drainage, irrigation systems, etc.). Fixed costs are added by using a dummy tillage that has only a cost and no other tillage characteristics. Variable inputs are defined as machinery repairs, fuel and other energy, machine lubricants, seed, fertilizer, pesticides, labor, and irrigation water. The variable costs are included either with the machinery; as a function of water, lime or fertilizer uses; or with the pesticide.

2.2. SWRRB

The components of SWRRB can be placed into eight major divisions--hydrology, weather, sedimentation, soil temperature, crop growth, nutrients, pesticide fate, and agricultural management. A detailed description of the SWRRB components was given by Arnold et al. (1990). A brief description is presented here. Since SWRRB and EPIC use essentially the same components for simulating several processes, descriptions are not repeated.

2.2.1. *Hydrology*

2.2.1.1. *Surface Runoff*. Same as EPIC.

2.2.1.2. *Percolation*. Same as EPIC.

2.2.1.3. *Lateral Subsurface Flow*. Lateral flow is defined as flow that travels laterally within the soil profile that returns to contribute to streamflow. SWRRB uses a kinematic storage routing model (Sloan and Moore, 1984) based on the mass continuity equation with each soil layer defining a control volume. The model is physically-based and accounts for slope, slope length, and saturated conductivity.

2.2.1.4. *Evaportranspiration*. Potential evapotranspiration is estimated with the Priestly-Taylor (1972) method or the Hargreaves method (Hargreaves and Samani, 1985). Required inputs are daily maximum and minimum air temperature and solar radiation. The model computes soil and plant evaporation separately as described by Ritchie (1972).

2.2.1.5. *Snow Melt*. Same as EPIC.

2.2.1.6. *Transmission Losses*. Many semiarid watersheds have alluvial channels that abstract large volumes of streamflow (Lane, 1982). The abstractions, or transmission losses, reduce runoff volumes as the flood wave travels downstream. SWRRB uses Lanes method described in Chapter 19 of the SCS Hydrology Handbook (USDA Soil Conservation Service, 1983) to estimate transmission losses. Channel losses are a function of channel width and length and flow duration. Both runoff volume and peak rate are adjusted when transmission losses occur.

2.2.1.7. *Ponds and Reservoirs*. Farm pond storage is simulated as a function of pond capacity, daily inflows and outflows, seepage, and evaporation. Ponds are assumed to have only emergency spillways. Required inputs are capacity and surface area. Surface area below capacity is estimated as a non-linear function of storage. Reservoirs are treated similarly except they have emergency and principal spillways. Thus, required inputs include volume and surface area at both spillway elevations and the principal spillway release rate.

2.2.2. *Weather*. The weather variables necessary for driving SWRRB are precipitation, air temperature, and solar radiation. If daily precipitation data are available, they can be input directly to SWRRB. If not, the weather generator can simulate daily rainfall and temperature. Solar radiation is always simulated. One set of weather variables may be simulated for the entire basin, or different weather may be simulated for each subbasin.

2.2.2.1. *Precipitation*. The SWRRB precipitation model developed by Nicks (1974) is a first-order Markov chain model. Thus, input to the model must include monthly proabilities of receiving precipitation if the previous day was dry and if the previous day was wet. Given the wet-dry state, the model determines stochastically if precipitation occurs or not. When a precipitation event occurs, the amount is determined by generating from a skewed normal daily precipitation distribution. The amount of daily precipitation is partitioned between rainfall and snowfall using average daily air temperature.

If precipitation is to be simulated for each subbasin, the amount generated from the skewed normal distribution is assumed to be the mean for all gages for the day. The storm center (location of maximum rainfall) is located in a rectangle with boundaries set at a distance of 100 Km from the basin's maximum and minimum x and y raingage coordinates. Thus, the storm center could be located in the basin or as much as 100 Km in any direction from the basin. Each storm center is defined by drawing two random numbers--one for the x scale and one for the y scale. Rainfall at each gage is computed using an area reduction function (Nicks and Igo, 1980). The function reduces rainfall as distance from the storm center increases and as rainfall duration decreases. Finally, the reduction factors are applied to the subbasin raingages and adjusted to insure that the mean rainfall agrees with the original skewed normal simulated mean.

2.2.2.2. *Air Temperature and Solar Radiation*. Same as EPIC.

2.2.3. *Sedimentation*

2.2.3.1. *Sediment Yield*. The Modified Universal Soil Loss Equation (MUSLE) (Williams, 1975a) and two newly developed MUSLE variations are provided for estimating subbasin sediment yield. There is also a provision for inputting four coefficients to a MUSLE-like equation. The user may choose which of the four equations interacts with other simulation processes. The hydrology model supplies estimates of runoff volume and peak runoff rate. The crop management factor is evaluated with a function of above-ground biomass, crop residue on the surface, and the minimum C factor for the crop. Other factors of the erosion equation are evaluated as described by Wischmeier and Smith (1978).

2.2.3.2. *Sediment Routing*.

1. Ponds and Reservoirs--Inflow sediment yield to ponds and reservoirs (P/R) is computed with MUSLE. The outflow from P/R is calculated as the product of outflow volume and sediment concentration. Outflow P/R concentration is estimated using a simple continuity equation based on volumes and concentrations of inflow, outflow, and pond storage. Initial pond concentration is input and between storm concentration decreases as a function of time and median particle size of inflow sediment.

2. Channel and Floodplain--The sediment routing model consists of two components operating simultaneously (deposition and degradation). The deposition component is based on fall velocity and the degradation component is based on Bagnold's stream power concept (Williams, 1980).

The particle size distribution of the detached sediment is estimated from the primary particle size distribution (Foster et al., 1980). The particle size of the sediment reaching a subbasin outlet is estimated using the MUSLE derived sediment deposition model (Williams, 1975b). The delivery ratio of each event is estimated as the ratio of the subbasin peak discharge rate to the peak rainfall excess rate raised to the 0.56 power. Since rainfall input is daily and runoff volume is estimated as a daily amount, rainfall excess rate is not simulated directly. However, the peak rainfall excess rate can be estimated as the difference between the peak rainfall rate and the average infiltration rate of the storm. The storm duration and the peak rainfall rate can be estimated by assuming that rainfall rates are exponentially distributed. Note, this rainfall rate assumption is not related to the time distribution of rainfall amount. The delivery ratio is used to estimate a routing coefficient that governs deposition as a function of particle size. Thus, as deposition occurs the particle size distribution is composed of finer material.

Deposition in the channel and floodplain from the subbasin to the basin outlet is based on sediment particle fall velocity. Fall velocity is calculated as a function of particle diameter squared using Stokes Law. The depth of fall through a routing reach is the product of fall velocity and reach travel time. The delivery ratio is estimated for each particle size as a linear function of fall velocity, travel time, and flow depth.

Stream power is used to predict degradation in the routing reaches. Bagnold (1977) defined stream power as the product of water density, flow rate, and water surface slope. Williams (1980) modified Bagnold's equation to place more weight on high values of stream power--stream power raised to 1.5. Also, since hydrographs are not simulated, one flow rate is calculated to represent the total hydrograph. The representative flow rate is determined from simulated peak discharge rate and runoff volume assuming a triangular hydrograph with two recession limbs.

Available stream power is used to reentrain loose and deposited material until all of the material is removed. Excess stream power causes bed degradation. Bed degradation is adjusted by the USLE soil erodibility and cover factors of the channel and floodplain.

2.3.4. *Soil Temperature*. Same as EPIC.

2.3.5. *Crop Growth Model*. A single model is used in SWRRB for simulating all crops. Energy interception is estimated as a function of solar radiation and the crop's leaf area index. The potential increase in biomass for a day is estimated as the product of intercepted energy and a crop parameter for converting energy to biomass. The leaf area index is simulated with equations dependent upon heat units.

Crop yield is estimated using the harvest index concept. Harvest index increases as a non-linear function of heat units from zero at planting to the optimal value at maturity. The harvest index may be reduced by water stress during critical crop stages (usually between 30 and 90% of maturity).

The fraction of daily biomass growth partitioned to roots is estimated to range linearly from 0.4 at emergence to 0.2 at maturity. Root weight in a soil layer is simulated as a function of plant water use within that layer. Root depth increases as a linear function of heat units and potential root zone depth.

The potential biomass is adjusted daily if one of the plant stress factors is less than 1.0 using the product of the minimum stress factor and the potential biomass. The water-stress factor is the ratio of actual to potential plant evaporation. The temperature stress factor is computed with a function dependent upon the daily average temperature, the optimal temperature, and the base temperature for the crop.

2.3.6. *Nutrients*.

2.3.6.1. *Nitrogen*. The amount of NO_3-N in runoff is estimated by considering the top soil layer only. The decrease in NO_3-N concentration caused by water flowing through a soil layer can be simulated satisfactorily using an exponential function. The average concentration for a day can be obtained by integrating the exponential function to give NO_3-N yield and dividing by volume of water leaving the layer (runoff, lateral flow, and percolation). Amounts of NO_3-N contained in runoff, lateral flow, and percolation are estimated as the products of the volume of water and the average concentration.

Leaching and lateral subsurface flow in lower layers are treated with the same approach used in the upper layer, except that surface runoff is not considered. When water is evaporated from the soil, NO_3-N is moved upward into the top soil layer by mass flow.

A loading function developed by McElroy et al. (1976) and modified by Williams and Hann (1978) for application to individual runoff events is used to estimate organic N loss. The loading function estimates the daily organic N runoff loss based on the concentration of organic N in the top soil layer, the sediment yield, and the enrichment ratio.

Crop use of N is estimated using a supply and demand approach. The daily crop N demand is estimated as the product of biomass growth and optimal N concentration in the plant. Optimal crop N concentration is a function of growth stage of the crop. Soil supply of N is limited by mass flow of NO_3-N to the roots. Actual N uptake is the minimum of supply and demand.

2.3.6.2. *Phosphorus*. The SWRRB approach to estimating soluble P loss in surface runoff is based on the concept of partitioning pesticides into the solution and sediment phases as described by Leonard and Wauchope (Knisel, 1980). Because P is mostly associated with the sediment phase, the soluble P runoff is predicted using labile P concentration in the top soil layer, runoff volume, and a partitioning factor. Sediment transport of P is simulated with a loading function as described in organic N transport.

Crop use of P is estimated with the supply and demand approach described in the N model. However, the P supply is predicted using an equation based on plant demand, labile P concentration, and root weight.

2.3.7. *Pesticides*. Same as EPIC.

2.3.8. *Agricultural Management*.

2.3.8.1. *Tillage and Residue*. The SWRRB tillage component was designed to partition the above-ground biomass at harvest. Part of the biomass is removed as yield, part is incorporated into the soil, and the remainder is left on the soil surface as residue. The model has no process interactions with incorporated residue. Also, tillage does not effect soil properties.

The fraction of above-ground biomass left on the surface is determined by the tillage strategy selected. Fall plow leaves 5% of the above-ground crop residue on the surface. Percentages assigned to surface residue for the three other tillage strategies are: 25% for spring plow; 50% for conservation tillage; and 95% for zero tillage. The residue decays at a rate determined by soil water content and soil temperature of the top soil layer.

2.3.8.2. *Irrigation*. The user has the option to simulate dryland or irrigated agriculture. If irrigation is selected, he must also specify the runoff ratio (volume of water leaving the field/volume applied) and a plant water stress level to trigger irrigation. The plant water stress factor ranges from 0 to 1.0 (1 means no stress and 0 means no growth). When the user-specified stress level is reached, enough water is applied to fill the root zone to field capacity.

3. Water Quality Model Applications

3.1. EPIC

3.1.1. *Water Quality Project Evaluation and Planning*. In 1992, many of the 16 Water Quality Demonstration Projects and the 74 Hydrologic Unit Area Projects began using one or more of the water quality models described here. Models are being used in conjunction with monitoring to help identify current water quality problem areas and to estimate the benefits of water quality improvement practices. Without the models, it is difficult to estimate benefits.

3.1.2. *European Economic Community Project*. A three-year European Community project involving French, Italian, English, Spanish, and Portugese scientists is using EPIC to simulate complex cropping systems in ten major European production regions. Dr. Guillerono Flichmann at the Mediterranean Agronomic Institute in Montpellier and Dr. Maurice Cabelguenne of the French National Institute of Agronomic Research in

Toulouse, are leading the project. EPIC's field-scale crop production and water quality predictions are being utilized in conjunction with farm-scale economic models to assess the effects of common European agricultural policies on the relative regional competitiveness and environmental impacts.

3.1.3. *EPA Nutrient Movement Study.* EPA is using EPIC as a tool to examine the impacts of tillage on nitrogen and phosphorus movement. Four tillage strategies were examined using 100-year EPIC simulations at 100 sites in Illinois. Simulated crop yields corresponded closely to expected yields. No-till corn/soybean rotations were found to reduce sediment bourne nutrients and nitrogen in runoff relative to conventional till continuous corn (Phillips et al.).

3.1.4. *Animal Waste Recycling Planning.* The EPIC model is currently being used to study animal waste disposal systems for the Upper North Bosque Hydrological Unit Project in Erath and Hamilton Counties, Texas. Since EPIC simulates the effects of manure application rates on crop growth, soil fertility, and water quality, it can help identify the levels of manure applications that could be recycled through forage production. Various forage production systems and soils were used to illustrate the interactions among crops, manure recycling, and nutrient losses.

3.1.5. *Crop and Tillage Impacts on Nutrient and Sediment Movement.* EPIC has been used to examine the water quality impacts of alternative crop tillage and conservation systems (Benson et al. 1990a). Nitrogen and phosphorus losses were simulated for two soils and five crop/tillage systems. The base system was continuous conventional tillage cotton. The simulated rotations of cotton/sorghum and cotton/sorghum/wheat reduced phosphorus losses for both conventional and conservation tillage. Also, the ratio of N loss to N fertilizer applied was reduced by the rotations, although total N loss increased slightly.

3.1.6. *Sustainable Agriculture's Impact on Production and Water Quality.* The Rodale Research Center in cooperation with the World Resource Institute used EPIC to study sustainable agricultural systems in the East and Midwest of the U.S. An expanded sustainable agriculture analysis for the U.S. is now underway. This analysis will examine the potential applicability of sustainable agricultural systems for all 10 production regions in the U.S.

3.1.7. *Evaluation of New Cropping Systems Influence on Water Quality.* Washington State University used EPIC to simulate a field experiment in integrated cropping for a 100-year period to extend the experimental data over time (Benson et al. 1990b). This study examined the feasibility of a spring pea and Medic/Medic/winter wheat rotation as a replacement for the current winter wheat/spring barley/winter wheat/spring pea rotation. The proposed system was studied using relatively short-term experiments that may not represent long-term system behavior. EPIC was used to simulate over 100 years of yields and nitrogen and phosphorus losses. The results indicated that not only were there some potential economic advantages (Goldstein and Young, 1987), but also some water quality enhancement and soil conservation. Both N and P losses were reduced.

3.1.8. *Cover Crop Impacts on Water Quality.* Meisinger et al. (1991) used EPIC to address the impacts of winter cover crops on water quality. This study compared EPIC simulations to historic studies to show the reasonableness of EPIC nutrient loss estimates. The EPIC simulations indicated that winter cover crops had the most potential for reducing nitrogen leaching in the humid Southeast and in irrigated areas.

3.1.9. *Erosion Impacts on Productivity and Water Quality.* Fribourg et al. examined the impact of erosion on nitrate loss by using EPIC to estimate N runoff and leaching for two eroded phases of two Tennessee soils, Memphis and Grenada. The EPIC mean corn yield estimates for both soils for the eroded and non-eroded phases were within 10 percent of the measured for three of the four soils and within 15 percent for the fourth. Estimated nitrogen losses for the eroded phase of the Grenada soil were greater than those of the non-eroded phase. This is not surprising considering the root restricting fragipan of the Grenada soil.

These examples of past and current EPIC applications serve to illustrate the tremendous potential of comprehensive interactive Water Quality models. These models help identify problems, examine systems for improving water quality, estimate potential benefits, and develop national farm programs which will lead to widespread adoption of the water quality measures.

3.2. SWRRB

The SWRRB model is currently being used throughout the world by consulting engineers, government agencies, universities, and chemical companies. Several applications follow to demonstrate the model's potential capabilities.

3.2.1. *Hydrology Unit Model of the United States (HUMUS).* The SCS is using the EPIC, SWRRB, and ROTO models as part of the 1997 Resource Conservation Assessment. The models will be linked to national economic models and used for national planning of water supply and quality on the 18 major river basins in the U.S. This system of models allows water, sediment, and attached pollutants to be tracked from their point or origin to major rivers, reservoirs, and coastal zones. Geographic information systems are utilized to integrate the models with national soils, land use, and digital elevation databases. The GIS automatically extracts model input from the map layers and display model output.

3.2.2. *Coastal Pollutant Discharge Inventory.* As a part of the National Coastal Pollutant Discharge Inventory, the National Oceanic and Atmospheric Administration (NOAA) is using SWRRB to estimate non-point source loadings from non-urban lands in all coastal counties of the U.S. (Singer et al., 1988). Site-specific data are obtained from the SCS's National Resources Inventory and Soils-5 data bases, NOAA weather stations, U.S. Geological Survey digital land use data tapes, and other local sources. Simulations have been run for cropland, rangeland, and forest land in approximately 770 subwatersheds comprising the Gulf Coast, Eastern, and Western coastal zones of the U.S. Results are compiled by season and added to a comprehensive data base containing pollutant loadings from all significant discharge sources.

3.2.3. *Pesticide Assessment.* SWRRB was modified to include simulation of pesticide concentration in the runoff and sediment. The EPA has adopted this version of SWRRB as its pesticide assessment model. They have prepared their own user's manual, "Pesticide Runoff Simulator" (Holst and Kutney, 1987). Many chemical companies and consulting firms are using this version of the model for environmental assessment.

3.2.4. *Effect of Urbanization on Reservoir Loadings.* White Rock Lake in Dallas was built in 1910. The SCS completed sediment survey on the lake in 1935, 1956, 1970, 1977, and 1984. Throughout this period, the percentage of urban area has increased from 0 to 77% in 1984. SWRRB was utilized to estimate the effects of urbanization on water and sediment delivery to White Rock Lake (Arnold et al., 1988).

Initially, simulated model results were compared with measured water yields, peak flow rates, and sediment yields. The comparisons showed that the model could do a reasonable job predicting the effect of urbanization on these variables. Also, the effect of urbanization on delivery ratios showed a positive linear correlation.

Once the model had been validated, future simulations, assuming no urbanization since 1935 and all urban since 1910, were compared with actual conditions. It was shown that surface runoff increased and sediment yield decreased as urban area increased. The weather generator in SWRRB was utilized to examine future scenarios. Prediction of reservoir capacity lost to sedimentation to the year 2050 was shown for three scenarios, thus showing the model's usefulness in planning and designing water resources projects.

3.2.5. *Supplemental Pond Irrigation*. A comprehensive simulation model was developed to optimize pond size for supplemental irrigation (Arnold and Stockle, 1991). The SWRRB model was chosen as the basis for the simulation model. It was modified to simulate crop yields, double cropping, and supplemental pond irrigation. A simple economic model was also added. SWRRB was then linked with a golden section search to determine the pond size that optimizes average annual return to management. The model also develops frequency distributions for risk assessment. Individual components such as water yield, ET, and crop yields were validated with measured data to insure proper model operation. Arnold and Stockle (1991) apply the model at two locations with extreme differences in climate and soils to demonstrate potential model capabilities.

3.2.6. *Water Rights*. SWRRB is currently being used in water rights disputes in New Mexico and Arizona. This is part of an effort being undertaken by the Bureau of Indian Affairs, and the Hopi and Navajo Indian Tribes, to quantify Indian rights to water in reservation basins. SWRRB was selected because the basins are ungaged (lacking measured rainfall, runoff, and diversion data) and because SWRRB is a continuous simulation model.

4. References

4.1. Arnold, J. G. 1990. 'ROTO-A continuous water and sediment routing model. ASCE Proc. of the Watershed Management Symposium'. Durango, CO. pp. 480-488.

4.2. Arnold, J. G. and C. O. Stockle. 1991. 'Simulation of supplemental irrigation from on-farm ponds'. ASCE J. Irrig. and Drainage 117(3):408-424.

4.3. Arnold, J. G. and J. R. Williams. 1987. 'Validation of SWRRB--Simulator for water resources in rural basins'. J. Water Resources Planning and Manage., ACSW, 113(2):243-256.

4.4. Arnold, J. G., M. D. Bircket, J. R. Williams, W. F. Smith, and H. N. McGill. 1988. 'Modeling the effects of urbanization on basin water yield and reservoir sedimentation'. Water Resources Bull. 23(6):1021-1029.

4.5. Arnold, J. G., J. R. Williams, A. D. Nicks, and N. B. Sammons. 1990. 'SWRRB-A basin scale simulation model for soil and water resources management'. Texas A&M University Press, College Station, TX. 255 pp.

4.6. Bagnold, R. A. 1977. 'Bedload transport by natural rivers'. Water Resources Res. 13(2):303-312.

4.7. Beasley, D. B. and L. F. Huggins. 1982. 'ANSWERS - Users manual'. EPA-905/9-82-001, USEPA, Region 5, Chicago, IL. 54 pp.

4.8. Benson, V. W., H. C. Bogusch, Jr., and J. R. Williams. 1990a. 'Evaluating alternative soil conservation and crop tillage practices with EPIC'. pp. 91-93 In P. W. Unger, T. V. Sneed, W. R. Jordan, R. Jenson (eds.) Proc. Intl. Conf. on Dryland Farming, Challenges in Dryland Agriculture - A Global Perspective, Aug. 1988, Amarillo/Bushland, TX. Texas Agri. Exp. Stn., 965 pp.

4.9. Benson, V. W., W. A. Goldstein, D. L. Young, J. R. Williams, C. A. Jones, and J. R. Kiniry. 1990b. 'Impacts of integrated cropping practices on nitrogen use and movement'. pp. 426-428 In P. W. Unger, T. V. Sneed, W. R. Jordan, R. Jenson (eds.) Proc. Intl. Conf. on Dryland Farming, Challenges in Dryland Agriculture - A Global Perspective, Aug. 1988, Amarillo-Bushland, TX. Texas Agric. Exp. Stn., 965 pp.

4.10. Carsel, R. F., C. N. Smith, L. A. Mulkey, J. D. Dean, and P. Jowise. 1984. 'User's manual for the pesticide root zone model (PRZM): Release 1'. EPA-600/3-84-109. U.S. Environmental Protection Agency. Environmental Research Laboratory, Athens, GA.

4.11. Cole, G.W., L. Lyles, and L.G. Hagen. 1982. 'A simulation model of daily wind erosion soil loss'. ASAE Paper #82-2575.

4.12. Dean, J. D., P. S. Huyakorn, A. S. Donigian, Jr., K. A. Voos, R. W. Schanz, Y. J. Meeks, and R. F. Carsel. 1989. 'Risk of unsaturated/saturated transport and transformation of chemical concentrations (RUSTIC)'. EPA/600/3-89/048a. U.S. Environmental Protection Agency. Environmental Research Laboratory, Athens, GA.

4.13. Donigian, A. S. and W. C. Huber. 1990. 'Modeling of nonpoint source water quality in urban and non-urban areas'. U.S. Environmental Protection Agency, Athens, GA. Contract No. 68-03-3513 (Draft).

4.14. Engel, B. A. and J. G. Arnold. 1991. 'Agricultural non-point source pollution control using spatial decision support systems'. (Draft).

4.15. Foster, G. R. and L. D. Meyer. 1972. 'A closed-form soil erosion equation for upland areas. In H. Shen (ed.), Sedimentation, Colorado State University, Fort Collins, CO, Chapter 12.

4.16. Foster, G. R., L. J. Lane, J. D. Nowlin, J. M. Laflen, and R. A. Young. 1980. 'A model to estimate sediment yield from field-sized areas: development of model'. In W. G. Knisel (ed.) CREAMS, A field scale model for chemicals, runoff, and erosion from agricultural manaaaagement systems. U.S. Dep. Agric. Conserv. Res. Report, Chapter 3, pp. 36-64.

4.17. Fribourg, H. A., D. D. Tyler, V. W. Benson, J. R. Williams, J. G. Graveel, J. Logan, and G. R. Wells. 'Environmentally sould agricultural soil use--prediction of yield, erosion and other off-site effects from corn production'. (Mimeo handout)

4.18. Godwin, D. C., C. A. Jones, J. T. Ritchie, P. L. G. Vlek, and L. G. Youngdahl. 1984. 'The water and nitrogen components of the CERES models'. In ICRISAT (International Crops Research Institute for the Semi-Arid Tropics). Proc. Internatl. Symp. on Minimum Data Sets for Agrotechnology Transfer. March 21-26, 1983. pp. 95-100. Patancheru, India: ICRISAT Center.

4.19. Goldstein, W. A. and D. L. Young. 1987. 'An agronomic and economic comparison of a conventional and low-input system in the Palouse'. American Journal of Alternative Agriculture. Spring 51-56.

4.20. Hargreaves, G.H. and Z.A. Samani. 1985. 'Reference crop evapotranspiration from temperature'. Appl. Engr. in Agric. 1:96-99.

4.21. Holst, R. W. and L. L. Kutney. 1987. 'U.S. EPA simulator for water resources in rural basins (EPA-SWRRB)'. (Draft)

4.22. Huggins, L. F. and E. J. Monke. 1966. 'The mathematical simulation of the hydrology of small watersheds'. Technical Report 1, Water Resources Research Center, Purdue University, West Lafayette, IN. 130 pp.

4.23. Johansen, N. B., J. C. Imhoff, J. L. Kittle, and A. S. Donigian. 1984. 'Hydrological Simulation Program--Fortran (HSPF): User's manual for release 8'. EPA-600/3-84-066. U.S. Environmental Protection Agency, Athens, GA.

4.24. Jones, C.A. 1983. 'Effect of soil texture on critical bulk densities for root growth'. Soil Sci. Soc. Amer. J. 47:1208-1211.

4.25. Jones, C.A., C.V. Cole, A.N. Sharpley, and J.R. Williams. 1984. 'A simplified soil and plant phosphorus model. I. Documentation'. Soil Sci. Soc. Am. J. 48:800-805.

4.26. Kletke, D.D. 1979. 'Operation of the enterprise budget generator'. Oklahoma State Univ., Agric. Exp. Sta. Res. Report P-790.

4.27. Knisel, W. G. (ed.). 1980. 'CREAMS: A field scale model for chemicals, runoff, and erosion from agricultural management systems'. USDA, Conservation Research Report No. 26. 643 pp.

4.28. Lane, L. J. 1982. 'Distributed model for small semi-arid watersheds'. J. Hydraulic Eng., ASCE, 109(HY10):1114-1131.

4.29. Leonard, R. A., W. G. Knisel, and D. A. Still. 1987. 'GLEAMS: Groundwater loading effects of agricultural management systems'. Trans. ASAE 30(5):1403-1428.

4.30. McElroy, A. D., S. Y. Chiu, J. W. Nebgen, A. Aleti, and F. W. Bennett. 1976. 'Loading functions for assessment of water pollution from nonpoint sources'. Environmental Protection Tech. Series, USEPA, EPA 600/2-76-151. 445 pp.

4.31. Meisinger, J. J., W. L. Hargrove, R. L. Mikkelsen, J. R. Williams, and V. W. Benson. 1991. 'Effects of cover crops on groundwater quality'. pp. 57-68 In W. L. Hargrove (ed.) Cover Crops for Clean Water. Proc. Intl. Conf., April 9-11, 1991, Jackson, TN. Soil Water Conserv. Soc.

4.32. Monteith, J.L. 1965. 'Evaporation and environment'. Symp. Soc. Exp. Biol. 19:205-234.

4.33. Nicks, A. D. 1974. 'Stochastic generation of the occurrence, pattern, and location of maximum amount of daily rainfall'. pp. 154-171 In Symp. on Statistical Hydrology, Aug.-Sept. 1971, Tucson, AZ. Mis. Publ. No. 1275.

4.34. Nicks, A. D. and F. A. Igo. 1980. 'A depth-area-duration model of storm rainfall in the Southern Plains'. Water Resources Res. 16(5):939-945.

4.35. Office of Technology Assessment. 1982. 'Use of model for water resources management, planning, and policy'. Summary, Six Chapters and Appendix. U.S. Gov't. Printing Office, Washington, DC.

4.36. Onstad, C.A. and G.R. Foster. 1975. 'Erosion modeling on a watershed'. Trans. ASAE 18:288-292.

4.37. Penman, H.L. 1948. 'Natural Evaporation from Open, Bare Soil and Grass'. Proc. Royal Soc. (London) A193:120-145.

4.38. Phillips, D. L., P. D. Hardin, V. W. Benson, and J. V. Baglio. 'Using the national resources inventory and the EPIC model to evaluate the impact of alternative agricultural management practices in Illinois'. Submitted to the Jour. Soil and Water Conserv. (Mimeo handout)

4.39. Priestley, C.H.B. and R.J. Taylor. 1972. 'On the assessment of surface heat flux and evaporation using large scale parameters'. Monthly Weather Review. 100:81-92.

4.40. Reddy, K. R., R. Khaleel, M. R. Overcash, and P. W. Westerman. 1979a. 'A nonpoint source model for land areas receiving animal wastes: I. Mineralization of organic nitrogen'. Trans. ASAE 22(4):863-872.

4.41. Reddy, K. R., R. Khaleel, M. R. Overcash, and P. W. Westerman. 1979b. 'A nonpoint source model for land areas receiving animal wastes: II. Ammonia volatilization'. Trans. ASAE 22(6):1398-1404.

4.42. Richardson, C. W. 1981. 'Stochastic simulation of daily precipitation, temperature, and solar radiation'. Water Resources Res. 17:182-190.

4.43. Richardson, C. W. 1982a. 'Dependence structure of daily temperature and solar radiation'. Trans. ASAE 25:735-739.

4.44. Richardson, C. W. 1982b. 'A wind simulation model for wind erosion estimation'. ASAE Paper No. 82-2576.

4.45. Ritchie, J.T. 1972. 'A model for predicting evaporation from a row crop with incomplete cover'. Water Resources Res. 8:1204-1213.

4.46. Seligman, N.G. and H. van Keulen. 1981. 'PAPRAN: A simulation model of annual pasture production limited by rainfall and nitrogen'. In Simulation of Nitrogen Behaviour of Soil-Plant Systems, M.J. Frissel and J.A. van Veen, eds. (Wageningen, The Netherlands, Jan. 28-Feb 1, 1980), 192-221.

4.47. Shaffer, M. J. 1991. 'Various chapters In R. F. Follett (ed.) Managing Nitrogen for Groundwater Quality and Farm Profitability'. ASAE Monograph. (in press)

4.48. Sharpley, A. N. and J. R. Williams (eds.) 1990. 'EPIC--Erosion/Productivity Impact Calculator: 1. Documentation'. USDA Tech. Bull. #1768. 235 pp.

4.49. Singer, M. P., F. D. Arnold, R. H. Cole, J. G. Arnold, and J. R. Williams. 1988. 'Use of SWRRB computer model for the national coastal pollutant discharge inventory'. In Proc. AWRA Symp. on Coastal Water Resources, 119-132.

4.50. Sloan, P. G. and I. D. Moore. 1984. 'Modeling subsurface stormflow on steeply sloping forested watersheds'. Water Resources Research 20(12):1815-1822.

4.51. U.S. Department of Agriculture, Soil Conservation Service. 1972. 'Hydrology'. Chapters 4-10. In National Engineering Handbook. USA Government Printing Office.

4.52. U.S. Department of Agriculture, Soil Conservation Service. 1983. 'Hydrology'. Section 4, Chapter 19. National Engineering Handbook. USA Government Printing Office.

4.53. Williams, J. R. 1975a. 'Sediment-yield prediction with universal equation using runoff energy factor'. In Present and Prospective Technology for Predicting Sediment Yield and Sources, USDA, ARS-S-40, pp. 224-252.

4.54. Williams, J. R. 1975b. 'Sediment routing for agricultural watersheds'. Water Resources Bulletin, AWRA 11(5):965-974.

4.55. Williams, J. R. 1980. 'SPNM, a model for predicting sediment, phosphorus, and nitrogen yields from agricultural basins'. Water Resources Bulletin, AWRA, 16(5):843-848.

4.56. Williams, J. R. and R. W. Haan. 1978. 'Optimal operation of large agricultural watersheds with water quality constraints'. Texas Water Resources Inst., Texas A&M Univ., TR-96. 152 pp.

4.57. Williams, J. R. and A. D. Nicks. 1982. 'CREAMS hydrology model--Option one'. pp. 69-86 In V. P. Singh (ed.) Applied Modeling Catchment Hydrology. Proc. Intl. Symp. Rainfall-Runoff Modeling, May 18-21, 1981, Mississippi State, MS.

4.58. Williams, J. R., C. A. Jones, and P. T. Dyke. 1984. 'A modeling approach to determining the relationship between erosion and soil productivity'. Trans. ASAE 27(1):129-144.

4.59. Williams, J. R., C. A. Jones, and P. T. Dyke. 1990. 'The EPIC model'. Chapter 2, pp. 3-02 In A. N. Sharpley and J. R. Williams (eds.) EPIC-Erosion/Productivity Impact Calculator: 1 Model Documentation. USDA Tech. Bull. No. 1768. 235 pp.

4.60. Williams, J. R., A. D. Nicks, and J. G. Arnold. 1985. 'SWRRB, a simulator for water resources in rural basins'. ASCE Hydraulics Journal, 111(6):970-986.

4.61. Wischmeier, W. H. and D. D. Smith. 1978. 'Predicting rainfall erosion losses'. Agriculture Handbook 537, USDA, SEA, 58 pp.

4.62. Woodruff, N.P. and F.H. Siddoway. 1965. 'A wind erosion equation'. Soil Sci. Soc. Amer. Proc. 29:602-608.

4.63. Young, R. A., C. A. Onstad, D. D. Bosch, and W. P. Anderson. 1987. 'AGNPS, agricultural non-point-source pollution model. A Watershed Analysis Tool'. 1987. U.S. Department of Agriculture, Conservation Research Report 35.

Assessment of Uncertainty in Non - Point Source Water Quality Models

A. Ghosh Bobba and Vijay P. Singh*

Aquatic Ecosystem Conservation Branch
National Water Research Institute
Burlington, Ontario, Canada, L7R 4A6

***Department of Civil Engineering**
Louisiana State University
Baton Rouge, Louisiana
U.S.A. 70803-6405

Abstract

To achieve effective environmental control, it is important develop methodologies for dealing with uncertainties in model simulation of contaminant behaviour and effects. Several procedures have been proposed to quantify uncertainties in modelling studies. This paper reviews the two methods that are widely applied, ie. functional analysis and Monte Carlo simulation ,

The first order part of the functional analysis method provides a measure of uncertainties in dependent variables in terms of uncertainties in independent variables. The procedure is based on first-order terms in the Taylor series expansion of the dependent variable about its mean value with respect to one or more independent variables. The major assumption in this procedure is that all independent and dependent variables are the second moment variables (SMV), which means that the behaviour of any SMV is completely described by its mean and standard deviation. The mathematically simplicity of the procedure allows application by simple input - output models. Consequently, it has been applied to many environmental simulators, e.g. hydrological models, stream water quality models, lake water quality models and ground water contamination models.

The Monte Carlo simulation (MCS) method uses a large number of repeated trials or simulations with the values for stochastic inputs or uncertain variables selected at random from their assumed parent probability distributions to establish an expected range of model uncertainty.

Introduction

Many mathematical water quality models have been developed to simulate contamination of and physical and biological processes occurring in stream waters. Applications range from identifying the processes affecting stream water quality to forecasting the quality for operational purposes.

It has been a common practice to describe the problems related to chemical and biological processes in river waters through deterministic partial differential equations **(PDE)**. These models are deterministic in that they provide a single response for each set of model parameters and initial conditions. However, there is always uncertainty, both in the evaluation of field data

V. P. Singh and B. Kumar (eds.), Water-Quality Hydrology, 243–264.

and in the use of mathematical models to predict the outcome of natural processes, since the processes are still poorly understood and full mathematical representation is usually too complicated and too costly to implement. In general, one must deal with four aspects of uncertainty when modelling water quality (Beck, 1987): (a) uncertainties about model structure, i.e., what basic processes are involved, how they interact, and how these processes and interactions can be mathematically characterized in an efficient and parsimonious manner; (b) uncertainty in the model parameters, i.e., parameter identification and calibration problems; (c) uncertainty associated with prediction of the future behaviour of the system, i.e., aggregation of model structure and model parameter uncertainties into overall prediction uncertainty; and (d) reduction of critical model uncertainties via carefully designed experiments and/or monitoring programs. There is also inherent variability and randomness in natural processes and their measurements. All these sources of uncertainty may be represented as input forcing terms in the balance equations. The initial conditions may also be random, either because knowledge of the real initial conditions is imperfect or because measurements are affected by random variations. The model coefficients (rate constants) are random either because our assessment is not perfect or because of random variations in measurements. Inputs may also be uncertain because estimates of future loadings, based on projections and future loadings, may be biased.

The above four types are strongly intertwined. For example, type (c) assumes that the model structure (type (a)) has been reasonably identified, and examines the propagation of model parameters (type (b)) and data uncertainties through the model and the resulting prediction uncertainty. Further, the techniques employed in type (c) may also identify the critical sources of model uncertainty that type (d) seeks to reduce. While each type is of great importance and their interactions cannot be ignored, type (c) is of primary importance from the point of view of the decision maker. The decision making processes often cannot wait for better models to be developed or new data to be obtained. Water quality standards must be set and management strategies implemented in light of existing models and data. Hence, assessment of prediction uncertainty has assumed a much larger role in water quality decision making in recent years. Fortunately, there are sound, well tested mathematical methods available for prediction - uncertainty estimation. These methods are first-order uncertainty and Monte Carlo simulation. This paper will describe the application of first-order error analysis (FOEA) and Monte Carlo simulation (MCS) analysis to a water quality model.

Functional Analysis

The title 'functional analysis' refers to any method which involves the use of an assumed function to approximate the mean, variance and/or higher order moments of an output variable (Y) as a function of one or more input variables (x). This grouping encompasses several closely related methods identified in the literature as error analysis, uncertainty analysis, or confidence interval development. A representative example of a functional analysis method is the method of Benjamin and Cornell (1970) which is based on using the first derivative terms in the Taylor series expansion of the dependent variable. If it is assumed that a functional relationship exists between one dependent variable, Y, and multiple independent random variables (x_1, x_2, ...,x_n) as:

$$Y = f(x) = f(x_1, x_2, \ldots. x_n) \tag{1}$$

Equation (1) can be linearized by writing the Taylor series expansion about the mean value of each independent variable; the mean and variance of Y can be obtained from the linearized equation as follows:

$$E(Y) = f(\underline{x}) \tag{2}$$

$$V(Y) = \sum_{i=1}^{n} \sum_{j=1}^{n} \cdot \frac{\partial f(x)}{\partial x_i} \cdot \frac{\partial f(x)}{\partial x_j} \cdot cov\ (x_i x_j) \tag{3}$$

where $\underline{x}$ denotes the mean vector of the independent random variable and E(Y) and V(Y) denote expected value and variance, respectively. All derivatives are evaluated at $\underline{x}$. Equation (2) then implies that the expected value of Y, E(Y), is the functional value of the mean values of all the independent variables.

If all input variables are mutually independent (i.e. uncorrelated), all covariances between any two different random input variables will disappear from Equation (3), and the variance of Y can be rewritten as:

$$V(Y) = \sum_{i=1}^{n} [\frac{\partial f(x)}{\partial x_i}]^2\ V(x_i) \tag{4}$$

Equation (4) can be used to estimate the effect of the uncertainty in each model parameter and input variable on the model output. The first order derivatives, $\partial f(x)/\partial x_i$, are also called sensitivity coefficients because they describe the variation in the dependent variable caused by a small change in each independent variable.

This approach has the advantage of only requiring estimates of the mean and variance (first and second moments) of the input parameters. In situations where limited information restricts analysis of the random components, the first-order analysis is a useful tool. In some cases,the first order-analysis gives answers comparable to those obtained by more complex analysis procedures. However, in more complicated situations, there are discrepancies between first-order analysis and nonlinearized methods such as a Monte Carlo technique (Scavia et al. 1981). Other functional analysis approaches, such as statistical estimation of moment generating equations (Karmeshu and Lara-Rosano 1987), point estimation techniques (Thorburn 1986), or numerical analysis (Dresnack and Dobbins 1968) are also usually limited to first and second moments.

Monte Carlo Simulation

Monte Carlo Simulation methods were developed in the 1940's as a numerical tool to solve complex sets of equations which were beyond the computational power available at the time (Ulam and von Neumann 1945; Metropolis and Ulam 1949). Monte Carlo simulation (MCS) is a conceptually simple process that has gained popularity with the development of high speed computers. For this method, a probability distribution type is identified or assumed for each uncertain characteristic in the model. This includes model equations, input variables, and parameters. The model is then run repeatedly for predetermined number of times, each time with the value for each uncertain characteristic being randomly selected from the probability distribution. The result is a distribution of values for the model prediction. This distribution can be regarded as the prediction error, representing the combined effects of all the uncertainties in characterizing the model.

Due to the characteristics of MCS, two important factors to be considered for its application are specifying the type of probability distribution of each random variable and selecting the total number of simulations to perform.

An accurate or appropriate description of the probability distribution of each random variable is necessary to properly utilize Monte Carlo simulation (MCS), but for environmental engineering problems, certain distributions are more meaningful and fit the data better than others; these include Normal, Log Normal, Gumbel, Log- Gumbel, Gamma, Log-Pearson Type -III, Beta, Weibull, and Uniform.

Originally, the method was used as a numerical technique to solve differential and integral equations (Meyer 1956). An excellent example of this application which elucidates the basic principles of Monte Carlo is given in a paper by Kahn (1950). Consider the problem of evaluating the following integral:

$$\int_0^\infty g(x)\, dx \qquad 0 < g(x) < 1 \tag{5}$$

The solution is to be determined using random sampling with the following rules: (a) x and y are selected at random from a population uniformly distributed between 0 and 1; and (b) if y is less than g(x), the game is success, otherwise it is failure. Figure 1 shows that the probability of successes is equal to the area under the curve g(x). Therefore an estimate for the solution of equation (5) can be obtained by r/N, where r is the number of successes and N is the total number of trials. This example highlights the two basic traits of Monte Carlo methods: random sampling from a specified distribution and the use of that sample in a specified equation.

Monte Carlo methods are based on random sampling of a specified set. When used as a numerical technique in integration, the specified set is set of possible solutions to the equations over the range being examined. In environmental modelling, Monte Carlo Simulation methods are used to randomly select input values or equation parameters. Because the Monte Carlo Simulation method provides a straight forward way of including stochasticity in environmental models, it is probably the most widely used stochastic method. The procedure can easily be extended to several variables simultaneously. Input variables and/or parameters can be linked by predicting the selection of one variable upon the random outcome of another. Monte Carlo Simulation methods have been used to model environmental systems ranging in complexity from first order decay of organic chemicals in a river (Kothandaraman and Ewing 1969; Burges and Lettenmair 1975) to highly diversified biological communities (Tiwari and Hobbie 1976). The method has also been used in lake models (Fedra 1979), predictor-prey models (Gardner et al. 1980), and ground water pollution models (Smith and Schwartz, 1981; Black and Freyburg 1987). Results of Monte Carlo Simulation are a useful standard to which other methods are compared (van de Kramer 1983; Malone et al. 1983).

Because each input variable is randomly selected for each simulation, its value changes from simulation to simulation. However, its frequency distribution will resemble its parent distribution only when a sufficient number of simulations have been performed. In other words, the greater the number of simulations, better resemblance between generated and parent distribution of each random variable. This in turn results in more reliable information on model output frequency distributions, provided the appropriate input probability distribution is known.

Several methods are available to determine an adequate number of simulations. Malone et al. (1983) cited two methods. One method consists of iterating the sample generating procedure until the sample statistic converges to its population or historical value. The other approach is based on the normality assumption of the mean value of any input variable. The expected error is kept less than the maximum desired error at a given confidence level as:

$$z_x = \frac{S}{\sqrt{n}} \le y \tag{6}$$

where z_x is a normal deviate at a confidence level, x; s is estimate of standard deviation of a selected variable; and y is the maximum desired error for a selected variable. Then the required number of simulations is found as:

$$n = \frac{(z_x * S)}{y} *2 \tag{7}$$

Another approach is to utilize the Kolmogorov - Simrnov (K-S) goodness-of-fit statistic(Spear,1970). Whitehead and Young (1979) used this sample in the Monte Carlo simulation of a dynamic, stochastic model for water quality to obtain distributions for BOD and DO which closely corresponded with the true distribution.

Non-Point Source Water Quality Model

SNSIM is a mathematical model which can be used to formulate a steady-state, one dimensional, simulation model of a stream network (Braster and Chapra 1973). SNSIM appears to be appropriate for application of FOEA and MCS error analysis methods. The reasons are that FOEA requires explicit expressions of mathematical models, i.e., analytical solutions, and SNSIM is capable of handling a variety of variables and parameters, such as non-point source waste loads, algal growth, and sediment oxygen demand. Using systems analysis, a mathematical model can easily be written to determine the levels of dissolved oxygen (DO), carbonaceous biochemical oxygen demand (CBOD) and nitrogen biochemical oxygen demand (NBOD) in a stream. It is designed to evaluate and or to predict the dissolved oxygen, and the carbonaceous and nitrogenous BOD profiles in a river or stream where the effects of dispersion can be assumed to be insignificant.

The stream network consists of a river and its tributaries which are segmented into sections of constant hydrologic, physical, chemical and biological parameters. Loads may be applied pointwise at the ends of the section or as distributed sources along its length. Also point sources of both BOD and DO deficit from minor tributaries can be input at the ends of a section and background loads of BOD and DO deficit can be introduced at the system's upstream ends. The mathematical equations for the calculations of the distribution of CBOD, NBOD and DO deficit have been explained by Braster and Chapra (1973).

An analytical expression is then applied to each section to determine its CBOD, NBOD and DO deficit response to the loadings. This expression is similar in form to the classical Streeter-Phelps equation. Mass balance equations are applied at the junction of sections as well as the more complex junction of the systems tributaries. In this simulation, the model generates results for the entire system.

A computer program was initially written for a single simulation of a river system (Braster and Chapra, 1973). For application of FOEA and MCS, the program was modified keeping most of features in the original program. The only modification made to the mathematical expressions in the carbonaceous BOD removal rate was that deoxygenation rate was assumed to be identical and that the (distance/velocity) term replaced with travel time. As an example , the restatement of the carbonaceous BOD equation :

$$L_{c,t} = L_{c,0} \cdot \exp(-k_c \cdot t) + \frac{L_{c,d}}{k_c} \cdot (1 - \exp(-k_c \cdot t))$$

where

$$t = \frac{x}{u} \tag{8}$$

Whenever the stream encounters different conditions, at any downstream point, say t_1 , such as the presence of point loads, tributaries, or different values of reaction coefficients, and others, the model needs to replace the initial value terms by combining all the effects from upper reaches, tributaries and point loads. At that point, the uncertainty associated with each initial variable contains not only the uncertainty at the headwaters, but all the components contributed by uncertainty from tributaries and point loads. Then, further downstream, say t_2, uncertainty in any process variable contributes to total uncertainty in the dependent variable in two ways: a) uncertainty of process variable from the head water t_1 is included in uncertainty terms of dependent variable at t_1, which then serves as the independent variable from t_1 to t_2 , directly contributes to the uncertainty in dependent variables; (b) the sum of these two components is the total contribution to the dependent variable at t_2 by a process variable.

In the modified model, FOEA is available as an option, and up to 16 variables can be randomized which includes: three reaction coefficients, (k_c, k_n, and k_a); algal oxygen production rate; sediment oxygen demand, headwater loads (HCOD, HNOD, HDOD); point source waste loads (WCOD, WNOD and WDOD); tributary load (TCOD, TNOD, TDOD); and distributed source of CBOD and CNOD. The only variable which is not allowed to be randomized is the travel time, t.

Application of FOEA and MCS techniques to A Non Point Source Water Quality Model

The one-dimensional, steady-state water quality model, SNSIM, is capable of simulating CBOD, NBOD and DO concentration, and evaluating the error term associated with each variable, using either FOEA or MCS technique, under the given condition of uncertainty in a multi-reach river system. To demonstrate the application of the methods, it is used for a hypothetical river basin Braster and Chapra (1973).

The river basin (Figure 2) consists of five headwaters, three point wasteloads from municipalities and industries, and three tributaries. A withdrawal is treated as a tributary with negative loads in the model. Three upper reaches receive substantial amount of non-point source loads of CBOD and NBOD and lower reaches have oxygen demands due to sediments and algal activity. Wasteload No. 1 has a 5000 pounds each per day of CBOD and NBOD.

Thirty three Monte Carlo runs were conducted with various assumptions of uncertainties in input variables and the discharge from wastewater treatment plant No. 1. Due to lack of actual data, all random input variables were assumed normally distributed with the standard deviations in the range from zero to 40% of the mean values. Each run consisted of 200 simulations.

The thirty three MCS runs required eight hundred ninty one sets of two hundred randomized values. The first three statistical moments were evaluated for each set of random values. The means and standard deviations of randomly generated values did not significantly deviate from input values to the randomization process. The average of eight hundred ninty one skewness coefficients was +0.022. Eight hundred sixteen values were within ± 0.3 and only 7 values were outside the range of ± 0.5. Based on this result, the random number generation process performed as desired.

Distribution of DO Values: The minimum DO concentration in the river system occurred between river miles twenty six and twenty eight depending, to some extent, on the waste loads from WWTP no. 1. The range of the minimum concentration was from 3.5 to 5.5 mg/l. The DO concentration at the end of the last reach was raised close to saturation concentration, in the range from 7.5 to 8.5 mg/l for twenty nine runs. The distributions of DO values at the sag point and at the river end were fitted to normal and log-normal distributions using five goodness-of-fit test (GOFT) statistics (Bobba et al. 1993).

At the sag point, the normal distribution hypothesis for the DO values was not rejected by any statistics at the 85% confidence level (CL) in 25 out of 33 runs, and rejected by all statistics at 97.5% CL in only 2 runs. The log-normal distribution was not rejected at the 85% CL in 17 runs and rejected at 97.5% CL in 2 runs.

At the stream end, the normal distribution was not rejected at the 85% CL in 19 out of 29 runs, and rejected at the 97.5% CL again in 2 runs. The log-normal distribution was not rejected at the 85% CL in all 9 runs analyzed.

The MCS performed GOFT tests for log-normal distributions only when the untransformed data are positively skewed and do not contain any negative values. At the DO sag point, either DO or DO deficit data were positively skewed and the GOFT test for log-normal distribution was performed on positively skewed data. However, at the river end, as DO values approach the saturation value, some DO deficit values were negative due to photosynthetic oxygen production. When DO data were negatively skewed and DO deficit was negative in any simulation, the GOFT test for log-normal distribution was not performed.

Based on this result, it can be concluded that both normal and log-normal distributions describe the distribution of DO values in the stream fairly well. However, because the normal distribution is more widely used and because it out performed the log-normal distribution in a small number of simulations, it would be the distribution of choice in this test.

Comparison of FOEA and MCS Results

Two series of computer runs were conducted to compare the output of FOEA and MCS. In one series, the CBOD and NBOD loads from the WWTPS were randomized. These two input variables are linearly related to the DO values in the river. In other series, three reaction rate coefficients, k_c, k_n, and k_a were randomized. These coefficients are non-linearly related to the DO values.

a) Randomization of Linear Input Variables: The predicted DO values and 95% confidence limits generated by FOEA and MCS along the main stream are shown in Figure 3. The standard deviations of the effluent load from WWTP is 20% of the mean value. For this series, WWTP No. 1 was assumed to discharge an average of 2500 pounds/day each of CBOD and NBOD. As expected for the linear input variables, the variations in DO due to the variations in waste loads were in close agreement as predicted by two methods. MCS appears to estimate a slightly larger variation than FOEA, but the maximum difference is less than 3% as shown in Figure 4. The maximum variation occurs beyond the DO sag point.

The GOFT test on the two hundred simulated DO values by MCS at the sag point did not reject the hypothesis of a normal distribution at 85% confidence level. In such a case, the probability of a DO value at the sag point being less than 5 mg/l was evaluated using three methods as follows:

1) Calculate the standardized value of the minimum required DO concentration (5.0 mg/l in this study) using the mean and the standard deviation of DO values at the sag point predicted by MCS as:

$$Z_5 \quad = \quad (5.0 - DO)/S_{DO} \tag{9}$$

then read the probability for Z_5 from the standard normal distribution table (parametric method).

2) Count the number of simulation which predicts DO value at the sag point less than 5.0 mg/l and divide it by the total number of simulations (nonparametric method).

3) Repeat (1) using the values predicted by FOEA.

Figure 5 depicts the probability of violation (DO < 5 mg/l) vs. the coefficient of variation of the effluent loads. MCS and FOEA produce essentially identical results in the range of variation presented in the Figure 5.

b) Randomization of Nonlinear Input Variables: In this series of MCS runs, three reaction rate coefficients, k_c,k_n, and k_a were assumed subject to variations. Figure 6 shows the FOEA predicts much wider confidence intervals than MCS, especially near the DO sag point. However, beyond the sag point, as the DO value approaches the saturation value, the errors predicted by two methods become more equal and both generally decrease. This can be more clearly seen in Figure 7 where the standard deviations of the output arrays, using both methods, are shown diverging to a maximum in the sag zone and the converging and decreasing beyond the sag zone.

c) Figure 8 shows the probability of DO values at the sag point being less than 5 mg/l. The two methods result in a large difference in predicting the probability of DO being less than 5 mg/l, over the entire uncertainty range of input variables, from 0.1 to 0.4 of the coefficient of variations. However, the difference in two methods does tend to decrease as the uncertainty in the reaction rate coefficients increases. Since the average DO value at the sag point is 5.2 mg/l (Figure 3), the maximum probability of DO being less than 5 mg/l can not exceed 0.5 with an estimated normal distribution, regardless of the uncertainty level in input variables.

Parameter Estimation

The sensitivity of a model to given input parameter or variable depends on the model structure as well as the probability distribution of the uncertainty in the parameter. In most of the simulation models, a few parameters usually exhibit major influence on the model output. The choice of a value is critical for these parameters, but not for others. Moreover, the degree of influence of a certain parameter may change with temporal and spatial variation. As a result, if an analyst is interested in the model prediction at a certain point, both in time and space, refining the values of one or two parameters which exhibit the most influence at that point may be needed.

One advantage in using FOEA is that it provides the components of the total uncertainty in the model output resulting from the uncertainty in each parameter and input variable. The uncertainty components can therefore be combined to give the total uncertainty, or several subtotals of partial uncertainties of similar types.

In the following example, sixteen input variables were exposed to uncertainty; the variables and their uncertainty levels are listed in Table 1. These variations might be much smaller than expected, but they still produce significant uncertainties in the predicted DO values.

The uncertainty components in the predicted DO deficiency, in mg/l, contain the combined effect of the uncertainties in the DO deficiency in headwaters, wasteload effluent, and tributary flows, together with the components for CBOD and NBOD. The minimum DO value in the stream is 5.15 mg/l at mile point 26 and the maximum total uncertainty in DO is 0.704 mg/l at the last point in the river. The extent of influence by each input variable on the uncertainty in the predicted DO changes with spatial variations depending on the characteristics of each variable. For example, a DO deficit in the headwaters has its maximum influence at mile point zero.

Figure 9 illustrates the variation of the uncertainty components in three reaction rate coefficients along the main stream. Note that the square of subtotal uncertainty in DO caused by these three variables is the sum of the squares of uncertainty in the DO caused by each variable, separately.

To present uncertainty components in the predicted DO graphically, the uncertainty components are combined into six subgroups. As shown in Figure 10, at zero mile point, the measurement error of DO (or equivalently DO deficit) in the head water is the only component. However, at the sag point (mile point 26), the uncertainty in CBOD reaction rate coefficient, k_c, is the most important factor. Finally, at the end point, the uncertainty in algal oxygen demand accounts for 71% of the total uncertainty in DO.

If the uncertainty in the algal term were reduced by half, total uncertainty in the DO forecast would decrease by 43% as shown in Figure 11. This example indicates that the method to reduce uncertainty should include increased survey and analysis to better define the more important parameters at the site of interest.

Wasteload Allocation

In the following example, WWTP No. 1 is the largest point-source load discharging 1 million gallons a day containing 5000 lbs/day each of CBOD and NBOD. To improve the downstream water quality, three alternatives are assumed possible for the plant's effluent. For each alternative, the probability of the DO value at the sag point less than the target value 5 mg/l is calculated for increase in the effluent load standard deviation from zero to 25% of the mean value. Figure 12 shows of this situation.

Although alternative B allows the discharge of more total BOD, it would cause less of a probable violation than alternative A at the same level of uncertainty. Alternative C significantly reduces the probability of a violation, but it is noteworthy that 25% variation in alternative C is as vulnerable to violation as only 7% variation in alternative A or B.

Limitations of methods
Functional Analysis

Functional analysis are in general, limited to generation of only the mean and variance of the output variable. Complete description of a distribution requires definition of all higher order moments. Thus, these methods do not entirely describe the distribution concept in the special case where the output is normally distributed. First order error analysis can be extended by truncating the Taylor Series expansion after the second or third order, but carrying higher order terms complicates the mathematics.

Monte Carlo Simulation Method

While the Monte Carlo method is straight forward, it is not without its weakness and limitations. Random number generation, the basis of the method, is not an easy process, and can be troublesome and even unreliable, especially on smaller computers. Random number generators usually produce a set of uniformly distributed values between 0 and 1. If some other distribution for the input variable is required, then the distribution of the random number must be transformed. Detailed discussion of the mathematics of the random variable transformations may be found in Benjamin and Cornell (1970) and Sobol (1974).

Another central issue in Monte Carlo simulation methods is determination of the number of times to run the model. Enough runs must be performed to obtain a statistically valid sample. The number required is usually a function of the output distribution, which is not known a priori. In practice, a certain number of trials are executed and the resulting statistics calculated. The process is then repeated with a larger sampling, new statistics calculated and compared to the previous run. If there is no 'significant' difference, then three hundred runs are sufficient. If there is a significant difference, the number of runs are increased and the new statistics compared to the run of six hundred. This process is repeated until no 'significant' difference is found between two successive runs. 'Statistical significance' can be determined by any one of several standard statistical tests for comparison of distributions (Benjamin and Cornell 1970).

Conclusions and Recommendations

In any environmental modelling, there are inevitable uncertainties in the model structure and parameter input values. The conventional deterministic modelling approach lacks mechanism for estimating the effects of uncertainties on the model response. Sensitivity analysis provides some insight, but is limited. In this paper two error analysis techniques applied in modelling process, first order error analysis (FOEA) and Monte Carlo simulation (MCS), are discussed in detail and some examples of their use are provided. The following is summary of the conclusions contained in this paper.

(a) The applicability of the first order error analysis (FOEA) is inherently restricted by three factors: (1) degree of nonlinearity in the model, (2) distribution of errors in each random variable, and (3) magnitude of errors. (b) one particular advantage of the first order error analysis is that it provides the components of total uncertainty in model output caused by each random input variable. When the model is highly nonlinear, first-order error analysis may need to be extended to some higher order-error analysis. (c) Monte Carlo simulation is conceptually simple, theoretically sound, and flexible. Two important factors to be considered in any application of Monte Carlo simulation are: (1) probability distribution for each random input variable needs to be selected, and (2) number of simulation requires specification. (d) These two error analysis procedures were incorporated, as options, in a BOD-DO model. This new model is capable of simulating water quality in a complex stream system as well as estimating uncertainties in predicted BOD and DO values which would result from uncertain input conditions. (e) The result of applying these procedures to the model is summarized as follows: (1) If the random input variables are linearly related to BOD and DO in the stream, such as wasteloads from treatment plants, and their distributions are normal, FOEA provides an exact solution for errors associated with predicted BOD, DO values. (2) If a limited number of nonlinear input variables are randomized, MCS provides the option of choice. (3) When a reasonably large number of input variables are randomized, the model output tends to be normally distributed. If moderate departures from normality are present in the random input variables, FOEA provides good estimates of uncertainties in model response. (4) The maximum uncertainty in predicted DO value occurs well beyond the DO sag point for the example selected. (5) The magnitude of influence of uncertainty in each random variable varies with the location along the stream. (6) One or two random variables exhibited a major influence on uncertainty in predicted DO value at any particular point in the stream for the example

Table 1: Uncertainties in input variables for test problem

Variables	Coefficient of Variation
CBOD reaction rate coefficient, k_c	0.1
NBOD reaction rate coefficient, k_n	0.1
Atmospheric reaeration rate coefficient, k_a	0.1
Headwater CBOD	0.1
Headwater NBOD	0.1
Headwater DO deficit	0.1
Effluent loads CBOD	0.1
Effluent loads NBOD	0.1
Effluent loads DO	0.1
Tributary loads CBOD	0.1
Tributary loads NBOD	0.1
Tributary loads DO deficit	0.1
Algal oxygen demand	0.2
Sediment oxygen demand	0.2
Non-point source CBOD	0.2
Non-point source NBOD	0.2

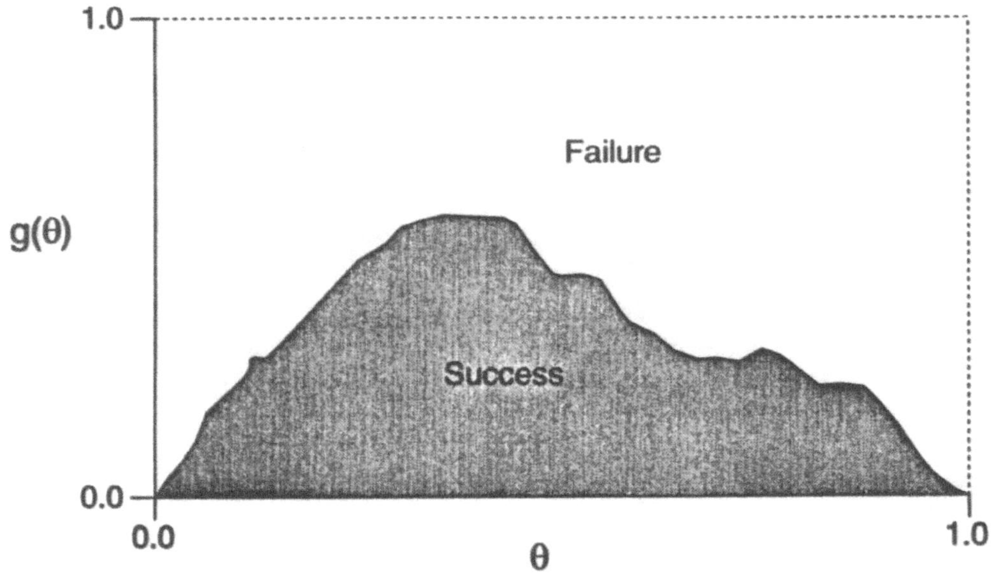

Figure 1 Application of Monte Carlo technique (From Khan 1950).

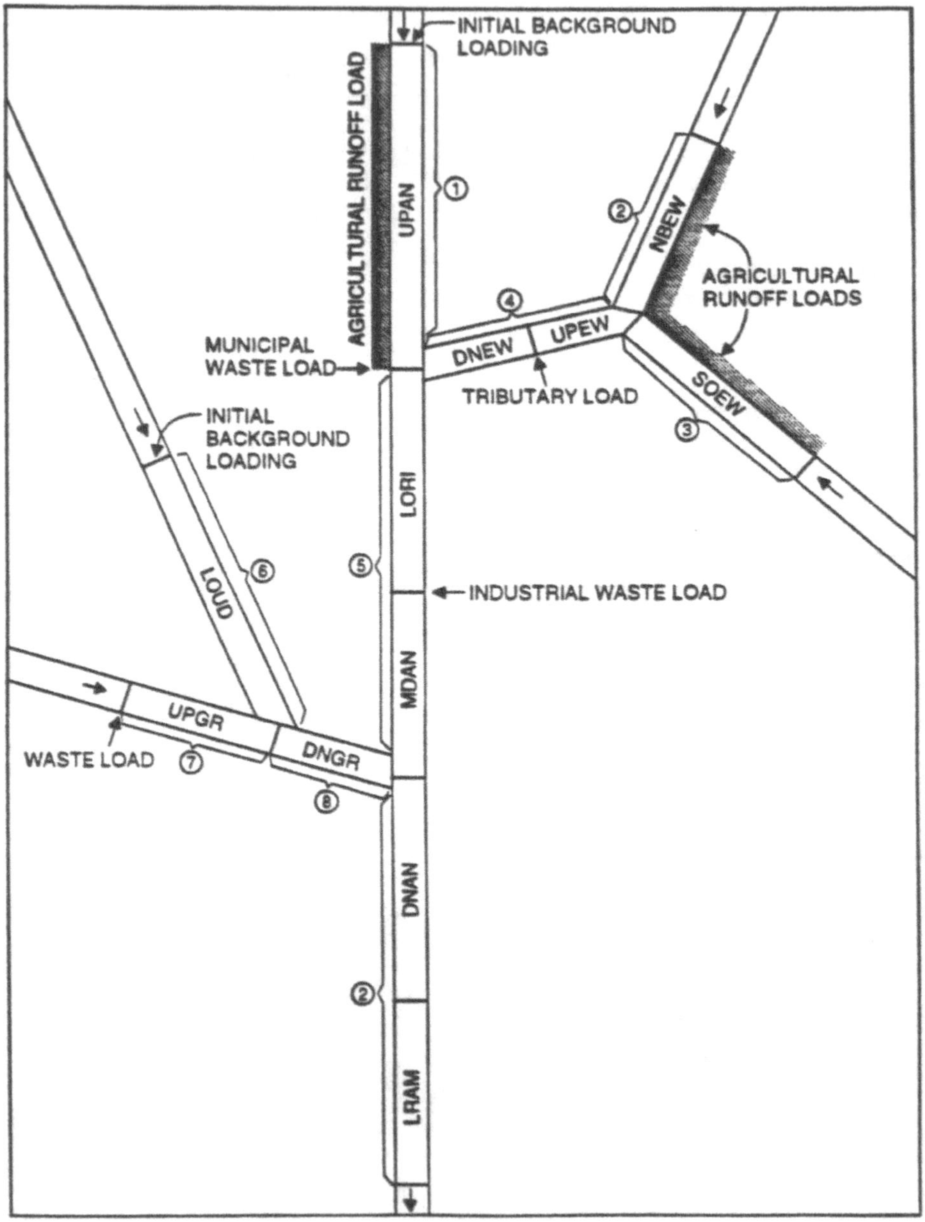

Figure 2 Schematics of river in watershed.

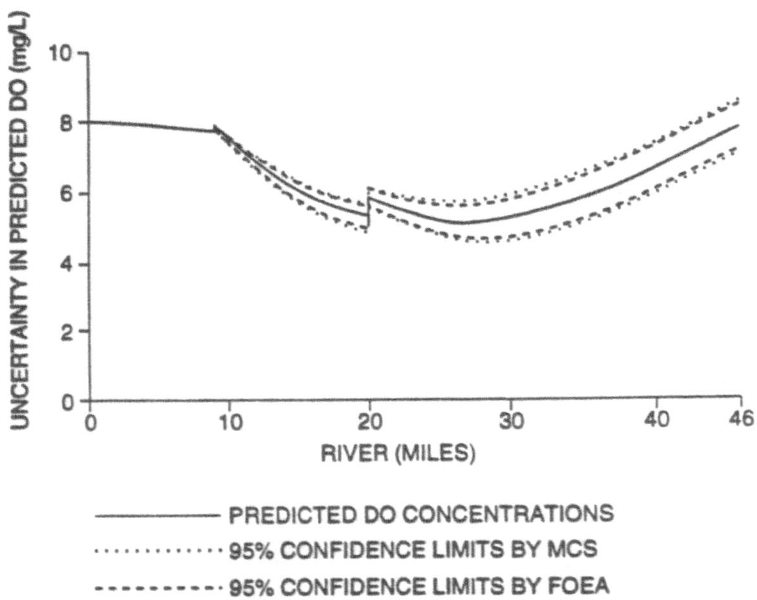

Figure 3 Predicted DO concentration and the 95% confidence limits along the main stream when point source wasteloads are randomized.

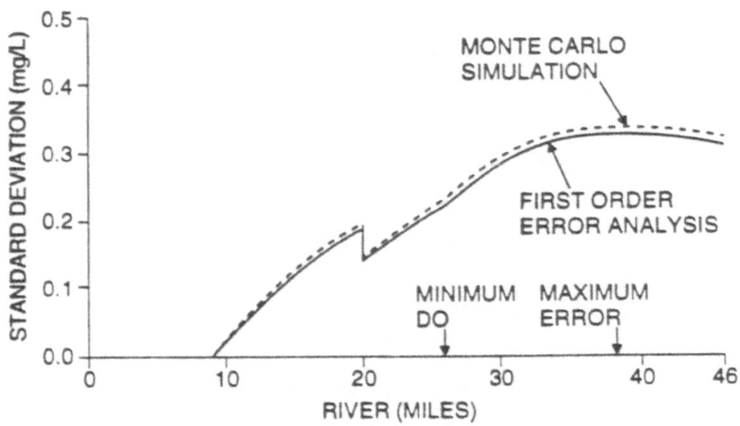

Figure 4 Uncertainty in predicted DO value along the main stream when point source wasteloads are randomized.

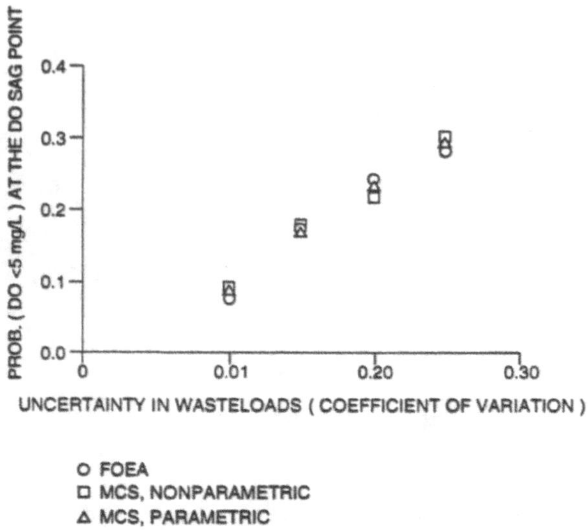

Figure 5 Probability of violation (DO_{sag} < 5 mg/l) when point source wasteloads are randomized.

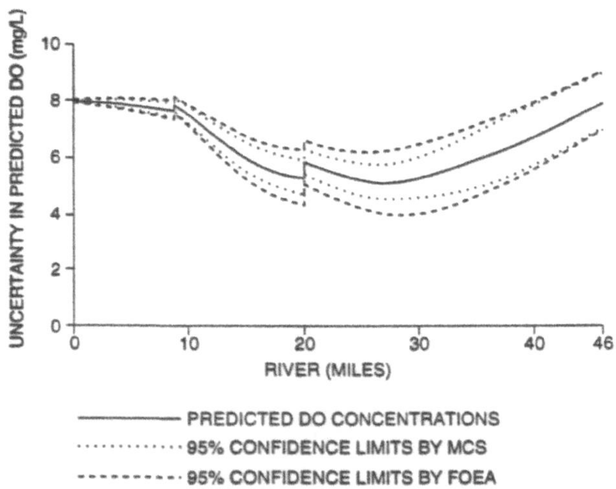

Figure 6 Predicted DO and the 95% confidence limits when k_c,k_n and k_a are randomized.

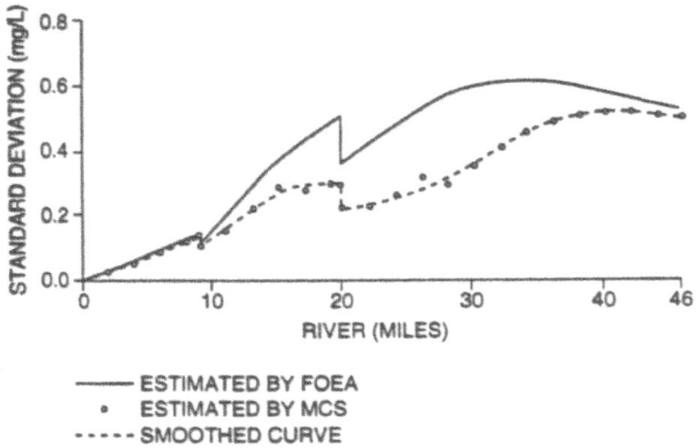

Figure 7 Uncertainty in predicted DO value along the main stream when k_c, k_n and k_a are randomized.

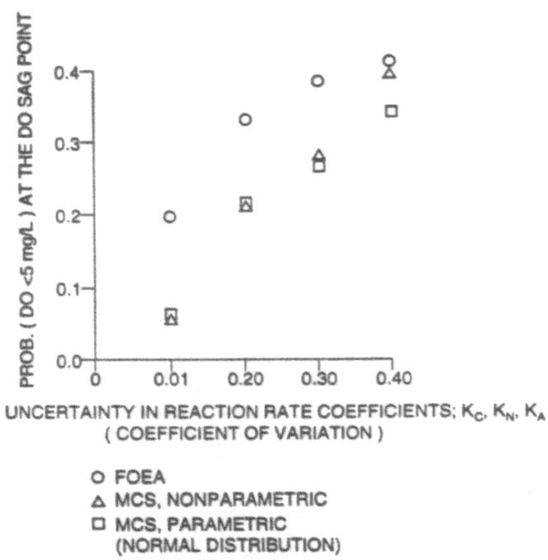

Figure 8 Probability of violation (DO_{sag} < 5mg/l) when k_c, k_n, and k_a are randomized.

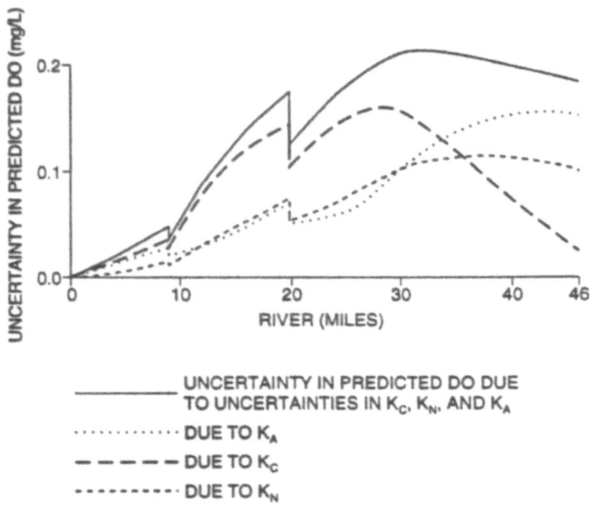

Figure 9 Uncertainty components in predicted DO due to uncertainties in k_c, k_n and k_a.

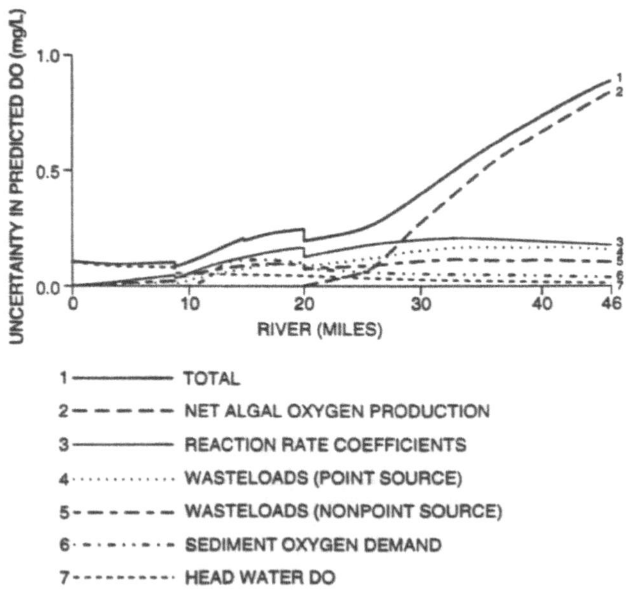

Figure 10 Uncertainty components in predicted DO when 16 input variables are randomized.

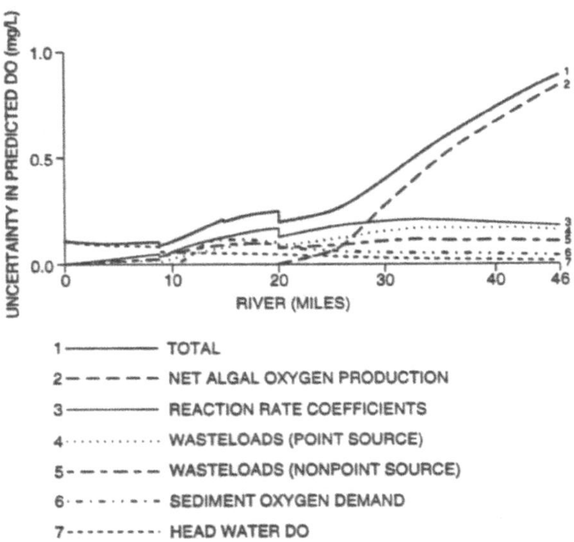

Figure 11 Variation of total uncertainty in predicted DO due to different level of uncertainty in algal oxygen production.

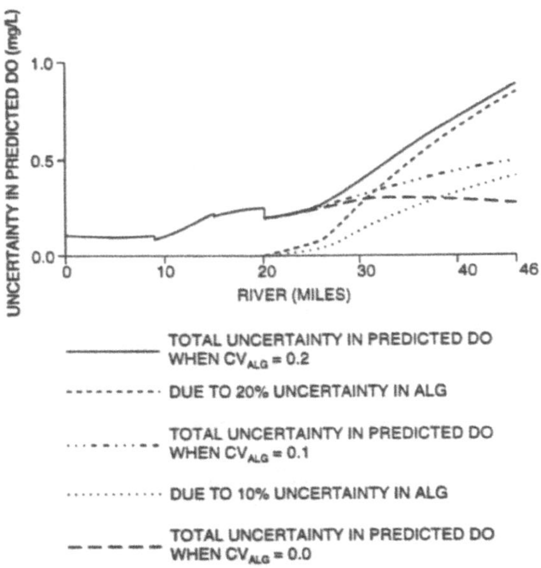

Figure 12 Probability of violation (DO < 5 mg/l) as a function of uncertainty levels in point source loads.

considered. (7) These error analysis procedures can be used to estimate the probability of violation of minimum DO requirements in stream. (8) The error analyses should be considered in effluent management programs.

Based on the findings in this simulation study, the following recommendations are made: (a) In this study, the modified model was applied to a fictious data set. For further verification, it is desirable to apply this model to data from actual river systems to further explore the utility of this modelling strategy. (b) This model may be further improved using procedures such as random number generations from 3-parameter gamma distribution, 4-parameter beta distribution, and data-defined and user defined distributions. (c) A general purpose goodness-of-fit test procedure is required to determine the distribution of input variables to available data. It should conclude extreme-value, 3-parameter gamma, and 4-parameter beta distributions.

Acknowledgments

We thank Dean Jeffries and G. Padmanabhan for helpful comments on early drafts of this paper. The encouragement and support from John Carey is also appreciated.

References

Beck, M.B. 1987. Water quality modeling: A review of the analysis uncertainty. Water Resources Research. Vol. 23, No. 5, pp. 1393-1441.

Benjamin, J.R., and Cornell, C.A. 1970. Probability, statistics and decision for Engineers.McGraw Hill Book Company, New York, N.Y.

Black, T.C. and Freyburg, D.L. 1987. Stochastic modeling of vertically averaged concentration uncertainty in a perfectly stratified aquifer. Water Resources Research. Vol. 23, No. 6., pp.997-1004.

Bobba, A.G. and Singh, V.P., 1994. Application of Monte Carlo analysis of saturated subsurface water contamination modelling. Proceedings of International Conference on hydrology and water resources and water resources (In press).

Braster, R.E and Chapra, S.C. 1973. Documentation for SNIM1/2, a computer program for the steady state water quality simulation of stream network. USEPA, Region II, 26 Federal Plaza, New York, NY 10007. pp 18.

Burges, S.J., and Lettenmaier, D.P. 1975. Probabilistic methods in stream water quality management.Water Resources Bulletin. Vol. 11. No. 1, pp. 115-130.

Cornell, C.A. 1972. First order analysis of model and parameter uncertainty. Proc. Int. Symp.on uncertainties in Hydrology and Water Resources Systems. University of Arizona, Vol. 2., pp. 1245-1272.

Dobbins, W.E. 1964. BOD and oxygen relationship in streams. Journal of the Sanitary Engineering Division. ASCE. Vol. 90., No. SA3, pp. 53-57.

Dresnack, R., and Dobbins, W.E. 1968. Numerical analysis of BOD and DO profiles, Journal of Sanitary Engineering. Vol. 94., SA5, pp. 789-807.

Fedra, K. 1979. A stochastic approach to model uncertainty. A lake modelling example.

Working paper No. WP-79-63, International Institute for Applied Systems Analysis, Vienna, Austria.

Gardner, R.H., Huff, D.D., O'Neill, R.V., Mankin, J.B., Carney, J., and Jones, J. 1980. Application of error analysis to a marsh hydrology model. Water Resources Research. Vol. 16., No. 4., pp. 659-664.

Gardner, R.H., O'Neill, R.V., Mankin, J.B., and Kumar, D. 1980. Comparative error analysis of six predictor prey models. Ecology vol. 61 No. 2, pp.323-332.

Kahn, Herman. 1950. Random sampling (Monte Car;o) techniques in neutron attenuation problems- PartI. Nucleonics. Vol. 6, No. 5. pp. 27-33.

Karmeshu and Lara-Rosano,F. 1987. Modelling data uncertainty in growth forecasts. Applied Mathematical Modelling. Vol. 11, No. 1, pp. 62-68.

Kothandaram, V., and Ewing, B.B. 1969. A probabilistic analysis of DO-BOD relationships in streams. Journal of the Water Pollution Control Federation. Vol. 41; No. 2; R73-R90.

van de Kramer, S. 1983. Monte Carlo simulation and first order error analysis: Two possible methods to cope with uncertainties in water quality modelling. Working paper No. WP-83-9, International Institute for Applied Systems Analysis, Vienna, Austria.

Malone, R., Bowles, D.S., Windham, M.P., Grenney, W.J. 1983. Comparison of techniques for assessing of loading uncertainty upon a longterm phosphorous model. Applied Mathematical Modelling. Vol # 7, No. 1, pp.11-18.

Melching, C.S. 1992. An improved first order reliability approach for assessing uncertainties in hydrologic modelling. J. of Hydrology, Vol. 132, pp. 157-177.

Melching, C.S., Yen, B.C., and Wenzel, H.G., Jr. 1990. A reliability estimation in modelling watershed runoff with uncertainties. Water resources Research. Vol. 25, No. 10, pp. 2275- 2286.

Metropolis, N and Ulam, S. 1949. The Montecaro Method. Journal of the American Statistical Association. Vol. 44, No. 247, pp. 335-341.

Meyer, H.A. 1956. Symposium on Monte Carlo Methods. Proceedings of Symposium held at the University of Florida. 1954. John Wiley and Sons, New York, N.Y., USA.

Rubinstein, R.Y. 1981. Simulation and Monte Carlo method. John Wiley and Sons, New York, N.Y.

Scavia, D., Canale, R.P., Powers, W.F., and Moody, J.L. 1981. Variance estimates for a dynamic eutrophication model of Saginaw Bay, Lake Huron. Water Resources Research. Vol. 17, No. 4, pp.1115-1124.

Scavia, D., Powers, W.F., Canale, R.P., and Moody, J.L. 1981. Comparison of first order error analysis and Monte Carlo simulation in time dependent lake eutrophication models. Water Resources Research. Vol. 17, No. 4, pp. 1051-1059.

Smith, L., and Schwartz, F.W. 1981. Mass transport. Part 2. Analysis of uncertainty in prediction. Water Resources Research. Vol.17, No. 2, PP. 351-369.

Smith, L., and Schwartz, F.W. 1981. Mass transport, part 3. Role of hydraulic conductivity data in prediction. Water Resources Research. Vol. 17, No.5, pp.1463-1479.

Sobal, I.M. 1974. The Monte Carlo Method. The University of Chicago Press. Chicago, MI. USA

Spear, R.C., 1970. The application of Kolmogorov-Renyi statistics to problems of parametric uncertainty in system design. International Journal of Control, Vol. 11, No. 5, pp. 771-

778.

Thoburn, D. 1986. A Baysian approach to density estimation. Biometrika, Vol. 73, No. 1, pp. 65-75.

Thomann, R.V. 1972. Systems analysis and water quality management. Environ. Res. Applications, New York., N.Y.

Thomann, R.U. 1982. Verification of water quality models. Journal of Environmental Engineering Division, ASCE. Vol. 108, No. EE5, pp. 923-940.

Tiwari, J.L. and Hobbie, J.E. 1976. Random differential equations as models of ecosystems: Monte Carlo simulation approach. Mathematical Bioscience, Vol. 28, No. 1, pp. 25-44.

Ualm, S.M., and von Neumann, J. 1945. Random ergodic theorems. Bulletin American Mathematical Society. vol. 51., No. 9., p660.

Whitehead, P., and Young,P. 1979. Water quality in river systems: Monte Carlo Analysis, Water resource Research, Vol. 19, No. 5, pp. 1266-1270.

ENVIRONMENTAL ASPECTS OF SUBSURFACE DRAINAGE PROJECTS

S.K. Kamra and K.V.G.K. Rao
Central Soil Salinity Research Institute,
Karnal- 132001, India.

ABSTRACT

Irrigation and drainage projects affect the physical, chemical, and biological components of the ecosystem and their eventual success depends on how adequately the environmental aspects are addressed at the planning and design phases. Drainage is generally required to combat the twin problems of waterlogging and soil salinity and to ensure sustained irrigated agriculture in the arid and semi-arid regions. While the benefits of drainage can be counted in terms of improved crop yields and increased economic gains, the cost to the environment may also be high. Environmental impacts may be positive or negative and may occur both on and off site. Their relative importance should be considered in the context of the socio- economic conditions and evaluated on both short and long term bases. The positive on site impacts of drainage projects, besides increased agricultural production and improved water regime and microclimate, also include reduction in water borne diseases and increased social, cultural and employmental opportunities. Among the negative offsite impacts, the concern for the disposal of saline drainage effluent, historically confined to major soluble salts and plant nutrients, now also includes toxic concentrations of several trace elements and pesticides. Reliable estimates of the volume and composition of drainage effluent in time are needed to envisage and plan environmentally acceptable strategies for its disposal or treatment. Public health issues also need to be addressed.

The paper reviews the environmental issues related to drainage projects and identifies the technical options for the management, reuse and disposal of drainage effluent. Improved on- farm water management practices for minimizing the volume and salt load of drainage water, relative merits of different modes of conjuntive use of fresh and saline water including drainage effluent, and water treatment options are discussed. An example on the application of a numerical model to compare the drainage salt load under alternate drainage designs during reclamation of a 2000 ha highly waterlogged saline area in Haryana is presented. Future research and monitoring needs are also discussed.

INTRODUCTION

It is estimated that currently 1 to 1.5 million ha of agricultural lands are damaged around the world each year through man- induced salinisation (Kovda, 1983). A major part of this problem can be attributed to a threefold

V. P. Singh and B. Kumar (eds.), Water-Quality Hydrology, 265–285.

expansion in the world's irrigated area between 1950 and 1986. Irrigation water supplies in arid and semi- arid regions usually exceed 10000 m^3/ha a year, significantly more than the annual precipitation and consequently have great influence on the environment. Postel (1990) estimated that India accounts for 20 million ha, or one third of the global 60 million ha salt affected soils. The social cost of waterlogging and soil salinisation is not easy to quantify. Salinisation causes occupational or geographical shifting of the population and reduction in aggregate regional income. WAPDA (1988) estimated annual loss of US$ 300 million to the economy of Pakistan due to decrease in productivity of soils affected with salinity in Punjab and North West Frontier Province. The agricultural losses from salinisation are projected at US$ 267 million a year in 2010 for Colorado River Basin of the US if controls are not instituted (El-Ashry et al., 1985). Joshi and Agnihotri (1984) estimated the annual loss due to soil salinity and waterlogging in eleven selected irrigation projects in India to be more than Rs. 6000 million (or US$ 200 million at current prices).

Much of the waterlogging and salinisation problems of irrigated lands can be attributed to the fact that adequate drainage measures were either not considered at the design and planning phases of irrigation projects or were not provided when the problem started to emerge or, when provided, were not properly designed or maintained. Lesaffre (1992) estimated that subsurface drains have been installed in more than 50 million ha area in the world (majority in developed countries). Biswas (1985) projected that about 45 million ha of the 92 million ha irrigated land in the developing world will require improvement by 1990 at US$ 22 billion cost at 1975 prices. Approximately 5.5 million ha irrigated land is expected to be drained in Egypt by 2000 while Pakistan plans to drain both by tubewell and tile drainage over 8.5 million ha by 1993. The large scale drainage projects are still at the take-off stage in India. Though sporadic pilot studies (see for example Rao et al., 1986) have been undertaken in the past, large scale drainage works in the states of Haryana, Punjab, Rajasthan, Gujarat, Karnataka and Maharashtra are being envisaged and planned only in the recent years.

Irrigation and drainage projects affect the physical, chemical and biological components of the ecosystem. The ICID's Environmental Impact Checklist (Mock and Bolton, 1991) describes the possible environmental impacts of these projects and also identifies the measures avoid or control the adverse effects. The drainage projects have several shortterm and longterm positive or negative on and offsite effects. The positive on site impacts of drainage projects, besides increased agricultural production and improved water regime and microclimate, also include reduction in water borne diseases and increased social, cultural, recreational and employmental opportunities. There is, however, a growing awareness of the hazards posed by saline drainage effluent to downstream environment. These waters generally contain high concentrations of soluble salts and plant nurients, and sometimes potentially toxic trace elements and pesticides. The disposal and storage of such waters can have remarkable impacts on the degradaion of surface and ground water quality, threat to aquatic organisms, wildlife and plants and potential risk to public health (Letey et al., 1986; Tanji, 1990). All these aspects need to be addressed during the planning and design phases of drainage systems.

The paper reviews the environmental issues which are likely to arise in drainage projects of India and summarizes the scope of technological options tried in other countries and in India for the management, reuse and disposal of drainage effluent. Improved on- farm water management practices for minimizing the volume and salt load of drainage water, relative merits of different modes of conjuntive use of fresh and saline water including drainage effluent, and water treatment options are discussed. An example on the application of a two- dimensional solute transport model to compare salt load of drainage effluent under alternate drainage designs in a 2000 ha highly waterlogged saline area in Haryana is presented. Future research and monitoring needs are also discussed. It is expected that the paper would provide management guidelines during planning and execution of large scale drainage projects in India to ensure their agricultural and environmental sustainability.

DISPOSAL, MANAGEMENT AND REUSE OF DRAINAGE WATER

If the drainage area is a small fraction of the total basin area, saline drainage water may be accomodated by disposal to the river system or by reuse. With more area coming under drainage, these modes of disposal are becoming increasingly inadequate. The options for managing drainage water, and strategies for reuse and disposal are summarized in Table 1 and will be discussed in the same framework.

1. Drainage Water Disposal

(a) River Discharge with and without Dilution

Disposal of the drainage water into rivers or canal can deal with large volumes and, in the medium term, may be the only practical solution at the inland sites. The acceptability of this mode of disposal depends largely on the extent of downstream water use and on the environmental susceptability of the river course and the flood plain. Discharge to the river system is inexpensive but it is likely that disposal of large volumes of saline drainage effluent through this mode will become increasingly difficult due to environmental restrictions, except perhaps when the discharge is very remote from the downstream user and higher dilution levels are possible.

Table 1 Options for the management, reuse and disposal of drainage water

1. Drainage water disposal

River Discharge with and without Dilution
Evaporate Drainage Water in Ponds
Deep Well Injection

2. Drainage water reuse

Agriculture
Industrial Use and for Cooling of Power Plants
Solar Ponds for Electric Energy
Salt Production in Evaporation Ponds
Aquaculture
Agroforestry

3. Source control for drainage water quality and quantity

Water Management: irrigation and leaching efficiencies
Drainage Management: depth and spacing of drains
Crop Management
Alternate Landuse

4. Drainage water treatment

Desalinization
Trace Element Treatment (Biological, Physical and Chemical Processes)

5. Human, aquatic and wildlife ecology

Protection, Restoration and Enhancement; removal of toxic elements
Provision of adequate fresh water

6. Institutional Changes

Water Pricing and Marketing
Modify or Eliminate Irrigation Subsidies
Modify Tax Structures and Rates
Impose Drainage Effluent Fees
Reallocate Water from Agriculture to other Uses

(Adapted from Tanji, 1990)

The maximum drain discharge generally takes place during monsoon (rainy season) when the irrigation demands are relatively less and consequently large volumes of saline drainage water can be discharged into the river system. However, there is a strong need to establish numeric water quality standards for various points along the river, monitoring and modelling of spatial and temporal water quality trends and provisions for diverting saline drainage waters away from the river. In a case study on the disposal of drain discharge into the Yamuna river during June to October, UNDP (1985) assumed the electrical conductivity (EC) of river and surface drainage water as 0.2 dS/m and 1.2 dS/m respectively, of subsurface drainage water (EC_a) as 6 and 10 dS/m and restricted the salinity of river water after mixing at 0.75 dS/m. Based on 80 % frequency discharges of the river at Wazirabad (Delhi) and of drain No. 2 and 8 of Haryana, additional allowable discharge of subsurface drainage water was calculated. Assuming a subsurface drainage surplus of 1.5m/ha/day during the monsoon period, the area that could evacuate its drain discharge into the river was calculated (Table 2). It is seen that about 0.1 million ha critical area in the command can evacuate even 10 dS/m drainage water in river Yamuna during July and August.

Table 2 Allowable subsurface drain discharge and drainable area into Yamuna river

Months	drain discharge m^3/s		Drainable area (ha)	
EC_a(dS/m)=	6	10	6	10
June	0.9	0.5	5,000	3,000
July	25.4	14.4	146,000	83,000
August	47.6	27.0	274,000	156,000
September	6.5	3.7	37,000	21,000
October	3.0	1.7	17,000	10,000

For sustainable irrigated agriculture in inland saline soils, drainage water needs to be disposed of in the ocean or in terminal lakes. The most intensively irrigated part of India, encompassing western portions of the states of Punjab, Haryana and Rajasthan, needs drainage but faces a serious long- term problem of effluent disposal. Barber (1985) estimated that about 1 million ha area in these states had water table less than 3 m from the surface and 0.25 million ha had water table at less than 1 m depth for long periods in the year. The ground water in most part of the region is saline and not suitable for irrigation. Among other engineering and management alternatives suggested by Barber (1985), the proposal of an outfall drain through Indian territory of 900 km, including 600 km in the irrigated area (in the states of Punjab, Haryana, Uttar Pradesh, Rajasthan and Gujarat), and requiring a lift of 80 m needs serious consideration and should not be dismissed only on the plea of huge investments involved. A similar 250 km long Left Bank Outfall Drain (LBOD), aligned along the left side of Indus river to dispose drainage effluent from sizeable areas of Punjab and Sind provinces into the Arabian sea, has already been commissioned in Pakistan.

(b) Evaporation Ponds

One means of drainage water disposal in inland basins with no outlet is its storage and possible evaporation in natural or artificially constructed ponds. This can be a convenient and economic option in areas which have high evaporative demand and abundance of natural depressions and salt lakes far from the cultivated lands. Accumulated salts and minerals may ultimately require disposal at some other facility or location. In India, many areas in Rajasthan can be ideal sites for construction of evaporation ponds.

Evaporation ponds can be constructed on individual farms or on a regional basin where drainage effluent of several farms can be collected. Each option has an unique set of economic, management, and environmental features influenced by geoographical and social considerations. Considering the size of average land- holding in India, it is foreseen that regional evaporation ponds will be more practical in our country. The practice can require assigning upto 10 % of the drainage area to the evaporation pond. Rao and Rao (1990) analysed from the observed and computed drainage discharge and evaporation that evaporation ponds will account for about 5 % of the drainage area in arid and semi- arid regions of Haryana (India). Kamra et al. (1993) report from initial results a reasonable sufficiency of a 1 ha surface area pond for storage of drainage effluent from 40 ha area i.e. 2.5 % of the drained area.

Environmental concerns related to evaporation ponds include possible groundwater pollution and exposure of wildlife to toxic elements in the pond. Installation of drain lines under the pond facilitates interception and recycling of seepage water back to the pond. Following practices can reduce hazard to the wildlife: (i) maintaining water to a depth of 0.6 m or more, (ii) discouraging vegetal growth, (iii) making steep slopes of waterside levees and (iv) providing good access to pond and regular monitoring of its water quality. Experiences in a number of countries indicate that the longterm problems of evaporation ponds may exist. Thus the solution should be seen only as a temporary and partial remedy.

(c) Deep Well Injection

Deep well injection, a liquid waste disposal method involving pressurized injection of wastes into suitable geological formation deep beneath the earth's surface, is a potential option for the disposal of agricultural drainage water. In oil and gas industry this technology consists of extraction of liquids (crude oil and waste brine) during drilling from the petroleum bearing strata which is simultaneously replenished with waste brine injected under high pressure. The injection of drainage water is different since it involves placement of liquids into a strata from which fluid has not been previously removed and may require greater injection pressures. Geohydrological conditions must be considered while selecting injection wells to ensure the protection of fresh groundwater supplies. Porous, saline water bearing sands confined above and below by impermeable rock strata are the most desirable receiving formations.

The treatment of drainage water may be required to ensure continued smooth operation of the injection well against clogging, chemical precipitation or biological growth of algae or fungi and fouling of the well due to decomposition of organics. The potential of seismicity induced by injection process must be considered. Deep injection of saline drainage water into isolated aquifers is a potential disposal option that prompts probing under Indian conditions.

2. Drainage Water Reuse

(a) Agriculture

The objective in irrigation and drainage projects should be to reuse as much of the drainage water as possible for irrigation and reclamation leaching. Large volume of effluents are utilized rather cheaply by reuse and if good quality water supplies are limited, saline drainage water can be used to irrigate crops of moderate to high salt tolerance at non-critical growth stages even if the yield is somewhat reduced.

There are two strategies to reuse drainage water for irrigation and reclamation. The cyclic strategy involves the use of relatively non saline water for pre- plant and early irrigations of crops, and of more saline drainage water during later growth stages when the plant is more salt tolerant. The reuse plan must consider the crop sensitivity to salt, growth stages, and salt concentrations of two water sources, and accordingly assess the need for surface storage of drainage water to coincide with the irrigation demands.

The other strategy for the reuse of drainage water is **blending**. This involves mixing water of two different salinities, to obtain irrigation water of suitable quality. The waters can be blended in the field or in the irrigation canal/channel. A major disadvantage of blending is that good quality water becomes unusable when it is blended with highly saline drainage water. Further, a blending facility is required and salt- sensitive crops cannot be included in the rotation plan. In most cases, the cyclic strategy is preferred over blending.

A modification of cyclic strategy (Rhoades, 1984), also known as **seasonal switching**, uses saline and nonsaline waters in a crop rotation that includes both salt-sensitive and tolerant crop species. Saline waters are used to irrigate tolerant crops; the resulting salt buildup in the soil is alleviated in the subsequent season when a sensitive crop is grown with nonsaline waters.

Although irrigation with saline drainage water looks promising, uncertainty still exists about the longterm effects of this practice on soil crusting, reduced water infiltration capacity, and accumualtion of toxic elements. If blending is used, the optimal mix of saline and nonsaline waters for different crops need to be worked out. Reuse is an accepted solution during the initial stages of drainage development when the volumes of saline drainage water are small. Irrigating with drainage water is likely to occur, not because it is a perfectly sound practice in the long run but because it may be the only immediate alternative avaialble.

(b) Industrial Use and for Cooling of Power Plants

Drainage water after pretreatment against corrosion or softening by ion exchange resins has been used in industry or for cooling of power plants. These procedures are unlikely to be adopted in India because of cost factor.

(c) Solar Ponds for Electrical Energy

Evaporation ponds and desalinisation technology (to be discussed later) results in end products which need disposal. Solar gradient ponds offer scope of using reject brine or salt to capture solar energy and convert it into electricity. Short wave solar rays penetrate through ponded water and heat lower levels of water. Ordinarily heated bottom water in the water bodies expands and rises to the surface to lose absorbed heat by convection and radiation. This is prevented in solar gradient ponds by storing dense saline water at the pond bottom and progressively less saline waters in upper layers. The less saline top layer is replenished with fresh water to compensate for evaporation.

Solar gradient ponds (Fig. 1) are 2- 5 m deep with three distinct zones. The top (upper convection) layer is 0.2m to 0.5m deep and contains uniform, low salinity water of upto 60,000 mg/l. The temperature and salinity increase with depth in the intermediate non convective gradient zone which is 0.75m to 1.5m thick and helps to insulate the underlying zone of heat storage. The lowest 1 to 3 m thick layer stores very hot and saline barine of about 150,000 to 250,000 mg/l. This zone is system's heat accumulating element from which heat can be extracted for potential conversion to electrical energy. Use of solar ponds for electrical power generation has been extensively researched in Israel. The solar pond area for a 5 MW plant in Israel is approximately 24 ha. Inital costs of these power plants are high, but operation costs are comparable to hydroelectric plants. Stringent requirements for controlling leakage of hazardous chemicals into groundwater add significantly to the cost. Solar gradient ponds do not offer solution of draiange water treatment at present, but technology may advance to make them feasible.

(d) Salt Production in Evaporation Ponds

The recovery of salts or minerals from concentrated drainage water in evaporation ponds has been discussed

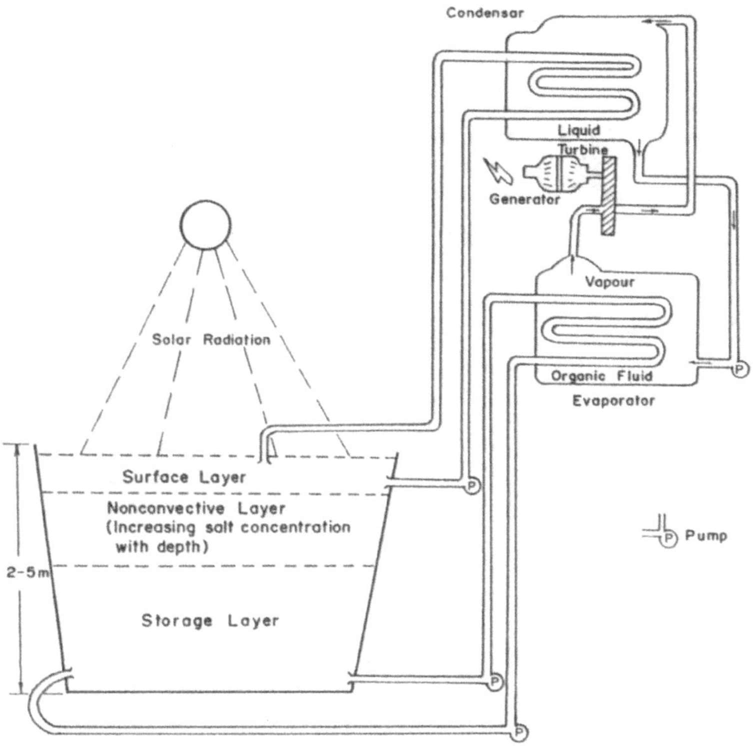

Fig. 1 Operation of a Solar Power Pond

above. Large scale salt production from evaporation ponds may have certain inherent problems like its possible contamination with toxic substances e.g. pesticides.

(e) Aquaculture

Though composite fish culture has been successfully tried in brackish waters of salinity upto 15 dS/m in inland areas (Garg, 1993), aquaculture in saline drainage water is still in conceptual phase because of the limited evidence of its success and also perhaps because of the apprehensions of trace elements in drain water.

(f) Agroforestry

Certain salt tolerant trees can be irrigated with saline water, including drainage effluent, after getting the seedlings established with fresh water during the first year. Once established, these trees draw water from the ground water for growth and ET. Similarly a few grasses can also be irrigated with saline water. Further aspects of agroforestry will be discussed in the next section.

3. Source control for drainage water quality and quantity

A consensus exists that source control for drainage water quality and quantity is its most important management option. For optimal management of water resources, water just sufficient to meet crop evapotranspiration and leaching demands should be applied to minimize drainage volumes. This requires structural and management changes in the existing delivery, distribution and application methods of irrigation water to improve irrigation efficiencies, and accurate estimation of the leaching requirements for a variety of soils, crop, and climatic conditions.

(a) Water Management: irrigation and leaching efficiencies

(i) Minimize leaching and deep percolation

Depending upon the ET of the crop and depth and salinity of irrigation water, an average of 0.3 to 32 tons of salt/ha can be added annually to irrigated soils (Rhoades, 1990). Irrigation must be adequate, over the long run, to prevent harmful accumulation of these salts in the root zone. The extent to which leaching of salts can be minimized is limited by the tolerance of plants to increased salinity in the lower parts of root zone. In most irrigation projects in India, current leaching fractions could be reduced appreciably without harming crops or soils. To avoid water table rise, leaching should take place during a fallow period or early in the growing period. Small, intermittent applications are generally more effective for leaching than continuous flood applications.

Very low leaching fractions are required for most crops after first reclamation leaching. Substantial leaching is required only for the most sensitive crops or when saline waters (including drainage water) are used for irrigation. Rainwater leaching can be enhanced by increasing the antecedent moisture of the soil profile by saline water irrigation prior to the onset of monsoon. Keeping in mind the existing irrigation practices, it is anticipated that changes in irrigation management and cropping patterns (rather than the leaching requirements) can reduce drainage requirements in India. Seepage from water distribution and collection systems may exceed field percolation losses.

(ii) Irrigation Scheduling

Irrigation scheduling is important when management includes salinity considerations. Scheduling of each irrigation allows the depth of leaching water to be accurately determined and prevents excessive deep percolation. ASAE (1981) presented methods for estimating rates of ET by irrigated crops and scheduling individual irrigations. Scheduling irrigation by traditional water balance methods is difficult if shallow ground water exists. Quantifying the groundwater contribution to crop ET and osmotic stress under saline conditions are the problems on which lot of work is currently going on.

Shainberg and Shalhevet (1984) reviewed the effect of frequency of saline water irrigation on crop yields and concluded that higher frequences result in higher yields. Frequent irrigations maintain low matric potential, which compensates for the osmotic stresses caused by saline water. If irrigations are applied frequently, each irrigation must be small which can seldom be applied as uniformly with surface methods as with sprinkler or drip irrigation.

(iii) Irrigation Methods

The volume of water needed for a given degree of leaching may be greater for furrow and other surface irrigation methods than for sprinkler and drip irrigation. Border irrigations require greater care to adjust flows to varying infiltration rates and application depths. Level basins facilitate uniform application of required depths of irrigation water over the entire area. However, high construction costs forbid their regular use in India in the near future. For improving irrigation uniformity in other surface methods, recommended specifications related to the length of water run, stream size, slope of soil and cut off ratio, if available, should be followed.

Drip and sprinkler systems help to maintain suitable water potentials in the root zone even with saline water. Sprinkler irrigation applies water uniformly and the application depths can be precisely controlled to meet the root zone moisture deficit and leaching demands. Frequent and small irrigations are possible with drip irrigation also. Drip irrigation enhances crop's threshold limits of salt tolerance by modifying the patterns of salt distribution and maintenance of constantly higher matric potentials. Though drip irrigation system is in its infancy in India, it has a great potential for irrigation with nonsaline or saline waters in the arid and semi- arid regions of the country.

(iv) Crop Use of Saline Subsurface Water

The depth and salinity of the water table, and the salt toleranece of plants determine the capillary contribution of ground water to a crop. Water table depths of 1.0 to 1.5 m generally provide maximum groundwater contribution for a wide range of soil types. The shallower water table depths apply for sandy, nonsaline soils.

(v) Improvement in Water Delivery Systems

An important part of salinity management is the prevention of watertable rise through reduction of seepage from canal, distributaries and water courses.The approximate quantitative estimates of seepage losses and utilization of irrigation water in some north Indian canals were computed by Gopinath (1976) and are presented in Table 3. Seepage can be reduced by lining canals with compacted earth, bentonite, concrete, and various geomembranes, or combinations of these materials. For smaller canals and laterals, closed conduits or concrete or PVC are practical alternatives. Good records of water deliveries are necessary to match water deliveries to crop requirements.

Table 3 Estimated seepage losses and utilization of irrigation water for north Indian canals

Particulars	Approximated losses (%)
Losses in main canal and branches	15- 20
Losses in major and minor distributaries	6- 8
Losses in field channels	20- 22
Losses during application including deep percolation, evaporation etc.	25- 27
Consumtive use of crops	28- 29

Source: Gopinath, 1976

(vi) Conjunctive Use

Conjunctive use of surface and ground water can be effectively used in certain cases to prevent salinisation or as management tool. Since there is a direct correlation between the cost of water and irrigation efficiency (Major, 1977), water pricing can be a means to increase irrigation efficiency for reduced watertable rise and salinisation. Transferable water entitlement is another mechanism for reducing waterlogging and salinisation. It allows water to be tranferred from a saline environment in a basin to non saline parts or vice versa. In Haryana, deep tubewells in fresh water areas are being used to augment canal water supplies for use in saline areas. Economic and political considerations have the major influence on implementation of such management options.

(b) Drainage Management

The volume of water collected by subsurface drains is a function of drain spacing and depth. There is a need to accurately determine the critical depth of various crops and soil types. Since deeper drains result in larger drainage volumes, minimum drain depth for optimal leaching and water table control is advantageous. An envisaged benefit of shallow drains is that subirrigation would be feasible for some soils and crops after reclamation leaching. Kamra et al. (1992), on the basis of numerical- modeling results and field experience, recommended that sub-surface drains installed at less than 1.75 m depth in the arid and semi-arid regions of India can maintain an adequate water and salt regime in the root zone for a good crop production in initially highly saline soils. Similarly the optimal drain spacing for crop growth and reclamation needs to be worked out for a variety of agro- climatic conditions. Methods for predicting and rapidly measuring concentrations of salts and trace elements and boron in drainage effluent are required.

(c) Crop Management

Crops vary in their tolerance to salinity of irrigation water. The appropriate choice of crops and cultivars is very important while irrigating with saline waters (including drainage water). Minhas and Gupta (1992) comprehensively presented the state, scope and management guidelines for using saline waters for crop production in different agro-climatic regions of India. They found that the salt tolerance of crops grown in different seasons follow the order: *rabi* > *kharif* > summer crops; legumes are more sensitive to salinity than the cereals and oilseeds; planting crops late in the season increases their sensitivity to salt stress, and comparatively higher salinity waters could be used in coarse- textured soils. Based on experimental evidence from different parts of the country, the germination and early seedling establishment stages have been identified to be salt sensitive in most crops. The other critical stages for crops are during the phase changes from vegetative to reproductive, i.e. heading and flowering to fruit setting. Breeding or selection of promising cultivars of crops for saline environment, besides having a high- yield potential and a high- salt tolerance, should also have the stability trait. Minhas and Gupta (1992) also present useful guidelines for using saline waters and selecting crops based upon clay percent in soil and the annual rainfall of the area (Table 4). Table 4 applies for irrigation waters of salinity EC_{iw} whose

Residual Sodium Carbonate, RSC (= carbonate + Bicarbonates- calcium - magnesium) in terms of me/l is less than 2.5.

Table 4 Guidelines for using saline waters (RSC< 2.5 me/l) for different crops

Soil texture (% clay)	Crop tolerance	Upper limit of EC_{iw} (dS/m) in annual rainfall regions (mm) < 350	350- 550	550- 750
Fine (>30)	S*	1.0	1.0	1.5
	ST	1.5	2.0	3.0
	T	2.0	3.0	4.5
Moderately fine (20- 30)	S	1.5	2.0	2.5
	ST	2.0	3.0	4.5
	T	4.0	6.0	8.0
Moderately coarse (10-20)	S	2.0	2.5	3.0
	ST	4.0	6.0	8.0
	T	6.0	8.0	10.0
Coarse (<10)	S	-	3.0	3.0
	ST	6.0	7.5	9.0
	T	8.0	10.0	12.5

* S, ST and T denote sensitive, semi- tolerant and tolerant crops respectively.

For arid and semi- arid regions, sensitive crops include beans, groundnut and berseem; bajra, sorghum and maize are semi tolerant crops while barley, sugarbeet, cotton, mustard, safflower and wheat are tolerant crops to saline water irrigations.

(d) Alternate Landuse

Salt tolerant trees have potential to lower watertable and reduce the volume of drainage through ET and consumptive use. A practical alternative to drainage under certain situations is to provide a permanent vegetative cover of trees and halophytic grasses. In waterlogged and/ or saline soils, trees in the agroforestry system can act as 'biological pumps' disposing water through foliage ET, while the companion halophytic shrubs/ grasses cover the soil to check upward flux of salts through evaporation. For canal command areas, Singh (1993) suggests planting of 100 m or more wide belt along the canals on both sides with high water demanding trees such as Eucalyptus, Populus, Laucaena and grasses such as Spartina, Panicum, Leptochloa, Brichairia for the interspaces. He also recommends field experiments and simulation studies for precise estimation of evapotranspiration of woody tree species to work out their potential for lowering shallow ground watertables.

4. Drainage Water Treatment

(a) Desalinization

Reverse Osmosis (RO) is the most promising desalinization technique for the treatment of agricultural drainage water. It involves application of high pressure on saline feedwater and forcing it through a semipermeable membrane to seperate water from dissolved salts and suspended materials. It results in increased salt concentration in the reject brine as relatively pure water passes through the membrane. Evaporation ponds may be required to dispose reject brine stream, if disposal to surface water is not possible. The cost of such ponds, which are to be abondoned or cleaned in the long run, contribute significantly to the treatment cost.

The feed waters to the RO system have strict quality criteria, necessitating sometimes the pretreatment of drainage waters. Sediments, silica and bacteria that clog or decompose the RO membranes need to be removed or destroyed. Waters rich in calcium and magnesium ions also need softening by substituting sodium for the calcium and magnesium. The conventional soda ash- lime treatment may not be practical for drainage waters because large quantities of soda ash and lime would be needed. Ion exchange appears to be a more feasible technique for softening drainage water. The drainage water is passed through a bed of resin beads with adsorbed sodium which gets exchanged with calcium and magnesium ions of the water. After the completion of ion exchange, the resin is regenerated by exposing it to waters having high sodium concentrations.

Another promising technique for desalinization is the vapor compression evaporation (VCE) which involves evaporation and condensation of the water to be treated. The reject brine, approximately 3 percent of the treated water, is highly concentrated and can be directly used in solar power ponds. It, however, is still more costly than the other salinization techniques.

(b) Trace Element Treatment

Serious efforts are being made in the developed countries for designing low cost methods for the treatment of trace elements. These techniques can be broadly divided into three categories namely biological, chemical and physical. Blue- green algae (cyanobacteria) are potentially useful for removing selenium, nitrogen, phosphorous and sulfur from drainage waters. The treatment process consists of biological reactors for bacteria- mediated reduction of selenium or other elements, microfiltration for removing fine solids and ion exchange for polishing the effluent stream. Microalgae and bacteria are fermented in the digestors to produce methane gas which powers the plant operation. The digested algal material reduces selenium in drainage water to an insoluble state in the reactor that can be seperated by microfiltration. The reactiom time in the biological reactors is less than an hour. There are evidences of biological reduction of selenium in drainage water from 330 µg/l- 550 µg/l to about 10- 40 µg/l after microfiltration and of boron from 10 mg/l to less than 1 mg/l after filtration and ion exchange.

The chemical processes, generally applied for the removal of trace elements from industrial effluent, can not be easily applied for treatment of drainage water because of its complex chemical nature. Selenium in the selenate form is reduced to elemental selenium and removed from aqueous phase by ferrous oxide process which consists of a chemical reactor and a solid gravity seperator. Not much work on this aspect can be reported at this stage. The physical processes involve the adsorption of ions on natural and synthetic surfaces of active materials, including ion exchange resins. There have been attempts for adsorption of selenium in the drainage water on iron filings. The cost is an important deterent for routine use of trace element treatments.

The fifth and sixth options listed in Table 1 involve ecological, socio- economical and institutional aspects that are greatly influenced by the political considerations. Serious debate is presently going on in India on key issues like water pricing and subsidies, but limited scope of this paper inhibits further discussion on these options vis- à-vis drainage.

Health Issues

The Joint WHO/ FAO/ UNEP Panel of Experts on Environmental Management for Vector Control (Birley, 1989; Tiffen, 1989) presented in good detail the water borne diseases that could emerge from irrigation and drainage projects. With respect to public health, drainage in principle reduces the incidence of diseases caused by mosquitoes, tropical blackflies and snails. Drainage is most effective against malaria mosquitoes because their favourite habitats of swamps or flooded areas are amenable to being drained. Blackflies do not usually breed in such areas (but in fast- flowing water) and snails can withstand dry conditions achieved with drainage. However, surface drains and drainage water ponds can be excellent habitats for mosquitoes and snails for spreading malaria and schistosomiasis. Clearing and maintenance of surface drains by dredging, weed removal, or drying and of pipe drains by periodic flushing to dislodge mosquito larvae and eggs is an important part of health effort. Drainage water and irrigation return flows may sometimes contain excessive concentrations of nitrates, pesticides, and other toxic compounds. Human health could be affected if such drainage effluents contaminate the drinking water, or are consumed domestically.

Hydro- Salinity Models

Hydro-salinity models, linking hydrological and salinity interactions of subsystems of an irrigation system, are effective tools to optimize regional management of water resources and for comparing the effectiveness of salinity control strategies. The hydro- salinity models of irrigation systems range from mass balance based models (Tanji, 1977; Oosterbaan and Pedroso-de-Lima, 1989) to complex salinity management models (Makin and Goldsmith, 1988; Gates and Grismer, 1989) which require exhaustive input data. Irrespective of scale, each model should be seen as a mean (and not an end in itself) to get better insight into the flow mechanisms of a system and to get better estimates of required variables than otherwise possible.

Several computer simulation models of water flow (Skaggs, 1978; Feddes et al., 1978; Belmans et al., 1983) and solute transport (Nour el- Din et al., 1987 a,b; Kamra et al., 1991 a,b) have been developed to relate drainage system design to soil, aquifer and climatic conditions. Kamra et al. (1991 a,b) developed and field validated a two dimensional finite element model that provides long-term predictions of the desalinization of a tile- drained soil, and of the associated changes in the quality of the groundwater and the drain effluent. This model was applied to estimate the salt loads of drainage effluent under alternate drainage designs from a 2000 ha intially highly saline area in Haryana. Five drain spacing (2S)- drain depth (d) combinations: 2S= 48m, d=1.0m; 2S= 67m, d=1.5m; 2S= 75m, d=1.8m; 2S= 77m,d=2.0m; 2S= 85m, d=2.5m) were considered. Table 5 presents the expected annual salt load in drainage effluent of these designs corresponding to two salinities, EC_{iw} (= 0.5 and 5.0 dS/m), of irrigation water.

Table 5 : Annual salt loading (10^4 x tonnes) of drainage effluent from 2000 ha area under alternate drain designs using different quality irrigation water

Drainage design 2S	d	T (yrs)	EC_{iw} = 0.5 dS/m 1	2	5	EC_{iw} = 5.0 dS/m 1	2	5
--(m)--								
48	1.0		4.6	4.1	2.7	4.6	4.3	3.1
67	1.5		4.6	4.4	2.9	4.6	4.4	3.3
75	1.8		4.6	4.3	2.9	4.6	4.4	3.2
77	2.0		4.9	4.6	2.8	5.0	4.7	3.3
85	2.5		5.8	5.0	2.9	5.9	5.1	3.0

It is observed that, irrespective of EC_{iw}, the annual amount of salts discharged during initial two years of reclamation from drains wider than 75m and deeper than 1.8m is significanly higher than the closer and shallower drains in sandy loam soils in Haryana. In the fifth year, the salt discharged from all design options is almost identical.

Future Drainage Research

Subsurface drainage criteria for arid and semi- arid regions of India needs modification to incorporate salinity control aspects. Further research is needed on the scope of controlled drainage for maximising the contribution of watertable to evapo-transpiration and for reducing the load of drainage effluent. Pilot small scale drainage trials need to be conducted in different waterlogged and saline commands before undertaking large scale projects. Similarly long- term evaluations and standards of synthetic envelopes for different soil conditions also need to be established. It is, however, foreseen that the environmental issues including the effects of saline drainage effluent on the water quality as well as surrounding natural environment will soon dominate the drainage scenario in India.

It is envisaged that hydro- salinity models are likely to play an increasing role in future for regional management of drainage systems. There is a great need to establish detailed data sets on the performance of irrigation and drainage projects which could be used for computer simulation studies. Towards this end, systematic monitoring and phase- wise evaluation of irrigation and drainage projects for basin level planning and execution cannot be over emphasized.

Acknowledgement

The authors are grateful to Dr. N. T. Singh, Director, Central Soil Salinity Research Institute, Karnal for his constant encouragement and interest in this study.

References

American Society of Agricultural Engineers (ASAE), 1981. Irrigation scheduling for water and energy conservation in the 80's. Proc. ASAE Irrigation Scheduling Conference, St. Joseph, Michigan, 231 pp.

Barber, W., 1985. The rising water table and development of waterlogging in Northwestern India. A Drainage Sub-Sector Report, World Bank, Washington, D.C.

Belmans, C., Wesseling, J.G. and Feddes, R.A., 1983. Simulation model of the water balance of cropped soil: SWATRE. J. Hydrology. 63: 271- 286.

Birley, M.H., 1989. Guidelines for forecasting the vector-borne disease implications of water resources development. WHO/FAO/UNEP Panel of Experts of Environmental Management for Vector Control.PEEM Guidelines Series n2. WHO, Geneva.

Biswas, A.K., 1985. Evaluating irrigation impact, guidelines for project management. Ceres, 106: 23- 26.

El-ashry, M.T., Schilfgaarde, J.V. and Schiffman, S. 1985. Salinity pollution from irrigated agriculture. J. Soil and Water cons. , Jan.- Feb.: 48- 52.

Feddes, R.A., Kowalik, P.J. and Zaradng, H., 1978. Simulation of water use and crop yield. Simulation monograph. PUDOC, Wageningen, The Netherlands, 189

Garg, S.K. 1993. Fish Culture. In: Salinity Research In H.A.U. (Ed. H.R.Manchanda, S.P.S. Karwasra, H.C. Sharma), C.C.S. Haryana Agricultural University, Hisar- 125004, India, 103- 109.

Gates, T. K. and Grismer, M.E., 1989. Irrigation and drainage strategies in salinity- affected regions. J. Irri. Drain. Div., Am. Soc. Civ. Eng.,115(2), 255- 283.

Gopinath, C. 1976. Problems and concepts in irrigation use. In: Role of irrigation in the development of India's agriculture, Indian Society of Agricultural Economics, Bombay, pp. 10- 15.

Joshi, P.K. and Agnihotri, A.K., 1984. An assessment od adverse effects of canal irrigation in India. Indian J. Agric. Econ., 38: 528- 536.

Kamra, S.K., Singh, S.R., Rao, K.V.G.K. and van Genuchten, M. Th., 1991 a. A semidiscrete model for water and solute movement in tile- drained soils: 1. Governing equations and solution. Water Resour. Res., 27(9), 2439- 2447.

Kamra, S.K., Singh, S.R., Rao, K.V.G.K. and van Genuchten, M. Th., 1991 b. A semidiscrete model for water and solute movement in tile- drained soils: 2. Field validation and applications. Water Resour. Res., 27(9), 2449- 2456.

Kamra, S. K., Rao, K. V. G. K., Singh, O. P. and Oosterbann, R. J., 1992. Effect of drain- depth on salinity control in irrigated lands of semi- arid regions. Proc. 5th Intern. Drainage Workshop, Lahore, Pakistan, vol. II, 2.78- 2.86.

Kamra, S.K., Rao, K. V. G. K. and Sharma, D.P., 1993. Design and evaluation of drainage measures for watertable and salinity control in arid soils. (Draft) Annual Report 1992- 1993, Central Soil Salinity Research Institute, Karnal, India.

Kovda, V.A., 1983. Loss of productive land due to salinisation. Ambio., 12(2): 91- 93.

Lesaffre, B., 1992. Overview of diagnosis and identification of problematic soils. Proc. 5th Intl. Drainage Workshop, Lahore, ICID- CIID, IWASRI, Vol.I, 17- 57. .

Letey, J., Roberts, C., Penberth, M. and Vasek, C., 1986. An agricultural dilemma: drainage water and toxics disposal in the San Joaquin Valley. Spec. Publ. 3319. Agric. Expt. Sta., Univ. California, Riverside, CA, U.S.A., 56pp.

Major, D.C., 1977. Multiobjective Water resources Planning. Water Resources Monograph 4, American Geophysical Union, Washington, D.C.

Makin, I.W. and Goldsmith, H., 1988. Selection of drainage and its phased implementation for salinity control. Irrigation and Drainage Systems, 2: 109- 121.

Minhas, P.S. and Gupta, R.K., 1992. Quality of Irrigation Water- Assessment and Management. Indian Council of Agricultural Research, New Delhi, 123pp.

Mock, J.F. and Bolton, P., 1991. Environmental effects of irrigation, drainage and flood control projects- checklist for environmental impact indication. Working Group on Environmental Impacts of Irrigation and Drainage, ICID, Draft Report OD/TN 50, HR Wallingford, U.K.

Nour el-Din, M.M., King, I. P. and Tanji, K. K., 1987a. A salinity management model: I. Development. J. Irri. Drain. Div., Am. Soc. Civ. Eng., 113(4), 440- 453.

Nour el-Din, M.M., King, I. P. and Tanji, K. K., 1987b. A salinity management model: II. 1 and 2- D applications. J. Irri. Drain. Div., Am. Soc. Civ. Eng., 113(4), 454- 468.

Oosterbaan, R.J. and Lima Pedroso de, 1989. SALTMOD Manual, International Institue for Land Reclamation and Improvement, Wageningen, The Netherlands.

Postel, S., 1990. Saving water for agriculture. In: State of the World 1989: a World Watch Institute Report on Progress Towards a Sustainable Society, World Watch Institute, Washington, D.C., 39- 58.

Rao, G.G.S.N. and Rao, K.V.G.K., 1990. Disposal of subsurface drainage waters through evaporation ponds in Haryana. Indian J. Agril. Sciences, 60(2): 97- 101.

Rao, K.V.G.K., Singh, O.P., Gupta, R.K., Kamra, S.K., Pandey, R. S., Kumbhare, P. S. and Abrol, I. P., 1986. Drainage Investigations for Salinity Control in Haryana. Bull. 10, Central Soil Salinity Research Institute, Karnal, India.

Rhoades, J.D., 1984. New strategy for using saline waters for irrigation. In: Water Today and Tomorrow, Proc. Specialty Conference of Irrigation and Drainage Div., ASCE, Flagstaff, Ariz. (USA), July 24- 26, 231- 236.

Rhoades, J.D., 1990. Overview: diagnosis of salinity problems and selection control practices. In: Agricultural Salinity Assessment and Management (ed. K.K. Tanji). Manual 71, Amer. Soc. Civil Eng., New York, 18- 41.

Shainberg, I. and Shalhevet, J., 1984. (Eds.) Soil Salinity Under Irrigation : Processes and Management. Springer-Verlag, New York, 349 pp.

Singh, Gurbachan, 1993. Agroforestry systems for optimal water use in salt affected soils. In: Sustainable Irrigation in Saline Environment (Ed. N.K. Tyagi, S.K. Kamra, P.S. Minhas and N.T.Singh), CSSRI, Karnal, India, 125- 136.

Skaggs, R. W., 1978. A water management model for shallow watertable soils . Tech. Rep. 134, Water Resour. Res. Inst., North Carolina State University, Raleigh, N.C.

Tanji, K.K., 1977. A conceptual hydro-salinity model for predicting salt load in irrigation return flow. Managing saline water for irrigation. Texas Tech. Univ., 49- 70.

Tanji, K. K. (Ed.), 1990. Agricultural Salinity Assessment and Management. Manual 71, Amer. Soc. Civil Eng., New York, 619 pp.

Tiffen, M., 1989. Guidelines for the incorporation of health safeguards in irrigation projects through Intersectoral co- operation. WHO/FAO/UNEP Panel of Experts on Environmental Management of Vector Control. PEEM Guidelines Series n1. WHO, Geneva

United Nations Development Programme (UNDP), 1985. Studies on the use of saline water in command areas of irrigation projects, Haryana, India: Interim technical report (draft) by Haryana Minor Irrigation and Tubewell Corporation.

Water and Power Development Authority (WAPDA), 1988. Private communication, 94, Ferozepur Road, Lahore, Pakistan.

Water Science and Technology Library

1. A.S. Eikum and R.W. Seabloom (eds.): *Alternative Wastewater Treatment.* Low-Cost Small Systems, Research and Development. Proceedings of the Conference held in Oslo, Norway (7–10 September 1981). 1982 ISBN 90-277-1430-4
2. W. Brutsaert and G.H. Jirka (eds.): *Gas Transfer at Water Surfaces.* 1984 ISBN 90-277-1697-8
3. D.A. Kraijenhoff and J.R. Moll (eds.): *River Flow Modelling and Forecasting.* 1986 ISBN 90-277-2082-7
4. World Meteorological Organization (ed.): *Microprocessors in Operational Hydrology.* Proceedings of a Conference held in Geneva (4–5 September 1984). 1986 ISBN 90-277-2156-4
5. J. Němec: *Hydrological Forecasting.* Design and Operation of Hydrological Forecasting Systems. 1986 ISBN 90-277-2259-5
6. V.K. Gupta, I. Rodríguez-Iturbe and E.F. Wood (eds.): *Scale Problems in Hydrology.* Runoff Generation and Basin Response. 1986 ISBN 90-277-2258-7
7. D.C. Major and H.E. Schwarz: *Large-Scale Regional Water Resources Planning.* The North Atlantic Regional Study. 1990 ISBN 0-7923-0711-9
8. W.H. Hager: *Energy Dissipators and Hydraulic Jump.* 1992 ISBN 0-7923-1508-1
9. V.P. Singh and M. Fiorentino (eds.): *Entropy and Energy Dissipation in Water Resources.* 1992 ISBN 0-7923-1696-7
10. K.W. Hipel (ed.): *Stochastic and Statistical Methods in Hydrology and Environmental Engineering.* A Four Volume Work Resulting from the International Conference in Honour of Professor T. E. Unny (21–23 June 1993). 1994
    10/1: Extreme values: floods and droughts ISBN 0-7923-2756-X
    10/2: Stochastic and statistical modelling with groundwater and surface water applications ISBN 0-7923-2757-8
    10/3: Time series analysis in hydrology and environmental engineering ISBN 0-7923-2758-6
    10/4: Effective environmental management for sustainable development ISBN 0-7923-2759-4
    Set 10/1–10/4: ISBN 0-7923-2760-8
11. S.N. Rodionov: *Global and Regional Climate Interaction: The Caspian Sea Experience.* 1994 ISBN 0-7923-2784-5
12. A. Peters, G. Wittum, B. Herrling, U. Meissner, C.A. Brebbia, W.G. Gray and G.F. Pinder (eds.): *Computational Methods in Water Resources X.* 1994 Set 12/1–12/2: ISBN 0-7923-2937-6
13. C.B. Vreugdenhil: *Numerical Methods for Shallow-Water Flow.* 1994 ISBN 0-7923-3164-8
14. E. Cabrera and A.F. Vela (eds.): *Improving Efficiency and Reliability in Water Distribution Systems.* 1995 ISBN 0-7923-3536-8

Water Science and Technology Library

15. V.P. Singh (ed.): *Environmental Hydrology.* 1995 ISBN 0-7923-3549-X
16. V.P. Singh and B. Kumar (eds.): *Proceedings of the International Conference on Hydrology and Water Resources* (New Delhi, 1993). 1996
    - 16/1: Surface-water hydrology ISBN 0-7923-3650-X
    - 16/2: Subsurface-water hydrology ISBN 0-7923-3651-8
    - 16/3: Water-quality hydrology ISBN 0-7923-3652-6
    - 16/4: Water resources planning and management ISBN 0-7923-3653-4
    - Set 16/1–16/4 ISBN 0-7923-3654-2

Kluwer Academic Publishers – Dordrecht / Boston / London

GPSR Compliance

The European Union's (EU) General Product Safety Regulation (GPSR) is a set of rules that requires consumer products to be safe and our obligations to ensure this.

If you have any concerns about our products, you can contact us on ProductSafety@springernature.com

In case Publisher is established outside the EU, the EU authorized representative is:

Springer Nature Customer Service Center GmbH
Europaplatz 3
69115 Heidelberg, Germany

Batch number: 10607320

Printed by Printforce, the Netherlands